T0072462

No Miracles Needed

The world needs to switch away from using fossil fuels to using clean, renewable sources of energy as soon as possible. Failure to do so will lead to accelerated and catastrophic climate damage, loss of biodiversity, and economic, social, and political instability. This book describes how to solve the climate crisis, and at the same time eliminate air pollution and safely secure energy supplies for all – without using "miracle" technologies. It explains how to use existing technologies to harness, store, and transmit energy from wind, water, and solar sources to ensure reliable electricity and heat supplies. It also discusses which technologies are not needed, including natural gas, carbon capture, direct air capture, blue hydrogen, bioenergy, and nuclear energy. Written for everyone, *No Miracles Needed* advises individuals, communities, and nations on what they can do to solve the problems, including the economic, health, climate, and land benefits of the solutions.

Mark Z. Jacobson is a professor of Civil and Environmental Engineering and Director of the Atmosphere/Energy Program at Stanford University. He has published six books and over 175 peer-reviewed papers. His work forms the scientific basis of the Green New Deal and many laws and commitments for cities, states, and countries to transition to 100 percent renewable electricity and heat generation. He received the 2018 Judi Friedman Lifetime Achievement Award, and in 2019 was selected as "one of the world's 100 most influential people in climate policy" by Apolitical. In 2022, he was chosen as the World Visionary CleanTech Influencer of the Year. He has served on an advisory committee to the U.S. Secretary of Energy, appeared in a TED talk, appeared on the David Letterman Show, and cofounded The Solutions Project.

"To those who wrongly insist we lack the tools to decarbonize our economy today, I say: read energy systems expert Mark Jacobson's amazing new book. In *No Miracles Needed*, Jacobson presents a comprehensive and detailed, yet highly accessible and readable blueprint for the options we have right now to address the climate crisis by taking advantage of existing renewable energy, storage, and smart grid technology combined with electrification of transportation systems, and efficiency measures. Read this book and be informed and engaged to help tackle the defining challenge of our time."

Michael Mann, Distinguished Professor of Atmospheric Science at
Penn State University and author of
The New Climate War

"Many people believe or fear that we can't solve the climate crisis, because we just don't have the technologies in hand to do so. This book should lay that fear to rest, once and for all."

Naomi Oreskes, co-author (with Erik M Conway) of
*The Big Myth: How American Business Taught Us to Loathe
Government and Love the Free Market*

"… shows impressively that numerous crises can be killed with one stone, without us having to wait for miracles: the energy, economic, health, and biodiversity crises can be solved by transitioning to a smart and complete supply of renewable energies. Let's not wait for miracles: let's simply implement it as soon as possible. Well worth reading!"

Claudia Kemfert, German Institute for Economic Research and
Professor of Energy Economics and Energy Policy at
Leuphana University

"… a highly compelling and accessible book laying out the best path for [our] energy future, one that is achievable with currently available technologies, with no need for some new miraculous breakthrough. This is a must read for all who care about the future of our society and our planet, written by the world's premier thinker on energy futures."

Bob Howarth, Cornell University

"… blends science, engineering and history into a readable cornucopia of information … Mark's style is to present approachable depth on dozens of major topics: everything you need to understand, and to join the fight against, the peril of our time."

Anthony R. Ingraffea, Cornell University

"Forget future miracle technologies promised by snake oil salespeople. This book offers a practical and real-world solution today. It is a must read for everyone concerned about climate change and air pollution and interested in the transition to a more sustainable all-purpose renewable energy future. It is sure to be one of the most important books that you will read this decade."

Peter Strachan, Aberdeen Business School,
Robert Gordon University

"Mark Jacobson's essential book, *No Miracles Needed*, offers clean, safe, and efficient solutions for our energy needs in this time of ever-growing climate chaos and disaster … The tools for producing, storing, and transmitting affordable and safe clean energy exist here and now with wind, water, and solar. *No miracles are needed*. A tireless and brilliant advocate for the environment, Professor Mark Jacobson's voice must be read, heard, and acted upon —now."

Heidi Hutner, Stony Brook University

No Miracles Needed

How Today's Technology Can Save Our Climate and Clean Our Air

Mark Z. Jacobson
Stanford University

CAMBRIDGE
UNIVERSITY PRESS

University Printing House, Cambridge CB2 8BS, United Kingdom

One Liberty Plaza, 20th Floor, New York, NY 10006, USA

477 Williamstown Road, Port Melbourne, VIC 3207, Australia

314–321, 3rd Floor, Plot 3, Splendor Forum, Jasola District Centre, New Delhi – 110025, India

103 Penang Road, #05–06/07, Visioncrest Commercial, Singapore 238467

Cambridge University Press is part of the University of Cambridge.

It furthers the University's mission by disseminating knowledge in the pursuit of education, learning, and research at the highest international levels of excellence.

www.cambridge.org
Information on this title: www.cambridge.org/9781009249546
https://doi.org/10.1017/9781009249553

First published 2023

A catalogue record for this publication is available from the British Library.

ISBN 978-1-009-24954-6 Paperback

To the young people of today, who will take us to the finish line
tomorrow

CONTENTS

FIGURES

FOREWORD

This is among the most important books you'll ever read, because it lays out in clear and frank terms the great problem of our age, and the great solution.

Burning things – coal, gas, oil, and biomass – has produced the prosperous world that we in the West inhabit. It has allowed us to heat and cool our buildings when the temperature is not to our liking, to light our spaces so as to extend our days, and to move ourselves and our stuff great distances with great ease. It has liberated us, that is, from many of the constraints that had traditionally governed human life.

But we now know that those liberations have come with unbearable cost. Breathing the smoky byproducts of all that burning kills more than 7 million of our brothers and sisters each year, far more than Covid, or HIV/AIDS, or malaria, or war. And that combustion has filled the air with invisible greenhouse gases that now threaten the very stability of our civilizations by raising the temperature and in the process melting the icecaps, destabilizing the jet stream and the Gulf Stream, raising the sea level, and sundry other catastrophes on a scale of destruction we'd previously imagined only in connection with atomic weapons.

So replace them we must – but with what? Mark Jacobson and his team have provided, after two decades of work, all the answers we need. Wind power, hydropower, and solar power – wind, water, and sun, or WWS to use his formulation – are sufficient to give us more than enough energy for our needs, and to do it at a cost that should allow for quick transition. This book lays out those essential facts in interesting, accessible, and readable fashion: it is a user's manual for a planet in transition, and one that should settle the panic in anyone who thinks we lack the resources to do what needs doing.

To state it plainly: there is no longer any technical or economic obstacle to the swift transition of our energy system to something far cleaner, cheaper, and more rational. We have the miracle technologies we require firmly in hand. You can point a sheet of glass at the sun and out the back will come light, air conditioning, information, mobility: all the requirements of modernity. Jacobson dutifully considers the possible drawbacks – will it use up minerals we don't possess in sufficient quantity, or occupy too much land – and comes back with mathematical assurances. He has the data.

But of course winning the argument is not the same as winning the fight. Shifting in the short time that climate science requires will mean overcoming both inertia and vested interest, which means that all of us, even if we are not engineers, have a role to play in getting the job done. Indeed, some of the most interesting sections of this volume describe Jacobson's own evolution into an activist of sorts, or at least someone trying to make the case for change. If he can overcome the sweaty panic that overtook him in the seconds before his nationwide interview with David Letterman, the rest of us can learn to make this case in letters to the editor and to our elected leaders.

In fact, it would be a dereliction of intellectual duty to read this book and then not take some actions to change the debate. If we had no readily available answer to the twin crises of climate change and air pollution, then I suppose we could in conscience ignore them. But the solutions are readily at hand. This book should empower you – and with not a moment to spare!

Bill McKibben

PREFACE

On July 11, 2011, I was invited to a dinner at the Axis Café and Gallery in San Francisco to discuss the potential of renewable energy as an alternative to natural gas hydrofracking in New York State. Little did I know it at the time, but that dinner would set off a chain reaction of events that turned a scientific theory, that the world has the technical and economic ability to run on 100 percent clean, renewable energy and storage for all purposes, into a mass popular movement to do just that. The movement catalyzed an explosion of worldwide country, state, and city laws and proposed laws, including the Green New Deal, and business commitments. Ten years after that meeting, critics were no longer mocking our ideas as pie-in-the-sky and tooth-fairy-esque. They were no longer claiming that transitioning to more than 20 percent renewables would cripple power grids. Instead, the discussion had changed to what is the cost of 100 percent renewables, how fast can we get there, and should we leave a few percent for non-renewables?

Why do we want to transition our worldwide energy system entirely to clean, renewable energy and storage for everything? This book first explores the three main reasons: to eliminate air pollution, global warming, and energy insecurity. Air pollution kills about 7 million people and injures hundreds of millions more each year worldwide. It is the second-leading cause of death. The impacts of global warming are accelerating as greenhouse gases and dark pollution particles in our atmosphere continue to increase. Such impacts include melting of glaciers and sea ice, rising sea levels, more droughts and floods, more intense hurricanes and wildfires, more air pollution and heat-related deaths and illnesses, agricultural shifts and famine, climate migration, species extinction, coral reef damage, and more. Lastly, fossil fuels are limited resources. As they run out, economic, social, and political instability will ensue. These three problems require an immediate and drastic solution.

Do we need *miracle technologies*? No. Then what is the solution? It is to transition the world's current combustion-based energy to 100 percent clean, renewable **wind, water, and solar (WWS)** and storage for all energy purposes and to eliminate non-energy emissions. The main idea behind the solution comes from the fact that air pollution health and climate problems arise from the same cause: combustion of fossil fuels, bioenergy fuels, and open biomass. **Fossil fuels** include coal, oil, natural gas, and all their derivatives, such as gasoline, diesel, kerosene, jet fuel, and liquefied natural gas. **Bioenergy fuels** include liquid biofuels, such as ethanol, biodiesel, and bio jet fuel, for transportation, and solid biomass, such as wood, wood pellets, dung, and vegetation, for electricity and heat. **Open biomass** includes forests, woodland, grassland, savannah, agricultural crops, and agricultural residues. Burning any of these leads to pollution that affects both health and climate.

To solve the problems, it is necessary to move away from combustion by electrifying and providing direct heat without combustion. For the electricity and heat to remain clean and available for millennia to come while not creating other risks, they need to originate from clean, renewable, and sustainable sources, namely WWS.

WWS includes energy from **wind** (onshore and offshore wind electricity), **water** (hydroelectricity, tidal and ocean current electricity, wave electricity, geothermal electricity, and geothermal heat), and **sunlight** (solar photovoltaic (PV) electricity, concentrated solar power (CSP) electricity and heat, and direct solar heat). WWS electricity and heat need to power all current energy sectors, which include the electricity, transportation, building heating and cooling, industrial, agriculture/forestry/fishing, and military sectors. Although human-designed energy systems cause about 90 to 95 percent of **anthropogenic** (human-produced) air pollution and 75 percent of anthropogenic greenhouse gas emissions, this book also discusses methods of eliminating non-energy anthropogenic emissions that damage air quality and warm the planet.

Many solutions to date that have focused on the climate problem have included some technologies that are less helpful than WWS technologies. This book describes such technologies, which raise costs to consumers and society, increase emissions relative to WWS sources, create substantial risks that WWS sources do not have, and/or delay the solution to pollution and global warming because of the long time they take to come online. Given our limited time and funding available to solve the pollution, climate, and energy security problems we face, it is essential to focus on known, effective solutions that can be implemented rapidly. Money spent on less-useful options will permit more health, climate, and energy insecurity damage to occur.

In fact, to solve the three problems posed here, we have 95 percent of the technologies that we need already commercially available. We also know how to build the rest, which include primarily long-distance aircraft and ships, powered by hydrogen fuel cells, and some industrial technologies. As such, we do not need *miracle technologies* to solve these problems. We need the collective willpower of people around the world to solve them.

Why 100 percent clean, renewable energy and storage for everything? Why not 50 percent, 80 percent, or 99 percent? First, the health plus climate damage of every bit of pollution that we allow to remain in the air is so enormous that it is important both morally and economically to eliminate 100 percent of emissions. Second, 99 percent is not an ambitious goal to shoot for. Did Magellan aspire to circumnavigate 99 percent of his way around the Earth? Did the Apollo 11 crew aspire to reach 99 percent of its way to the moon? No. One hundred percent is the goal because that is the best society can do and will result in the cleanest air and most stable climate possible for future generations. Societies often strive for the best and safest.

How fast do we need to transition? In order to avoid more than 1.5 degrees Celsius global warming compared with temperatures between 1850 and 1900, we need to eliminate at least 80 percent of all emissions by 2030 and 100 percent no later than 2050, but ideally by 2035. In order to avoid tens of millions more air pollution deaths, we need to eliminate all emissions even faster.

Can we reach the goal of 100 percent WWS across all energy sectors and eliminate non-energy emissions at that speed? This book examines this question, including the data and scientific studies that say we can. It concludes that a transition among all energy and non-energy sectors worldwide is economically possible with technology that is almost all existing. The main obstacles are social and political.

This book is for lay-readers concerned about the massive air pollution, climate, and energy security problems the world faces. To summarize, it discusses why no *miracle technologies* are needed to solve these problems in the short period we have left to do so. The solution is to use existing and known technologies to harness, store, and transmit energy in the wind, the water, and the sun, and to ensure reliable electricity and heat supplies worldwide. The book also discusses what technologies are not helpful or needed but are being pursued vigorously. "Transition highlights" throughout the text offer examples of changes to renewable energy somewhere in the world. Finally, the book gives information about what individuals, communities, and nations can do to solve the problems, as well as the cost, health, climate, and land benefits of the solution.

1 WHAT PROBLEMS ARE WE TRYING TO SOLVE?

Why do we want to transition all of our energy to clean, renewable energy? Why don't we just continue burning fossil fuels until they run out, which may be in 50 to 150 years? For three major reasons. Namely, fossil fuels today cause massive air-pollution health damage, climate damage, and risks to our energy security. These three problems, which have the same root cause, require immediate and drastic solutions. The longer we wait to solve these problems, the more the accumulated damage. This chapter examines each problem, in turn.

1.1 The Air Pollution Tragedy

Today, air pollution is the second-leading cause of human death and illness worldwide. It also kills and injures animals; impedes visibility; and harms plants, trees, crops, structures, tires, and art. Because air pollution causes such enormous loss and cost, controlling it is one of the greatest challenges of our time.

What is air pollution? **Air pollution** occurs when

gases or aerosol particles in the air build up in concentration sufficiently high to cause direct or indirect damage to humans, plants, animals, other life forms, ecosystems, structures, or works of art.

What are gases and aerosol particles? A **gas** is a group of atoms or molecules that are not bonded to each other. Whereas a liquid occupies a fixed volume and a solid has a fixed shape, a gas is unconfined and freely expands with no fixed volume or shape.

An **aerosol particle** consists of 15 or more gas atoms or molecules, suspended in the air, that have bonded together and changed phase to become a liquid or solid. An aerosol particle can contain one chemical or a mixture of many different chemicals. An **aerosol** is an ensemble, or cloud, of aerosol particles.[1] Aerosol particles are distinguished from cloud drops, drizzle drops, raindrops, ice crystals, snowflakes, and hailstones, in that the latter all start as an aerosol particle but grow far more water on them than the former.

Gases and aerosol particles may be emitted into the air naturally or by humans (**anthropogenically**). They may also be produced chemically in the air from other gases or aerosol particles. Natural air pollution problems on the Earth are as old as the planet itself. Volcanos, natural fires, lightning, desert dust, sea spray, plant debris, pollen, spores, viruses, bacteria, and bacterial metabolism have all contributed to natural air pollution.

Humans first emitted air pollutants when we burned wood for heating and cooking. Today, anthropogenic air pollution arises primarily from the burning of fossil fuels and bioenergy fuels used for energy, and from the burning of open biomass for land clearing or ritual, or due to arson or carelessness. Air pollutants also arise from the release of chemicals to the air, such as from industrial processes or leaks.

The main **fossil fuels** burned today are coal, natural gas, and crude oil. Crude oil is refined into multiple products, including gasoline, diesel, kerosene, heating oil, naphtha, liquefied petroleum gas, jet fuel, and bunker fuel. **Bioenergy** fuels burned are either solid fuels, such as wood, vegetation, or dung, or liquid fuels, such as ethanol or biodiesel. **Open biomass** includes forests, woodland, grassland, savannah, and agricultural residues. Anthropogenic emissions have contributed not only to indoor and outdoor air pollution, but also to acid rain, the Antarctic ozone hole, global stratospheric ozone loss, and global warming.

In 2019, 55.4 million people died from all causes worldwide.[2] Air pollution enabled about 7 million (12.6 percent) of the deaths, making it the second-leading cause of death after heart disease.[3] Of the air pollution deaths, about 4.4 million were due to outdoor air pollution and about 2.6 million were due to indoor air pollution.[3] Indoor air pollution arises because 2.6 billion people burn solid fuels (wood, dung, crop waste, coal) and kerosene indoors for cooking and heating.[4] Air pollution also causes hundreds of millions of illnesses each year.

The deaths and illnesses arise when air pollution particles (mostly) and gases trigger or exacerbate heart disease, stroke, chronic obstruction pulmonary disease (chronic bronchitis and emphysema), lower respiratory tract infection (flu, bronchitis, and pneumonia), lung cancer, and asthma.

Almost half of all pneumonia deaths worldwide among children aged five and younger are due to air pollution.[4] Many children who die live in homes in which solid fuel or kerosene is burned for home heating and cooking. Their little lungs absorb a high concentration of aerosol particles in the air that result from fuel burning. They die of pneumonia because their immune systems weaken owing to the assault of air pollutants on their respiratory systems. Most of the casualties are in developing countries, where indoor burning often still occurs on a large scale. These deaths and illnesses not only devastate families, but also incur tremendous cost. The worldwide cost of all air pollution death and illness is estimated to be over US$30 trillion per year today.[5]

Transition highlight
In 2019, 7 million people died from air pollution worldwide. China and India absorbed the brunt of mortalities, with a combined total of 3.6 million deaths (52 percent of the total). Nigeria, Pakistan, Indonesia, Bangladesh, the Philippines, and Russia all suffered more than 100,000 air pollution deaths that year. The highest per capita air pollution death rates were in North Korea, Georgia, Chad, Nigeria, Bosnia and Herzegovina, and Somalia, respectively.

Around half the mass of aerosol particles emitted worldwide is in natural particles. However, natural particles are mostly large and thus do not penetrate deep into people's lungs. On the other hand, combustion particles, which are almost all from human sources today, are mostly small and penetrate deep into the lungs. Most combustion particles are also emitted near where people live, so people breathe in these particles. As a result, about 90 to 95 percent of air pollution deaths today are caused by anthropogenic air pollution. Of these deaths, about 90 percent are due to air pollution particles; the rest are due to air pollution gases, primarily ozone.

Because combustion during energy production is the world's major source of air pollution, changing the world's energy infrastructure to eliminate combustion will largely eliminate air pollution death and illness worldwide. This goal can be accomplished by transitioning to 100 percent clean, renewable energy and storage for everything.

1.2   Global Warming

1.2.1   The Natural Greenhouse Effect

Global warming is the human-caused increase in the average temperature of the Earth's lower atmosphere since the Industrial Revolution above and beyond the temperature due to the natural greenhouse effect. The natural greenhouse effect is the increase in the Earth's average temperature above its temperature without an atmosphere. The natural greenhouse effect is due to the build-up of natural greenhouse gases in the atmosphere since the formation of the Earth. Greenhouse gases are gases that are mostly transparent to sunlight but that absorb some of the heat emitted by the surface of the Earth. All objects in the universe, including the Earth, emit heat.

The Earth has three main sources of heat. The first and, by far, the most important, is sunlight, also called solar radiation. The Earth absorbs sunlight and converts it to heat, also called infrared radiation. About 99.97 percent of the heat emitted by the surface of the Earth originates from sunlight. The remaining 0.03 percent of heat originates from the interior of the Earth from two sources, each in relatively equal proportions. One is heat left over from the formation of the Earth, called primordial heat. Owing to gravitational compression of the Earth's interior during its formation, and despite heat loss over time, the temperature at the center of the Earth is still about 4,300 degrees Celsius. This heat transfers slowly to the surface of the Earth by conduction, which is the process by which molecules transfer energy to each other when they collide. Primordial heat also gets to the surface by volcanic activity. The other source of interior heat is heat released during the decay of radioactive elements in the Earth's interior. The main elements that decay are uranium, thorium, and a small fraction of potassium. The decay products of these elements decay further as well. The resulting heat transfers slowly to the surface, also by conduction.

Greenhouse gases in the Earth's atmosphere are transparent to sunlight, allowing it to penetrate to the Earth's surface. However, the same gases trap a portion of the Earth's outgoing heat, warming the ground and air near the ground. The more greenhouse gases present, the greater the trapping of heat and warming of the air. When the greenhouse gases are natural, the resulting warming is called the natural greenhouse effect.

The primary natural greenhouse gases in the Earth's atmosphere are water vapor, carbon dioxide (CO_2), ozone, nitrous oxide, methane, and oxygen gas. Oxygen gas is a weak greenhouse gas, but it is so abundant in the air (20.95 percent of all air molecules) that it has a non-trivial natural warning impact. Nitrogen gas, which comprises 78.08 percent of the molecules in the Earth's atmosphere, is not a greenhouse gas.

If the Earth had no atmosphere, thus no natural greenhouse effect, its average surface temperature would be about minus 18 degrees Celsius (zero degrees Celsius is the freezing temperature of water). At that temperature, little life would exist on Earth's surface.

During Earth's 4.6-billion-year history, several processes released to the air all of the Earth's natural greenhouse gases, except for ozone. These processes included emissions (through volcanos, fumaroles, and geysers) of greenhouse gases from the Earth's interior, bacterial metabolism, bacterial photosynthesis, and green-plant photosynthesis. Ultraviolet sunlight cooked some oxygen to produce ozone, most of which formed high above the ground, in what is now called the stratospheric ozone layer. The formation of the ozone layer was critical for protecting the surface of the Earth from harmful ultraviolet sunlight, permitting life to move from underwater and underground to above the ground.

Natural greenhouse gases raised the temperature of the Earth substantially compared with the Earth without an atmosphere, permitting life to flourish on the Earth. Just before the start of the Industrial Revolution, around 1760, Earth's average temperature was about 15 degrees Celsius. That is 33 degrees Celsius higher than Earth's temperature without greenhouse gases (minus 18 degrees Celsius). This warming was due to the natural greenhouse effect. Of this temperature rise, about 66 percent was due to water vapor, about 25 percent was due to background carbon dioxide, and about 6.2 percent was due to background ozone, most of which is in the upper atmosphere.[6]

1.2.2   Global Warming

Global warming is the rise in the Earth's globally averaged ground and near-surface air temperature above and beyond that due to the natural greenhouse effect, due to human activity. The Earth's average global warming in the period 2011 to 2020 compared with the period 1850 to 1900 was about 1.09 degrees Celsius.[7] Since this is an average value, some places on Earth have warmed more, whereas others have warmed less or cooled. For example, the Arctic has warmed by over 5 degrees Celsius. Many other high-latitude locations (parts of Canada, Northern Europe, and Russia) have warmed by 2 to 5 degrees Celsius. The North Atlantic Ocean has cooled slightly.

1.2.3   Causes of Global Warming

Global warming is due to four major warming processes partially offset by one major cooling process (Figure 1.1). The four major warming processes are anthropogenic greenhouse gas emissions, anthropogenic warming particle emissions, anthropogenic heat emissions, and the urban heat island effect. The cooling process is anthropogenic cooling particle emissions.

1.2.3.1 Anthropogenic Greenhouse Gas Emissions

The primary anthropogenic greenhouse gases contributing to global warming are carbon dioxide, methane, halogens, ozone, nitrous oxide, and anthropogenic water vapor.

The primary anthropogenic sources of **carbon dioxide** are fossil-fuel combustion, bioenergy combustion, open biomass burning, and chemical reaction during industrial processes, such as cement manufacturing, steel production, and silicon extraction. Owing to these emissions, carbon dioxide in the air has increased from about 275 parts per million (ppm) to 420 ppm, or by 53 percent, between 1750 and 2021. One part per million of carbon dioxide means that, for every million molecules of total air, one molecule is carbon dioxide. Carbon dioxide has been increasing in the air, not only owing to its emissions from human activity, but also because it stays in the air a long time. The major removal methods of carbon dioxide from the air are its dissolution into the oceans and other water bodies and green-plant

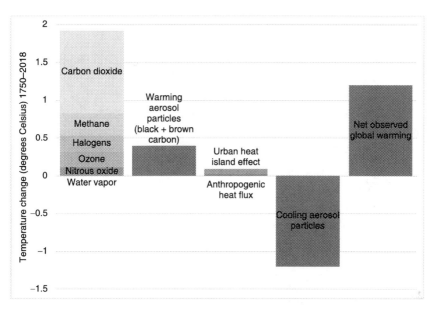

Figure 1.1 Estimated primary contributors to net observed global warming from 1750 to 2018. Warming aerosol particles include black and brown carbon from fossil-fuel burning, biofuel burning, and open biomass burning. Cooling aerosol particle components include sulfate, nitrate, chloride, ammonium, sodium, potassium, calcium, magnesium, non-brown organic carbon, and water. Of the gross warming (warming before cooling is subtracted out), 45.7 percent is due to carbon dioxide, 16.3 percent is due to black plus brown carbon, 12 percent is due to methane, 9 percent is due to halogens, 8.8 percent is due to ozone, 4.3 percent is due to nitrous oxide, 3 percent is due to the urban heat island effect, 0.7 percent is due to anthropogenic heat flux, and 0.23 percent is due to anthropogenic water vapor. Source: Jacobson, *100% Clean, Renewable Energy*.[8]

photosynthesis (the conversion of carbon dioxide and water vapor into oxygen and cell material by plants and trees). However, these sinks remove carbon dioxide very slowly over many decades.

The primary anthropogenic sources of **methane** are natural gas, coal, and oil mining leakage; fossil-fuel combustion, bioenergy combustion; open biomass burning; and leakage from landfills, rice paddies, livestock, and manure. Methane is removed from the air primarily by chemical reaction in the air itself and by bacterial metabolism at the surface of the Earth.

Halogens are a series of synthetic chemicals whose main uses are as refrigerants, solvents, degreasing agents, blowing agents, fire extinguishants, and fumigants. The first halogen was invented in 1928. Halogens enter the atmosphere when appliances or tubes sealing them in liquid form leak or are drained, and the liquid evaporates.

Most halogens are **halocarbons,** which are chemicals that contain carbon and possibly hydrogen, but also either chlorine, bromine, fluorine, or iodine. The main types of halocarbons are the following. **Chlorofluorocarbons** (CFCs) are halocarbons containing carbon, chlorine, and fluorine. **Halons** are halocarbons containing carbon and bromine. **Perfluorocarbons** are halocarbons containing carbon and fluorine. **Hydrofluorocarbons** are halocarbons containing carbon, fluorine, and hydrogen. Some halogens, such as sulfur hexafluoride, have no carbon, so are not halocarbons.

Because chlorofluorocarbons and halons contain stratospheric-ozone-destroying chlorine and bromine, most countries outlawed them through international agreement starting with the 1987 Montreal Protocol. Hydrofluorocarbons and perfluorocarbons were developed as ozone-layer-friendly replacements. However, because many of them are greenhouse gases with long lifetimes in the air, such chemicals, while not directly damaging to the ozone layer, have the unintended consequence of enhancing global warming.

Ozone is the only greenhouse gas with no emission source. It forms chemically in the air. About 90 percent of ozone resides in the upper atmosphere (stratosphere), and the rest resides in the lower atmosphere (troposphere). The troposphere is the layer of air between the ground and 8 kilometers above sea level at the North and South Poles and between the ground and 18 kilometers above sea level at the equator. The stratosphere is the layer of air just above the troposphere and extends to about 48 kilometers above sea level. Because of the substantial abundance of ozone in the stratosphere, the stratosphere is also called the ozone layer.

In the stratosphere, ozone (which has three oxygen atoms) forms chemically following the breakdown of oxygen gas (made of two oxygen atoms bonded together) into two unbonded oxygen atoms, by ultraviolet sunlight. Atomic oxygen then combines with oxygen gas to form ozone.

In the troposphere, ozone is produced chemically following the breakdown, by ultraviolet sunlight, of nitrogen dioxide into atomic oxygen. The atomic oxygen then combines with the oxygen gas that we breathe (molecular oxygen) to form ozone. The nitrogen dioxide comes either from direct emissions or from chemical reaction between nitric oxide and certain reactive organic gases. Most emissions of nitric oxide, nitrogen dioxide, and reactive organic gases result from the burning of fuels by humans. Some comes from natural forest burning and bacterial metabolism. Some nitric oxide comes from lightning.

Since the Industrial Revolution, the mass of tropospheric ozone has increased by about 43 percent because of the worldwide increase in air pollution (the anthropogenic emissions of nitric oxide, nitrogen dioxide, and reactive organic gases). Since the late 1970s, stratospheric ozone has declined by about 5 percent owing to the increased presence of chlorofluorocarbons and halons within the stratosphere.

Ozone has a relatively short lifetime in the air. Most of its loss is due to chemical reaction. Just as its concentration has grown rapidly in the troposphere owing to increases in air pollution, its tropospheric concentration and warming impact can decrease rapidly if air pollution levels decrease. This is one reason that a strategy to eliminate air pollution can help to decrease global warming as well.

Nitrous oxide (laughing gas) is a colorless gas emitted naturally by bacteria in soils and in the oceans. Because it is long-lived, nitrous oxide stays in the air for up to hundreds of years once emitted. It is a powerful greenhouse gas, so it causes substantial warming per molecule during this period. Humans have increased the abundance of nitrous oxide in the air through fertilizer use, agricultural waste, sewage, legumes (plants in the pea family), bioenergy burning, biomass burning, jet-fuel burning, nylon manufacturing, and aerosol spray can manufacturing. Agriculture (fertilizers, agricultural waste, and legumes) is the largest source of human-emitted nitrous oxide today.

Anthropogenic water vapor comes from two main sources. The first is evaporation of water that is used to cool power plants and industrial facilities that run on coal, natural gas, oil, biomass, or uranium. The second is emission of water vapor during the burning of fuels for energy. Water vapor emitted annually from these sources is only about 1/8,800 of the 500 million metric tonnes of water vapor emitted per year from natural sources. Nevertheless, this relatively small anthropogenic emission rate of water vapor contributes a modest 0.23 percent of global warming.[6]

1.2.3.2 Anthropogenic Warming Particle Emissions

Dark aerosol particles may contribute more to today's global warming than any other chemical aside from carbon dioxide (Figure 1.1).[9,10,11,12,13] Dark particles, also called **warming particles,** contain primarily black and brown carbon.

Black carbon is an agglomerate of solid spherules made of pure carbon and attached to each other in an amorphous shape. The source of black carbon is incomplete combustion of diesel, gasoline, jet fuel, bunker fuel, kerosene, natural gas, biogas, solid biomass, and liquid biofuels. Black carbon is often visible to the eye and appears black because it absorbs all wavelengths of sunlight, transmitting none to the eye. Black carbon particles convert the absorbed light to heat, raising the temperature of the particles and causing them to re-radiate some of the heat to the surrounding air.

Black carbon and greenhouse gases warm the air in different ways from each other. Greenhouse gases are mostly transparent to sunlight. They warm the air by absorbing heat emitted by the surface of the Earth. They then re-emit half of that heat upward and half downward, raising the ground and near-surface air temperatures.

Black carbon particles, on the other hand, heat the air primarily by absorbing sunlight, converting the sunlight to heat, then re-emitting the heat upward and downward, like with greenhouse gases. Black carbon particles also absorb and re-emit heat itself, but that process is important for them only at night and when black carbon concentrations are high.

When other aerosol material, such as sulfuric acid, nitric acid, water, or brown carbon, coats the outside of a black carbon particle, the black carbon heats the air 2 to 3 times faster than without a coating because more light hits the larger particle, thus more light bends (refracts) into the particle. Inside the particle, this light bounces around until it hits and is absorbed by the black carbon core.

Black carbon not only warms the air but also evaporates clouds and melts snow. When black carbon enters a cloud, it absorbs sunlight that bounces around in the cloud, converts the sunlight to heat, then emits the heat to the cloud, warming the cloud. If a sufficient number of black carbon particles is present, this warming can cause the cloud to evaporate completely. When black carbon falls on snow or sea ice, it similarly absorbs sunlight, converts the sunlight to heat, then emits the heat to the ice or snow, melting it.

Thus, for four reasons (its strong absorption when pure, its stronger absorption when coated, its ability to evaporate clouds, and its ability to melt snow and sea ice), black carbon is the second-leading cause of global warming after carbon dioxide. In fact, per molecule in the air, black carbon causes over a million times more warming than does carbon dioxide.[11] However, because black carbon particles last only days to weeks in the air, their concentrations are much lower than are those of carbon dioxide, which lasts decades in the air. Nevertheless, because black carbon is continuously emitted, it always causes a strong warming.

Brown carbon is also a particle component that increases global warming and causes health problems. Whereas brown carbon is generally more abundant than is black carbon, brown carbon is much less effective per unit mass at causing warming than is black carbon. As such, black carbon causes more overall warming than does brown carbon.

Whereas black carbon contains pure carbon, brown carbon contains carbon, hydrogen, and possibly oxygen, nitrogen, and or other atoms. In other words, brown carbon is a type of **organic carbon** (which is a chemical containing carbon, hydrogen, and other atoms). Not all organic carbon is brown carbon. Brown carbon is the subset of organic carbon that absorbs short (blue) and some medium (green) wavelengths of visible light. The remaining long wavelengths (red) and some of the green are transmitted to the viewer's eye, making the particle haze appear brown. The more green light that is transmitted (the less that is absorbed), the more yellow the particles appear. Other organic carbon particles are often white or grey because they do not absorb much or any visible light.

The sources of brown carbon, the combustion of fossil fuels, bioenergy, and biomass, are also the sources of black carbon. However, the relative amount of brown or black carbon from a combustion source depends largely on the temperature of the flame. Hotter flames favor black carbon, whereas cooler flames favor brown carbon. For example, in smoldering biomass (a low-temperature flame), the ratio of brown to black carbon is about 8 to 1. In diesel combustion (a high-temperature flame), the ratio is about 1 to 1.

Because black and brown carbon particles together cause such a large warming per molecule and have such short lifetimes in the air, reducing their emissions is the fastest way of slowing global warming.[11]

Because such particles both cause substantial human death and illness, reducing their emissions not only slows global warming but also immediately improves human health. Thus, two major reasons exist to eliminate black and brown carbon particle emissions: to slow global warming rapidly and to improve human health rapidly.

1.2.3.3 Anthropogenic Heat Emissions

Anthropogenic heat emissions are emissions of heat from the use of electricity; friction created by vehicle tires on the road; the combustion of fossil fuels, biofuels, and biomass for energy; nuclear reaction; and anthropogenic biomass burning. Such heat emissions warm the air directly. Much of the hot air eventually rises, converting the heat energy into **gravitational potential energy**, which is energy embodied in air lifted to a certain height against gravity. Differences in gravitational potential energy between one location and another create winds, which carry with them kinetic energy. Thus, some anthropogenic heat emissions are converted to energy in the wind. The increases in both temperature and wind speeds due to heat emissions cause liquid water to evaporate. Since water vapor is a greenhouse gas, the production of water vapor accelerates the impact of the original heat emissions.

In sum, much of the heat from anthropogenic heat emissions converts to other forms of energy. Since energy is conserved, the different forms of energy persist in the atmosphere (or oceans and land). Overall, though, the impacts of anthropogenic heat emissions are much less than are those of greenhouse gases, which persist for decades to centuries and cause greater overall warming than do anthropogenic heat emissions. Anthropogenic heat may contribute to about 0.7 percent of global warming to date (Figure 1.1).[6] As such, eliminating combustion and nuclear reaction, which a WWS system does, substantially reduces anthropogenic heat emissions as well as the emissions of air pollutants, greenhouse gases, and warming particles.

1.2.3.4 The Urban Heat Island Effect

The urban heat island effect is the temperature increase in urban areas due to the covering of soil and replacing of vegetation with impervious surfaces, such as concrete and asphalt. Covering surfaces reduces evaporation of water from soil and plants. Because evaporation is a cooling process, eliminating it warms the surface. Built-up areas also have sufficiently different properties of construction materials that they enhance urban warming relative to surrounding vegetated areas. Worldwide, the urban heat island effect may be responsible for about 3 percent of gross global warming (warming before cooling is subtracted out) (Figure 1.1).

1.2.3.5 Cooling Particle Emissions

Cooling particles are light-colored aerosol particles that cool the Earth's surface by reflecting sunlight to space and by thickening clouds, which are largely reflective. Cooling particles contain primarily sulfate, nitrate, chloride, ammonium, sodium, potassium, calcium, magnesium, non-brown organic carbon, and water. Because cooling particles tend to be more soluble in water than are warming particles, cooling particles allow water vapor to condense readily on them, enhancing cloudiness, thereby cooling the climate. Warming particles, on the other hand, tend to heat clouds, helping to burn them off. Like with warming particles, cooling particles last only days to weeks in the air and cause major air pollution health damage. Like with warming particles, eliminating cooling particle emissions will improve human health dramatically. However, eliminating cooling particles will raise global temperatures. This is why a strategy of eliminating all greenhouse gases, warming particles, and cooling particles simultaneously through a transition to WWS is necessary to solve both air pollution and global warming problems together.

1.2.4 Impacts of Global Warming

Global warming has already caused the world significant financial loss, and the cost is expected to grow to over $30 trillion per year by 2050.[5] Losses arise due to coastline erosion (from sea level rise); fishery and coral reef damage; species extinction; illness and death due to heat stress and heat stroke; agricultural loss; more famine and drought; more wildfires and air pollution; increased climate migration; and more severe weather and storminess (e.g., hurricanes, tornados, and hot spells).

Higher temperatures increase air pollution in cities where the pollution is already severe.[14,15] Higher temperatures also increase the risk of wildfires, which themselves cause air pollution, loss of life, and structural damage. For example, during November 2018, three major wildfires in California, enhanced by drought and unusually high November temperatures, killed dozens of people, displaced hundreds of thousands more, rendered several thousand people homeless, and produced dangerous levels of air pollution throughout the state for over 2 weeks.

Similarly, global warming has already caused a lot of damage by increasing hurricane duration, size, wind speed, and storm surge. Global warming has also caused agriculture crops to fail in many parts of the world, triggering mass migrations. Such migrations are already occurring from the Middle East and North Africa to Europe, and from Central America to the United States, for example.

1.2.5   Strategies for Reducing Air Pollution and Global Warming Together

Because all aerosol particles together are the leading cause of air pollution mortality, reducing both cooling and warming particles is desirable from a public health perspective. However, Figure 1.1 indicates that cooling particles cause more cooling than warming particles cause warming globally. As such, if emissions of all warming and cooling particles are eliminated together without eliminating other sources of heat, global warming will worsen.

Similarly, since cooling particles mask half of global warming, eliminating only cooling particles will roughly double net global warming.

One strategy to address global warming and human health simultaneously is to eliminate only warming particles. The downside of this strategy is that it permits most global warming and air pollution to continue.

Thus, Figure 1.1 suggests that the best strategy for addressing human health and climate simultaneously is to eliminate greenhouse gases, cooling particles, and warming particles simultaneously. This will also reduce most anthropogenic heat and water vapor emissions.

This book is about understanding and implementing that strategy – eliminating all anthropogenic emissions of greenhouse gases, warming particles, and cooling particles at the same time. This strategy will be accomplished by transitioning the world's energy to 100 percent wind, water, and solar plus storage for all energy and by eliminating non-energy emissions.

1.3 Energy Insecurity

Energy insecurity is a third major problem that needs to be addressed on a global scale. Several types of energy insecurity are of concern.

1.3.1 Energy Insecurity Due to Diminishing Availability of Fossil Fuels and Uranium

One type of energy insecurity is the economic, social, and political instability that results from the long-term depletion of non-renewable energy supplies. Fossil fuels and uranium are limited resources and will run out at some point. As fossil-fuel supplies dwindle, their prices will rise. Such price increases will first hit people who can least afford them – those with little or no income. These people will suffer, since they cannot warm their homes sufficiently during the winter, cool their homes sufficiently during the summer, or pay for vehicle fuel easily.

Higher energy prices will also increase the cost of food and ultimately lead to economic, social, and political instability. The end result may be chaos and civil war.

A solution to this problem is to transition to an energy system that is sustainable – one in which energy is at less risk of being in long-term short supply. Such a system is one that consists of **clean, renewable energy**, which is energy that is replenished by the wind, the water, and the sun. Solutions that do not solve this problem are fossil-fuel power plants, with or without carbon capture, and almost all nuclear power plants, because they rely on fuels that will disappear over time.

1.3.2 Energy Insecurity Due to Reliance on Centralized Power Plants and Oil Refineries

A second type of energy insecurity is the risk of power loss due to a reliance on large, centralized electric power plants and oil refineries. If a city or an island relies on centralized power plants, and one or more plants or the transmission system goes down, power to a large portion of the city or island may be unavailable for an indeterminate period. Such an event can result from severe weather, a power-plant failure, or terrorism. An accidental fire or act of terrorism at an oil refinery or gas storage facility can similarly cause a disruption in local and regional oil and gas supplies.

For example, a September 14, 2019, terrorist attack on two Saudi Arabian oil processing facilities knocked out the production of 5 million barrels of oil per day, or 5 percent of the world's and half of Saudi Arabia's daily oil production. Oil and gas refineries and storage facilities worldwide are continuously at risk of being attacked, and many become targets during conflict. Although decentralized power generation and storage facilities provided by WWS do not decrease the risk of attack to zero, they decrease the risk significantly because of the difficulty in taking down hundreds to thousands of smaller individual units rather than one or two larger ones.

> **Transition highlight**
>
> On September 18, 2017, Hurricane Maria hit Puerto Rico and knocked out power to its 1.5 million people for almost 11 months. The hurricane toppled 80 percent of the island's utility poles and transmission lines. With ten oil-fired power plants, two natural gas plants, and one coal plant, the island's energy supply was all but wiped out by the loss of transmission. The long delay in restoring power to individual homes and businesses occurred because of the need to rebuild most of the transmission system. A more distributed energy system with rooftop solar photovoltaics (PV), distributed onshore and offshore wind turbines, and local battery storage would have allowed hospitals, fire stations, and homes to maintain at least partial power during the entire blackout period and would have reduced the time required to restore power to most customers. In fact, in early 2019, the main utility in Puerto Rico proposed to divide the island into eight interconnected microgrids dominated by solar and batteries. If one microgrid goes down, the other seven will still function. On April 11, 2019, Puerto Rico went even further and passed a law to go to 100 percent renewable electricity by 2050.

Another problem with large, centralized power plants is that they do not serve the 940 million people worldwide without access to electricity,[16] and they poorly serve another 2.6 billion people who have access to only dirty solid fuels (dung, wood, crop residues, charcoal, and coal) for home cooking and heating.[3] Burning solid fuels fills homes with smoke that causes short- and long-term illness to hundreds of millions of people and death to 2.6 million people worldwide each year.[3] Similarly, centralized power plants cannot provide power to remote military bases. Those bases obtain their electricity from diesel transported long distance and used in diesel generators. For example, in 2009, 7 liters of diesel fuel were burned during the transport of each liter of diesel used to produce electricity in U.S. military bases in Afghanistan.[17] Many soldiers died during the transport of the fuel.

Because WWS technologies are largely **distributed** (decentralized), it is possible to use them in microgrids to reduce this lack of access to electricity. A **microgrid** is an isolated grid that provides power to an individual building, hospital complex, community, or military base. A microgrid may either be far from a larger grid or wired to a larger grid but disconnected from it. A WWS microgrid consists of any combination of solar PV panels, wind turbines, batteries, other types of electricity storage, heat pumps, hydrogen fuel cells for electricity and heat, vehicle chargers, and energy-efficient appliances. Electricity in a microgrid may also be used to purify wastewater, desalinate salty water, and/or grow food in a container farm or a greenhouse.[18] When used in a microgrid, WWS can bring electricity to people without previous access to it.

In sum, a transition to WWS facilitates the creation of microgrids and results in the use of more distributed energy sources. Both factors reduce the chance that severe weather, power-plant failure, or terrorism will deny people energy. Fossil-fuel power plants, with or without carbon capture, and nuclear power plants do not solve this insecurity problem because these plants are large and centralized. In addition, fossil fuels almost always require the import of fuel to a region. With a clean, renewable energy microgrid, this problem is eliminated since all energy is produced locally from natural sources, namely wind, water, and sunlight.

1.3.3   Energy Insecurity Due to Reliance on Fuel Supplies Subject to Human Intervention

A third type of energy insecurity is the risk associated with fuel supplies that can be manipulated or fluctuate substantially in price. Such risks often arise when one country relies on another country to supply its energy. For example, many countries, particularly island countries, must import coal, oil, and/or natural gas to run their energy system. Similarly, prior to the 2022 war in the

Ukraine, over 40 percent of the European Union's natural gas was imported from Russia. During the war, bans placed on Russian fuel decreased the flow substantially. Japan imports over 75 percent of its oil, primarily from the Middle East. Israel imports over 90 percent of its oil, primarily from Azerbaijan and Kazakhstan. Importing fuel not only results in higher fuel prices, but also creates reliance of one country on another. This reliance may be tested in times of international conflict. In some cases, a country that controls the energy may withhold it through a ban, an embargo, or price manipulation, or just may not be able to supply it anymore. Similarly, fossil-fuel and uranium fuel supplies, even within a country, can be held up by a labor dispute or civil war.[19]

Fossil-fuel power plants, with or without carbon capture, and nuclear power plants are particularly prone to this problem because they rely on fuels that must be supplied continuously, either from across country borders or from within the country. In many cases, especially for island countries, the fuels must be transported long distance.

A clean, renewable WWS energy system built within a country avoids this type of energy insecurity. This is mainly because WWS requires no mined fuels (oil, natural gas, coal, or uranium) to run. Instead, WWS relies only on natural energy sources. Eliminating mined fuels eliminates the energy insecurity associated with them.

Although a country that supplies 100 percent of its own energy with WWS minimizes the risk of energy insecurity due to international conflict and price manipulation, a benefit arises when adjacent countries trade WWS electricity between each other. Such trading, in the absence of conflict, reduces the overall cost of energy and improves the reliability of the overall energy system.

1.3.4 Energy Insecurity Due to Fuels That Have Mining, Pollution, or Catastrophic Risk

A fourth type of energy insecurity is the risk associated with byproducts of energy use. The perpetual mining of fossil fuels and uranium causes health damage to miners and major environmental degradation. For example, underground coal mining results in black lung disease to many miners. Underground uranium mining results in high cancer rates from the decay products of radon. In addition, plants and vehicles that burn fossil fuels produce air pollution that kills millions of people worldwide each year. Nuclear power plants produce radioactive waste that must be stored for hundreds of thousands of years. Nuclear plants also run the risk of a reactor core meltdown. The historic spread of nuclear energy to dozens of countries has also contributed to the proliferation of nuclear weapons in several of these countries.

A transition to clean, renewable energy avoids these risks to health, the environment, and public safety. The continued use of fossil fuels, with or without carbon capture, and of nuclear power, prolongs these energy security problems.

2 WWS SOLUTIONS FOR ELECTRICITY GENERATION

The solution to air pollution, global warming, and energy insecurity is, in theory, simple and straightforward: Electrify or provide direct heat for all energy; obtain the electricity and heat from only wind, water, and solar sources; store energy, transmit electricity over long distance; and reduce energy use. This chapter first explores the main components of a wind–water–solar system, then focuses on the WWS electricity-generating technologies that will replace traditional energy sources, thereby eliminating all global anthropogenic emissions from such energy sources.

2.1   Components of a WWS System

Figure 2.1 summarizes the main components of a 100 percent wind–water–solar energy, storage, transmission, and equipment system that maintains grid stability. It includes WWS electricity and heat generation; hydrogen generation; electricity, heat, cold, and hydrogen storage; transmission and distribution; energy efficiency; and appliances and machines that use WWS electricity.

What is meant by electrifying or providing direct heat for everything? Most all energy worldwide is currently used for electricity, transportation, heating and cooling of buildings, and industry. In a 100 percent WWS world, all modes of transportation will be converted

WWS Generation

WWS electricity generation
Onshore/offshore wind
Rooftop/utility photovoltaics
Concentrated solar power
Geothermal electricity
Hydroelectricity
Tidal and wave electricity

WWS heat generation
Solar heat/CSP steam
Geothermal heat

WWS Grid

Transmission/distribution
AC/HVAC/HVDC lines
Distribution lines
Grid management
Software
Demand response

WWS Storage

Electricity storage
Batteries
CSP storage
Pumped hydro storage
Hydropower reservoirs
Flywheels
Compressed air
Gravitational storage

District heat storage
Water tanks
Boreholes
Water pits
Aquifers

District cold storage
Water tanks
Ice
Aquifers

Building heat storage
Water tanks
Thermal mass

Hydrogen storage
Hydrogen storage tanks

WWS Equipment

Building and district air/water heating
Electric heat pumps

Building and district cooling
Electric heat pumps

Industrial heat
Arc/induction/resistance furnaces
Dielectric and electron beam heaters
Heat pumps/CSP steam

Hydrogen generation/compression
Electrolyzers/compressors

Transportation vehicles
Battery-electric
Hydrogen fuel-cell

Some appliances/machines
Induction cooktops
Electric leaf blowers/lawnmowers
Heat pump dryers

Efficiency/reduced energy use
Insulate/weatherize buildings
LED lights/efficient appliances
Telecommute/public transport

Figure 2.1 Main generation, transmission, storage, and use components of a 100 percent WWS system to power the world for all purposes. CSP is concentrated solar power, AC is alternating current electricity, HVAC is high-voltage alternating current electricity, HVDC is high-voltage direct current electricity, and LED is light-emitting diode.

to either battery-electric vehicles or hydrogen fuel-cell vehicles, where the hydrogen is produced from WWS electricity (green hydrogen). Electric heat pumps will be used for most air and water heating and air conditioning in buildings. Heat from geothermal reservoirs and sunlight will provide additional air and water heat for buildings. A portion of heating and cooling will come from centralized facilities and be distributed through water pipes to buildings. The remaining heating and cooling will be produced in buildings themselves.

For high-temperature industrial processes, existing electricity-based technologies, such as arc furnaces, induction furnaces, resistance furnaces, and dielectric heaters, will be used to create high-temperature heat.

Energy use in general will be reduced by capturing and recycling waste heat and cold, improving insulation, using more energy-efficient appliances, and creating more pedestrian- and bike-friendly cities.

All the electricity and direct heat in this new paradigm will be powered with WWS sources. Energy not used right away will be stored as electricity, heat, cold, or hydrogen. Electricity will also be transmitted from where it is produced to where it is needed through short- and long-distance electricity transmission lines. In cities, some heat and cold will be transported by hot and cold water pipes. The WWS energy generation technologies for each city, state, and country will include a combination of onshore and offshore wind turbines, solar photovoltaics on rooftops and in power plants, concentrated solar power plants, geothermal plants, conventional and run-of-the-river hydroelectric power plants, tidal and ocean current devices, and wave devices.

Types of storage will include electricity, heat, cold, and hydrogen storage. Major electricity storage options include pumped hydropower storage, existing hydroelectric dams, CSP coupled with thermal energy storage, batteries, flywheels, compressed air storage, and gravitational storage with solid masses. Major heat storage media will include water, soil, and heat-absorbing materials. Major cold storage media will include water and ice. Hydrogen, a form of electricity storage, will be used primarily for long-distance, heavy transport; steel production; and microgrids. Hydrogen will be produced by splitting water with WWS electricity (electrolysis). In some systems, storage will be co-located with energy generation to reduce cost. For example, batteries will often be co-located with residential rooftop solar PV systems to reduce the use of grid electricity when electricity prices are high. Reducing the use of grid electricity also reduces the occurrence of wildfires, which are often caused by transmission line sparks.

WWS electricity-generating technologies are generally defined in terms of their nameplate capacity. **Nameplate capacity** (also called rated capacity, generating capacity, or plant capacity) is the maximum instantaneous discharge rate of electricity from an electricity-producing machine's generator, as determined by the manufacturer of the machine. Whereas a **motor** converts electricity to mechanical motion, a **generator** is just a motor running in reverse, converting mechanical motion to electricity. Nameplate capacities are given in units of power. The base unit of power is the **watt**. When 1 watt of power is produced, 1 **joule** of energy is created (discharged) by an electricity generator per second. Thus, the nameplate capacity of a wind turbine is the rate of energy discharge from the turbine's generator. In other words, a wind turbine that has a nameplate capacity of 1 kilowatt (1,000 watts), can discharge no more than 1,000 joules of energy per second in the form of electricity from its generator.

Energy storage is similarly defined in terms of power and energy. The **peak charge or discharge rate of storage** is the maximum power (rate of change of energy) into or out of storage, respectively. The **peak storage capacity** is the maximum energy that can be stored and equals the peak discharge rate multiplied by the number of hours of storage at the peak discharge rate. Thus, energy stored in a battery is akin to water stored in a reservoir.

2.2   Onshore and Offshore Wind Electricity

Wind turbines convert the energy in the wind into electricity. The energy that arises due to the movement of air or water is **kinetic energy**. In most wind turbines, a slow-turning turbine blade spins a shaft connected to a gearbox. Progressively smaller gears in the gearbox convert the slow-spinning motion (3 to 20 rotations per minute for modern turbines) to faster-spinning motion (750 to 3,600 rotations per minute), just like shifting from the big gear to smaller gears on a bicycle allows one to pedal faster. A fast spinning motion is needed to convert mechanical energy to electrical energy in a wind turbine's generator.

Some modern wind turbines, called **direct-drive** turbines, are gearless, with the shaft connected directly to the generator. To compensate for the slow spin rate of a gearless turbine's shaft within the generator, the generator must be larger and heavier than it is with a geared turbine. However, because direct-drive turbines avoid the use of a gearbox, they are simpler, require less maintenance, and produce less noise than do geared turbines. Because each has advantages, both direct-drive and geared turbines are still manufactured today.

The **hub height** of a wind turbine is the height above the ground or ocean surface of the axis that the turbine spins around. The power output of a wind turbine increases with increasing turbine hub height because wind speeds generally increase with increasing height above the ground or ocean surface in the lower atmosphere. As such, taller turbines capture faster winds.

Wind farms are often located on flat open land, within mountain passes, on ridges, and offshore. Individual turbines to date have ranged in nameplate capacity from less than 1 kilowatt (1,000 watts) to 15 megawatts (million watts).

Small individual wind turbines, with nameplate capacities of 1 to 10 kilowatts, are often used to produce electricity in the backyard of an individual home or within a city street canyon. These local turbines do not produce much total electricity but, depending on wind speed, the amount is often sufficient to offset much of a homeowner's electricity usage.

Onshore wind farms usually contain a few to dozens of mid-sized wind turbines (1 to 8 megawatts in size) to power a part of a town or a city.

Offshore wind farms usually contain a few to dozens of mid- to large-sized turbines (3 to 15 megawatts in size). One particular 12-megawatt turbine, for example, designed for offshore use, has a 150-meter hub height above the ocean surface. Its blade diameter is 220 meters. Thus, the height of its furthest vertical extent is 260 meters, or 80 percent of the height of the Eiffel Tower. It can provide electricity for up to 16,000 households.[20]

Offshore wind turbines have either bottom-fixed foundations or floating foundations. Bottom-fixed foundations are used primarily in water depths down to 50 meters. However, a new design allows bottom-fixed foundations down to a depth of 90 meters.[21] Floating wind turbines avoid the need for a foundation that extends from the water surface to the ocean floor. They have a floating platform secured to the sea floor by cables and can be placed in water of any depth.

High-altitude wind energy capture has also been pursued, although it has not been commercialized to date.

Because the wind does not always blow and, when it does blow, its speed changes uncontrollably over time, winds are variable in nature. As such, wind turbine electricity output also varies with time, and wind is called a **variable WWS resource**. Another term commonly used to describe variability is **intermittency**. However, all energy resources are intermittent owing to scheduled and unscheduled maintenance. Variable resources are those whose energy outputs vary with the weather in addition to being affected by maintenance. Because wind output is variable, combining wind with batteries and other types of electricity storage helps to match variable demand for electricity with supply. This is particularly necessary when wind produces a high percentage of the total electricity on the grid but less so when it produces a low percentage.

2.3   Wave Electricity

Winds passing over water create surface waves. The faster the average wind speed, the longer the wave is sustained, the greater the distance the wave travels, and the taller the wave becomes. **Wave power devices** capture energy from ocean waves and convert that energy into electricity.

Because wave power output varies with time, wave power is also considered a variable energy source. However, wave power output is less variable than is offshore wind power output, even at the same location. The reason is that waves form from winds dragging the water over a long distance. The variability in wind speed at a given location along a wave's path has little impact on the wave, which already has momentum due to upstream winds. As such, waves at a given point represent the impact of winds accumulated over a long distance, whereas winds at the same point are instantaneous and more variable.

One type of wave power device is a free-floating device that bobs up and down with a wave, creating mechanical energy that is converted to electricity in a generator housed inside the device. The electricity is then sent through an underwater transmission cable to shore. Most of the body of the device is submerged under water. With some designs, the entire device is just below the water line so that it cannot be seen from the coast.

Another type of wave device has arms connected to a pier, jetty, breakwater, floating platform, or fixed platform. The other ends of the arms are connected to floaters that rise and fall with the motion of a wave. Because the device is connected to a structure, it is not free-floating in the open ocean. The pumping motion due to the rising and falling motion of the floaters is transmitted, through fluid pressure, to a power station on the platform that it is connected to. The fluid pressure is used to spin a hydraulic turbine, whose rotating motion is converted to electricity in a generator. Because the turbine and generator are on land, this type of wave device is easier to maintain than one completely offshore. However, a shore-based wave device can be placed in fewer locations than can a free-floating device, which can be placed anywhere over the ocean where waves occur.

Despite the development of a variety of wave energy technologies, the wave energy industry is still less mature, in 2022, than are other WWS industries. As a result, wave energy costs are still higher than are costs of other WWS technologies. Nevertheless, like with solar and wind costs, wave energy costs are expected to decline as some prototypes are deployed in increasing quantities.

2.4   Geothermal Electricity and Heat

Geothermal energy is energy extracted from hot water or steam residing below the Earth's surface. In both cases, the heat originates from hot rocks or soil. Rocks and soil are both heated in four ways: by the conductive (atom-by-atom or molecule-by-molecule) transfer of energy from the center of the Earth (which has a temperature of 4,300 degrees Celsius); the conduction of heat from volcanos; the decay of radioactive elements in the Earth's crust; and the downward conduction of sunlight that hits the Earth's surface. Most high-temperature rocks are found near volcanic activity, which generally occurs near tectonic plates. Low-temperature warm rocks and soil can be found anywhere underground. Even in locations where the ground surface is frozen, the soil temperature 6 meters deep is still high enough to warm buildings.

Depending on the temperature of the heat, geothermal energy is used to provide heat for buildings or electricity. Low-temperature heat (below 120 degrees Celsius) is used to heat buildings. It is extracted by ground-source heat pumps; pipes that capture native (naturally occurring) steam or hot water from hot springs; and absorption of soil heat by pipes circulating water or air.

High-temperature heat (120 to 400 degrees Celsius) is used to generate electricity. Geothermal electricity generating plants produce a steady supply of electricity over time. As such, they are baseload plants, which are electricity-generating plants that produce a constant supply of electricity for an extended period to meet loads on the grid. Loads are demands for electricity, such as for powering light bulbs or refrigerators.

Prior to the 1900s, steam and hot water from the Earth were used only to provide heat for buildings, industrial processes, and domestic water. The first use of geothermal heat to produce electricity was by Prince Piero Conti of Larderello, Tuscany, Italy. In 1904, he lit four light bulbs by using high-temperature steam from a geothermal field near his palace to drive a steam engine attached to a generator. A steam engine uses steam to push and pull a piston up and down. This piston is attached to a rotating cylinder. A generator converts the rotating motion to electricity.

In 1911, Conti installed the first geothermal power plant, which had a nameplate capacity of 250 kilowatts. This plant grew to 405 megawatts by 1975. The second electricity-producing plant was built at the Geysers Resort Hotel, California, in 1922. This plant was originally used only to generate electricity for the resort, but it has since been developed to produce a portion of electricity for the state of California.

Today, the three major types of geothermal plants for electricity production are dry steam, flash steam, and binary. Dry and flash steam geothermal plants operate when the geothermal reservoir temperature is 180 to 370 degrees Celsius or higher. In both cases, two boreholes are drilled – one for steam alone (in the case of dry steam) or liquid water plus steam (in the case of flash steam) to flow up, and the second for cool, liquid water to return after it passes through the plant.

In a **dry steam plant**, the pressure of the steam rising up the first borehole powers a turbine, which drives a generator to produce electricity. About 70 percent of the steam recondenses after it passes through a condenser, and the rest is released to the air as water vapor. Because several chemicals, including carbon dioxide, nitric oxide, sulfur dioxide, and hydrogen sulfide, in the geothermal reservoir steam do not recondense along with water vapor, these gases are emitted to the air as well.

In a **flash steam plant**, the liquid water plus steam from the geothermal reservoir enters a water tank held at low pressure, causing some of the water to vaporize ("flash"). The vapor then drives a turbine. About 70 percent of this vapor is recondensed. Again, the remainder escapes with carbon dioxide and other gases. The liquid water is injected back into the ground.

Binary geothermal plants are developed when the geothermal reservoir temperature is 120 to 180 degrees Celsius. Water rising up a borehole is enclosed in a pipe and heats, through a heat exchanger, a low-boiling-point organic fluid, such as isobutane or isopentane. The evaporated organic turns a turbine that powers a generator to produce electricity. Because the water from the reservoir remains in an enclosed pipe when it passes through the power plant, and is reinjected to the reservoir, binary systems emit virtually no carbon dioxide or other pollutants. About 15 percent of geothermal plants today are binary plants.

2.5 Hydroelectricity

Hydroelectricity (hydropower) is produced by water flowing downhill through a water turbine connected to a generator. Most hydropower is produced from water held in a reservoir behind a large dam. This type of hydropower is referred to as large, or conventional, hydropower. Hydropower dams require a reservoir behind them, which results in the flooding of large areas of land. The largest conventional hydropower plant in the world is the Three Gorges Dam on the Yangtze River in China. Its nameplate capacity is 22.5 gigawatts (billion watts). Some other large hydropower plants include the 14-gigawatt Itaipu plant on the Parana River bordering Brazil and Paraguay, and the 10.2-gigawatt Guri plant on the Caroni River in Venezuela.

A growing portion of hydroelectricity is produced by water flowing down a river directly through a turbine or by water that is diverted through pipes and a turbine near the edge of a river before returning to the river. This type of hydroelectricity is **run-of-the-river hydropower**. The advantage of run-of-the-river hydropower over conventional hydropower is that large amounts of land are not flooded behind a dam with the former. As a result, run-of-the-river hydropower is less useful for storage. However, run-of-the river hydro, with a modest storage pond behind it, like with conventional hydropower, can provide electricity within 15 to 30 seconds of a need. The largest run-of-the-river hydropower dam worldwide is the 11.2-gigawatt Belo Monte Dam in Brazil.

A conventional hydropower plant consists of a dam, a water storage reservoir behind the dam, penstocks, sluice gates, a powerhouse, and a downstream water outlet. A **penstock** is a pipe, channel, or tunnel through which water flows from the storage reservoir to a water turbine. A **sluice gate** is a gate to stop or control water flow between the storage reservoir and the penstock. A **powerhouse** is a building containing water turbines, generators, and power transmission cables. When water passes through a **water turbine** in the powerhouse, the turbine's blades spin, rotating a metal shaft connected to the generator, which converts the mechanical rotating motion into electricity.

A run-of-the-river hydropower facility consists of a water turbine and generator, but no reservoir, except for a small holding pond in some cases, and no pipes, except when the water is diverted a short distance away from the river, through the turbine, and back to the river.

Only a fraction of the many dams with reservoirs worldwide contain hydroelectric equipment. For example, the United States has about 84,000 dams, but only 2,500 (about 3 percent) of these have hydroelectric power associated with them.

A hydropower plant can be run as a baseload plant, a load-following plant, or a peaking power plant. Whereas a baseload power plant produces a constant supply of electricity for an extended time, a **load-following plant** runs continuously but ramps its power production up and down to meet 5- to 15-minute average changes in demand on the grid. Hydropower plants, when run as load-following plants, run continuously and adjust their output every 5 to 15 minutes to meet such changes. A **peaking power plant** generates electricity within seconds to a few minutes to meet specific peaks in electricity demand that other electricity sources, such as wind turbines and solar panels, cannot meet immediately. Some power plants, such as some natural gas plants, serve no other purpose but to meet peaks in demand. A hydropower plant, on the other hand, is flexible enough that it can run as a baseload, load-following, or peaking plant as needed. However, a hydropower plant's output is often limited by competing uses for the water that it holds and by how much water it can release downstream at a given time.

In a hydropower plant, neither the powerhouse nor the penstock needs to be located inside the dam. The penstock can be built to channel water around the dam to a powerhouse in front of or to the side of the dam. This may be desirable in cases where turbines are added to a hydropower plant, years after the dam is built, in order to increase the plant's peak discharge rate (nameplate capacity) while keeping the annual average water stored in the reservoir constant. Adding turbines is useful for plants that are run in load-following or peaking mode. The addition of turbines to a hydropower plant is called **uprating**. Uprating a hydropower plant may be "*one of the most immediate, cost-effective, and environmentally acceptable means of developing additional electric power*".[22]

The average power output from a hydropower facility is limited not only by the nameplate capacity of its generators but also by the annual average amount of water available in the reservoir to run through the turbines. A measure of the practical average power output of a hydropower facility that accounts for both the water availability and the turbine nameplate capacity is the installed capacity.

Installed capacity for hydropower is the smaller of the average power produced by available water in a hydropower reservoir and the nameplate capacity of turbines in the hydropower plant itself.[23,24] For other types of power-producing technologies, such as wind turbines and solar panels, the installed capacity equals the nameplate capacity of the technology.

2.6 Tidal and Ocean Current Electricity

Tidal currents (tides) are back-and-forth currents in the ocean caused by the rise and fall of the ocean surface due to the gravitational attraction among the Earth, the moon, and the sun. The rising and sinking motion of the ocean surface forces water below the surface to move horizontally as a current. Because tides run about 6 hours in one direction before switching directions for another 6 hours, they are fairly predictable.

An ocean current is a continuous flow (in one direction) of seawater driven primarily by winds or by temperature and salinity gradients in the ocean. The Gulf Stream current, for example, is a warm, fast-flowing current driven by winds. It runs from the Gulf of Mexico, past the tip of Florida, up the U.S. coast, to the Newfoundland coast, then to the North Atlantic ocean, where it splits into two other currents. Ocean currents, in fact, run along all major ocean coastlines in the North and South Pacific Oceans, the North and South Atlantic Oceans, and the Indian Ocean. For example, the current running along the west coast of North America from Washington State down to Southern California is the California Current. The current running northward along the west coast of South America is the Humboldt (Peru) Current.

A tidal turbine captures the kinetic energy of an ebbing and flowing tidal current in the ocean, the continuous flow of an ocean current, or a river current. Because both tidal and ocean currents are steady, tidal turbines usually supply baseload (constant) electricity to the grid when they are not undergoing maintenance.

Tidal turbines can be mounted on the sea floor or hang under a floating platform. They can also be placed in rivers to capture the energy of continuously flowing river water or of tidal water that ebbs and flows in a river. For example, tides from the Atlantic Ocean flow through Chesapeake Bay, near Virginia, then up the Potomac River, past Washington D.C. and Maryland, all the way up to Harper's Ferry, West Virginia, a land-locked U.S. state.

Like a wind turbine, a tidal turbine consists of a blade that spins a turbine, which provides rotational energy to a generator. The generator converts the rotational energy to electrical energy that is transmitted to shore. A tidal turbine's rotor, which lies under water, may be fully exposed to the water or placed within a narrowing duct that directs water toward it.

One type of tidal turbine, called a tidal kite, has 12-meter-wide wings and is shaped like an airplane. It is tethered to the sea floor and steered autonomously in a predetermined pattern. Water passes through its blades 40 meters underwater, causing the blades to spin. The rotation produces up to 1.5 megawatts of electricity in a generator housed in the device. The electricity is sent through a cable to shore, where its first use will be to power part of the Faroe Islands.[25]

2.7 Solar Photovoltaic Electricity

A **solar photovoltaic** panel consists of an array of cells containing a material that converts sunlight into direct current (DC) electricity. An inverter then converts the DC electricity into alternating current (AC) electricity, which may be used immediately in a building, or put in local storage, or sent to the electricity grid for others to use. Photovoltaics are often mounted just above the ground in large (utility-scale) power plants. They are also mounted on or built into the structure of roofs or walls of buildings. They can further be mounted on carports, parking lots, and parking structures, and built into the sides of cars, trucks, and buses. Transparent PVs have been developed to cover windows, but these have not yet been commercialized.[26] Solar PV arrays that are thin and flexible enough to be rolled up and transported have been commercialized.[27] Such panels are useful for disaster relief, for when grid electricity is not available, for remote military bases that must be moved regularly, and for communities that need a rapid but non-permanent source of electricity. Flexible panels can also be installed on a vehicle and used to provide electricity for it.[28] In utility-scale plants, PV panels are mounted either at a fixed tilt or on trackers that rotate to follow the sun. For most other applications, they are mounted at a fixed tilt relative to the ground.

Transition highlight
Solar PV systems have been built to float on lake or ocean surfaces or to rest over canals. During August 2021, Indonesia began construction of southeast Asia's largest floating solar farm (145-megawatt nameplate capacity). The country has potential for 28 gigawatts of floating solar PV among 375 lakes and reservoirs. A 3.6-kilometer stretch of the Narmada irrigation canal in Vadodara, India, for example, is covered with 33,816 PV panels, generating 10 megawatts of peak power. The system not only saves cropland from being covered with solar, but it also reduces water evaporation from the canal.[29]

The presence of clouds, shadow-casting buildings or trees, snow, desert dust, air pollution, or very high temperatures can reduce the efficiency of a solar PV panel. Rainfall naturally cleans panels. However, during long periods of no rain, PV efficiencies in moderately polluted locations can drop by 10 to 20 percent. In areas with substantial wind-blown desert dust, losses can be up to 40 percent.[30]

When the sun is behind a tree, a thin cloud can enhance sunlight incident upon a panel by bouncing light from the cloud, around the tree, onto the panel. Similarly, light can bounce off of snow to a panel, increasing panel output. Low temperatures can also increase panel output compared with high temperatures.

Because the sun disappears during the night and is often blocked by clouds, air pollution, or obstacles during the day, sunlight is a variable energy source, and solar PV electricity is an intermittent source of electricity. However, solar PV can be combined with batteries or other types of electricity storage to provide steadier (baseload) output and, in many situations, load-following or peaking output.

2.8   Concentrated Solar Power

With **concentrated solar power**, mirrors or reflective lenses focus (concentrate) sunlight onto a collector containing a fluid to heat the fluid to a high temperature. The heated fluid (usually synthetic oil or molten salt) is then used to boil water. The resulting steam passes through a steam turbine connected to a generator, which produces electricity. Up to 30 percent of the energy in the heat is converted to electricity.

Like solar PV, CSP is a variable and intermittent electricity source. However, some types of CSP generators allow the heat to be stored for many hours so that CSP electricity can be produced during the night or when heavy clouds are present. Thus, CSP combined with storage can be used to provide baseload, load-following, or peaking power. Solar PV must be connected to batteries or another type of electricity storage to accomplish the same.

A disadvantage of CSP compared with solar PV, though, is that CSP requires direct sunlight to operate effectively, whereas solar PV can use either direct or diffuse sunlight. **Direct sunlight** is the beam of light that comes directly from the sun and hits a solar PV panel or CSP mirror. **Diffuse sunlight** is light that is bounced out of the direct solar beam by gases and particles (including by cloud drops) in the air, but is then redirected by other gases or particles to the solar panel or mirror. Thus, light under a thick blanket of cloud that blocks the direct beam of the sun is almost all diffuse light.

Deserts, which have few clouds, are usually the best places for direct sunlight, thus CSP. Solar photovoltaics can operate anywhere on Earth, including at the North and South Poles, and even if skies are almost always cloudy, because they can convert diffuse and direct light to electricity.

One type of CSP collector is a set of long **parabolic trough** (U-shaped) mirror reflectors, which focus light onto a pipe containing oil. The hot oil flows to a chamber, where the heat is transferred to water, which boils to produce steam that is sent to a steam turbine connected to a generator to produce electricity.

A second type of collector is a **central tower receiver** with a field of mirrors surrounding it. In the central tower, the focused light heats a circulating thermal storage fluid, such as a **molten salt** mix, to 500 to 600 degrees Celsius. The molten mix often consists of sodium nitrate and potassium nitrate. The hot salt flows to a heat exchanger, where the heat is transferred to water, which boils to produce steam that is sent to a steam turbine connected to a generator to produce electricity. After passing through the steam turbine, the steam is piped to a condenser, where it recondenses to a liquid. From there, the cool liquid water is piped back to the heat exchanger to be boiled again. Thus, the water circulates in a closed loop. Meanwhile, after exchanging its heat with water to boil it in the heat exchanger, the molten salt temperature is now down to about 260 degrees Celsius. The salt is then moved to a "cold" holding tank until it is recirculated to the central tower. Thus, the molten salt also circulates, but in a separate closed loop.

In many parabolic trough and central tower CSP plants, some of the hot fluid (oil or molten salt mix) is stored in an insulated thermal storage tank before it is used to boil water in the heat exchanger. The purpose is to delay the boiling of water, and thus electricity production, until nighttime, when no immediate solar electricity from the plant is available.

A third type of CSP technology is a **parabolic dish-shaped** (like a satellite dish) reflector that reflects light onto a receiver while rotating to track the sun. The receiver transfers the heat to hydrogen in a closed loop. The expansion of hydrogen against a piston or turbine produces mechanical power used to run a generator to produce electricity. The power conversion unit is air-cooled; thus, water-cooling is not needed. Parabolic dish CSP is not coupled with thermal energy storage.

CSP plants require either air- or water-cooling. The use of air-cooling, which is desirable in water-constrained locations, reduces overall CSP plant water requirements by 90 percent at a cost of only about 1 to 5 percent less electric power production.[31]

Because the components of CSP plants are made of abundant raw materials, material shortages are not expected to limit the mass production of CSP plants. For example, CSP plants consist primarily of mirrors, receivers, and thermal storage fluid. Mirrors are made mostly of glass with a reflective silver layer on the back of the glass. Receivers are stainless steel tubes with an outer surface that absorbs sunlight and that is surrounded by an outer antireflective glass tube. None of these materials has limited availability.

3 WWS SOLUTIONS FOR ELECTRICITY STORAGE

Hydropower, geothermal, and tidal electricity generators can serve as baseload (constant-output) generators because their output can be held steady for long periods. However, wind, solar PV, and wave outputs vary during a day and seasonally. As such, these electricity sources provide variable output. Given that solar and wind may end up supplying 90 percent or more of all WWS energy generation worldwide, on average, it is important to have electricity storage technologies available to provide backup power when both solar and wind are unavailable. Major electricity storage options include existing hydroelectric dams, pumped hydroelectric storage, batteries, concentrated solar power coupled with thermal energy storage, flywheels, compressed air energy storage, and gravitational storage with solid masses. These technologies are discussed herein.

3.1   Hydroelectric Power Reservoir Storage

Conventional hydroelectric power (hydropower) plants have built-in storage since the water that generates the electricity for them is stored in a reservoir behind a dam. Whereas the water can be drained through a turbine to create electricity as needed, the reservoir can only be charged naturally through rainfall and runoff. Thus, the electricity storage associated with a conventional hydropower plant differs from that of battery storage. A battery can be charged when excess electricity is available and discharged when electricity is needed. A hydropower plant can produce electricity precisely when it is needed, but it cannot control when rain and runoff will refill the reservoir. The only option for controlling recharge is to turn the hydropower plant into a pumped hydropower storage facility.

Hydropower plants can be run as baseload plants, load-following plants, or peaking plants. When they are used for load following and peaking, hydropower plants can meet minute-by-minute changes in electricity demand on the grid faster than can natural gas, coal, or nuclear power plants. The ramp rate of a power plant is the speed at which it ramps up from zero to maximum power. The ramp rate of a hydropower plant is fast. For example, a hydropower turbine can generate full power from no power within 15 to 30 seconds. That is the time required for the water to flow from the sluice gate, after it opens, through the penstock, and to the turbine. Once the water reaches the turbine, the turbine spins and immediately produces electricity in an attached generator. As such, the ramp rate of a hydropower plant is 100 percent in 15 to 30 seconds. In comparison, the ramp rate of a natural gas open cycle turbine is 100 percent in 5 minutes; that of a natural gas combined cycle turbine is 100 percent in 10 to 20 minutes; and those of coal and nuclear plants are both 100 percent in 20 to 100 minutes.[32]

Because hydropower plants can produce electricity within 15 to 30 seconds of a need, they are often used to provide peaking power when no other electricity is available to match power demand with supply on the electricity grid. At a high continuous electricity discharge rate, hydropower plants can meet a high continuous demand for hours to days, depending on the size of the reservoir and the nameplate capacity of installed turbines in the plant. When used only to meet gaps in supply, such plants may sit idle for days to weeks before being used several days in a row for 2 to 3 hours per day.

3.2   Pumped Hydropower Storage

A related type of storage that can discharge its electricity quickly is **pumped hydropower storage**. Pumped hydropower storage is used primarily to fill in short-term gaps in electricity supply, usually on the order of minutes to a day.

Pumped hydropower storage consists of two reservoirs – an upper one and a lower one. The lower one can be the ocean, a natural lake, a human-made lake or reservoir, or a continuously running river. The upper one can be a natural lake or a human-made lake or reservoir. One newly proposed pumped hydropower storage system consists of Lake Mead, the reservoir behind Hoover Dam, and a pumping station 32 kilometers downstream along the Colorado River.[33]

The idea behind pumped hydropower storage is that, when excess electricity is available or when electricity prices are low, a motor pumps water uphill through pipes, from a pumping station in the lower reservoir to the upper reservoir. When electricity is needed, water drains downhill through a water turbine connected to the motor running in reverse as a generator, producing electricity.

The overall efficiency of a pumped hydropower storage system (ratio of electricity delivered to the sum of electricity delivered and electricity used to pump the water uphill) is around 80 percent. The ramp rate for pumped hydropower storage is the same as for conventional hydropower, from zero to 100 percent power in 15 to 30 seconds. As such, pumped hydropower storage ramps up faster than do natural gas plants, which ramp to 100 percent power in 5 to 20 minutes, or nuclear plants, which ramp to 100 percent power in 20 to 100 minutes.

Transition highlight
Excluding hydropower reservoirs, 97 percent of all electricity storage built on Earth prior to 2021 was pumped hydropower storage. Worldwide, about 530,000 potential pumped hydropower sites exist. These sites can store an estimated 22 million gigawatt-hours of energy for the electricity grid. This represents 100 times the energy needed to back up a 100 percent renewable electricity system worldwide.[34]

The largest pumped hydropower facility in the world is the 3.6-gigawatt facility in China's Hebei province. It can store up to 50 gigawatt-hours of producible electricity. At the peak discharge rate of 3.6 gigawatts, that stored electricity can be extracted continuously for about 14 hours.

3.3   Stationary Batteries

Batteries in a 100 percent WWS world will be used primarily for vehicles; electronic devices (e.g., phones, watches, computers, flashlights), moveable equipment (e.g., leaf blowers, lawnmowers, chain-saws), and stationary electricity storage. Batteries are useful because they generate electricity almost immediately on demand. As such, an electric vehicle accelerates to a high speed faster than does an internal combustion engine vehicle. Similarly, batteries can provide needed electricity to the grid much faster than can coal, natural gas, nuclear, or even hydroelectric facilities. Stationary batteries can provide load-following or peaking power for a large electricity grid or an isolated microgrid. Wall-mounted or floor-mounted battery packs are commercially available for homes, commercial buildings, microgrids, and city electric grids.

A **battery** is an electrochemical cell that converts chemical energy into electricity. A **battery pack** is a collection of such cells. Each cell consists of two half-cells divided by a separator. Each half-cell consists of a current collector, an electrode, and an electrolyte solution.

The **current collector** is a metal cap in each half-cell through which electrons move either forward or backward. In both cells, the current collector is attached to an **electrode**. The current collector in one cell is called the positive current collector. In a **lithium battery**, the electrode attached to the positive current collector is often made of lithium cobalt oxide, which is the source of lithium ions (Li^+) that move back and forth between the half-cells. The current collector in the other cell is called the negative current collector, and the electrode attached to it is often made of pure carbon (graphite). The purpose of the graphite is to receive and bond to lithium ions and electrons.

The **electrolyte solution** contains a lithium salt dissolved in an organic solvent. The liquid solution facilitates the passage of lithium ions between the two half-cells. The **separator** allows lithium ions, but not electrons, to pass between the two half-cells.

During battery charging, a charger plugged into the wall is wired to both current collectors. Electrons first flow from the charger to the negative current collector. The build-up of negative charges there induces lithium cobalt oxide in the other half-cell to break apart into a lithium ion, an electron, and cobalt oxide. The lithium ion is then drawn like a magnet from the positive current collector, through the electrolyte solution, and through the separator, to the electrons accumulating at the negative current collector, neutralizing the charge there.

Electrons released at the positive current collector cannot pass through the separator. Instead, they take the path of least resistance by flowing from the positive current collector, through a wire, to the negative current collector, completing the circuit and adding more electrons to that side. At the negative current collector, the electrons combine with lithium ions and graphite to form lithium-graphite, neutralizing the positive charge of the lithium ions. When no more lithium ions flow from the positive to negative current collectors, the battery is fully charged.

During battery discharging to provide power for an electronic device, the device draws electrons from the negative current collector. The electrons pass through the wire to the device. The device consumes some of the electrons. Remaining electrons continue to flow to the positive current collector. The negative charge build-up at the positive current collector induces positively charged lithium ions obtained from the dissociation of lithium-graphite to flow from the negative current collector, through the separator in the electrolyte solution, to the positive current collector, where they recombine with an electron and cobalt oxide to reform lithium cobalt oxide. When all lithium ions have moved from the negative current collector to the positive current collector, the battery is depleted and needs recharging.

Two parameters important for determining the use of batteries in different applications are their **round-trip efficiency** (ratio of electricity delivered to energy stored in the battery) and the number of times they **cycle** (fully discharge plus charge) before the battery is no longer useful. Battery round-trip efficiencies are generally 80 to 90 percent. A typical car battery goes through 500 to 1,000 cycles before needing to be replaced. However, some newer lithium batteries degrade by only 35 percent after 28,000 full charge/discharge cycles over 8 years.[35] If these cycles occurred once daily, the battery would theoretically last over 77 years. However, batteries, even if idle, also degrade, so it is more likely that those cycles need to be performed over 20 to 40 years without significant battery degradation.

Batteries aside from the standard lithium-ion ones just described have been developed. These include the following.

Iron–air batteries. Iron–air batteries are also called reversible rust batteries. These batteries have one cell containing iron oxide (rust) and another containing pure metallic iron. During charging, an electric current is used to break down iron oxide into metallic iron and oxygen. The oxygen is released to the air. During discharging, metallic iron combines with oxygen from the air to reform iron oxide and produce electricity. The main advantage of an iron–air battery is the low cost of the main material (iron) and the system as a whole. The estimated battery pack cost in 2021 is $20 per kilowatt-hour of energy storage,[36] which compares with $100 to $200 per kilowatt-hour for lithium-ion batteries. A **kilowatt-hour** is a unit of energy, just as the joule is a unit of energy. There are 3.6 million joules in a kilowatt-hour of energy. Good things about iron–air batteries are that iron is a safe and durable element, and the batteries can be scaled and recycled easily. However, because iron–air batteries are much larger, for the same storage capacity, than lithium-ion batteries, iron–air batteries are not useful for transportation, only for stationary electricity storage.

Basalt stone batteries. These batteries consist of two insulated steel tanks, each filled with pea-sized basalt stones. One tank is cold and the other, hot. Excess renewable electricity from the grid is used to compress, thus heat, air from the cold tank. The lower the temperature of the cold tank, the less the energy needed to raise the temperature by a fixed amount. This is because compressing hot air takes more energy than compressing cold air. The compressed, hot air is then transferred to the hot tank with an electric heat pump, which then expels cold air to the cold tank. The hot tank heats to 600 degrees Celsius, whereas the cold tank cools to –30 degrees Celsius. Owing to the insulation, the hot and cold tanks can maintain their temperatures for several days. To store more energy, more stones are added to each tank. When electricity is needed, more cold air from the cold tank is compressed and sent to the hot tank, forcing hot air from the hot tank to move to an expander. As air expands and cools in the expander, it flows through an expansion turbine to generate electricity. The cold air is then returned to the cold tank. Because basalt is inexpensive, the cost of this battery is also anticipated to be low. The round-trip efficiency of the battery is 60 percent. However, if some of the heat lost is sent to a district heating system, then the efficiency increases to 90 percent. Currently, a 10-megawatt-hour plant based on this concept is being developed in Denmark.[37,38]

Sodium sulfur batteries. Sodium sulfur batteries contain sodium and sulfur in different half-cells. During discharge, sodium in the sodium half-cell gives up an electron. The electron travels through an external circuit that has an electricity-using device (such as a light bulb) connected to it and then to the sulfur side of the battery. The positive sodium ion simultaneously passes through a membrane to the sulfur side, where it reacts with the sulfur and the electron to form a sodium–sulfur compound. During charging, the reverse process occurs. Sodium sulfur batteries have high energy densities, high charge and discharge efficiencies, and long cycle lives. Because they are made of inexpensive material, such batteries are also economical at large sizes, which makes them ideal for stationary electricity storage. They can also be used for vehicles. In fact, one of the first applications of the sodium sulfur battery was in the 1991 Ford Ecostar demonstration electric vehicle, which never went into commercial production. Sodium sulfur batteries operate at high temperature (300 to 350 degrees Celsius). Research is currently being carried out to allow the batteries to be used at medium temperatures (120 to 300 degrees Celsius) and room temperature.[39,40]

Aluminum-ion batteries. These batteries use aluminum instead of lithium to flow between cells. An aluminum ion, Al^{3+}, carries 3 times as much the charge as does a lithium ion, Li^+. This allows aluminum-ion batteries to be smaller than lithium-ion batteries. However, the higher charge also results in greater interaction of aluminum ions than lithium ions with other chemicals in the battery, reducing the efficiency of aluminum-ion versus lithium-ion batteries. Aluminum-ion batteries are less flammable than are lithium-ion batteries. Recently, an aluminum-ion battery paired with a graphene electrode was found to charge up to 60 times faster than a lithium-ion battery and hold up to 3 times the charge of other aluminum-based battery cells.[41]

Salt water batteries. In these batteries, a concentrated saline solution (salt water or sodium sulfate mixed with water) is used as the electrolyte to conduct ions. These batteries are similar to lithium-ion batteries, except that with salt water batteries, the sodium ion instead of the lithium ion migrates. However, the voltages are one-third those in lithium-ion batteries. Salt water batteries are suitable for stationary power storage and can be run for 10,000 cycles or more. Salt water batteries are also non-flammable, non-explosive, and have no toxic chemicals, so they are easily recycled.

Vanadium flow batteries. These batteries consist of two separated tanks of vanadium dissolved in a liquid electrolyte. The fluid in each tank is pumped through a pipe, past a membrane that separates the pipes. Ions are exchanged from one side of the membrane to the other, and a current simultaneously flows between the two tanks through an external wire. During electricity charging, ions flow in one direction through the pipe. During discharging, they flow in the opposite direction. A flow battery is much larger than is a lithium-ion battery. As such, flow batteries are used only for stationary electric power storage.

3.4 Concentrated Solar Power with Storage

Parabolic trough and central tower CSP plants work by heating a fluid (either oil for a parabolic trough or a molten salt mixture for a central tower), which passes by water in a heat exchanger to boil the water. The resulting steam runs a steam turbine to generate electricity. For both the parabolic trough and central tower plants, the hot fluid can first be stored in an insulated tank before it is sent to the heat exchanger. This allows electricity production to be delayed until nighttime, or when heavy clouds are present, electricity demand is high, or no other WWS electricity source is available. After the steam passes through the turbine, it is routed to a condenser, which liquefies the steam and sends it back to the heat exchanger in a closed loop to be reheated. After the molten salt exchanges its heat with water in the heat exchanger, the molten salt is also cold and is sent to a holding tank before it is passed through either the parabolic trough or the central tower receiver again to reheat.

Although CSP heat can be stored in a tank containing molten salt, the heat can alternatively be stored in a phase-change material. A phase-change material is a material that absorbs a large amount of heat when it melts and releases a large amount of heat when it solidifies. Phase-change materials require less volume, thus they cost less to store heat, than does molten salt.[42]

The storage associated with CSP is usually sized to last up to 15 hours at the peak discharge rate (nameplate capacity) of the CSP generator before the storage is depleted. Storage of up to 15 hours allows for 24 hours per day of electricity production from CSP plus storage.

Transition highlight
The Gemasolar CSP plant in Seville, Spain, first demonstrated 24 hours per day of CSP-with-storage electricity production during July of 2011. The plant has provided electricity continuously (24 hours per day) for stretches of up to 36 days. During cloudy and winter days, a CSP-with-storage plant also produces electricity at night, but for fewer hours.

The capacity factor of a CSP plant is the actual annually averaged power produced by the generator in the plant divided by the maximum possible power produced by the generator (its nameplate capacity). With no storage, the capacity factor of a CSP plant is limited by the instantaneous electricity generated, averaged over a year. Excess heat not used immediately is wasted because the generator can produce only a limited amount of electricity. A typical capacity factor of a CSP plant without storage in a sunny location is around 25 percent. With storage, however, the capacity factor increases to around 65 percent or higher. This is accomplished first by adding more mirrors and molten salt (for a central tower CSP plant) and storing the excess salt after it is heated. During the night, or when sunlight is weak, the stored hot salt is used to boil water. The additional steam is used to produce electricity in the absence of sunlight, increasing the plant's annual electricity output, thus its annual average capacity factor.

The capacity factor of a CSP plant can be increased even further if wasted heat that is not turned into electricity is captured and used to produce heat for industrial processes or used as a source of heat for heat pumps.

Finally, CSP plants with storage can help to keep the electric grid stable. With storage, CSP plants can ramp from zero to 100 percent power production in about 10 minutes,[43] which is faster than the ramp rate of a coal or nuclear plant (20 to 100 minutes) and similar to that of natural gas plants (5 to 20 minutes).[32] As such, CSP with storage can help meet peaks in electricity demand with 100 percent WWS.

3.5   Flywheels

A flywheel is a spinning wheel or disk, usually made of steel or carbon fiber, that rotates around an axis. The axis is perpendicular to the ground, like with a spinning top, so that gravity acts equally on all sides of the spinning wheel.

A flywheel stores energy as rotational energy, then converts that energy to electricity as needed. A flywheel is an electric motor, an energy storage device, and a generator, all in one. When excess electricity is available, it powers an electric motor to rotate the flywheel up to a high speed. Of the energy added to a flywheel, a small portion is used to keep the flywheel rotating. The rest is maintained as stored energy. If the flywheel is run in a vacuum, air resistance is zero, so frictional losses are minimized. Another way to minimize frictional loss is to use an electromagnetic bearing or permanent magnet. This allows the spinning rotor to float.

When electricity is needed, the motor turns into a generator, which converts rotational energy into electricity, with built-in electronics. When electricity is produced, the flywheel slows down but does not stop.

A flywheel can store more and more rotational energy until its rotor shatters. Steel flywheels are limited to around 3,000 rotations per minute. High-energy-density carbon fiber flywheels can spin up to 60,000 rotations per minute. Aside from possible breakage, flywheels require little maintenance and have a long lifetime (about 20 years) with no impact on the environment past the use of materials to build them.

A flywheel accumulates and stores kinetic energy produced by either steady or intermittent electricity over any period of time and releases that energy as electricity at a fast rate over a short period. As such, flywheels are ideal for storing excess electricity from intermittent solar and wind energy on the electric power grid. However, flywheels developed to date have stored only small amounts of energy (e.g., 3 to 25 kilowatt-hours) and have relatively high loss rates (3 percent per hour). As such, the stored electricity must be used quickly.

On the other hand, flywheels can begin to discharge quickly (within 4 milliseconds) and can discharge at a high rate (10 to 100 kilowatts). Thus, two useful applications of a flywheel are to provide short-term peaking power for the electric grid and to charge electric vehicles quickly.[44] For example, a flywheel that has a peak discharge rate of 100 kilowatts can add 20 kilowatt-hours to an electric vehicle in 12 minutes.

3.6 Compressed Air Energy Storage

Compressed air energy storage (CAES) is another technology that can be used to accumulate intermittent renewable electricity in storage over a long period, then to resupply that electricity to the grid when needed. With compressed air storage, excess intermittent electricity is used to compress air. When electricity is needed, the compressed air is expanded in an expansion turbine. The turbine's rotating shaft is connected to a generator, which converts the rotational energy to electricity.

Compressed air can be stored in an underground cavern, a salt dome, an aquifer, or a closed vessel. Although compressed air storage has been studied extensively, only a few large-scale facilities have been built, including one in Germany and two in the United States (Alabama and Texas). The locations for large-scale underground storage facilities are limited geographically.

However, a small, compressed air storage system can be connected to an electricity-producing device, such as a wind turbine. The wind turbine operates normally to produce electricity in a generator. Electricity from the generator at hub height is then used to power a motor that compresses air that is stored in a storage tank. When electricity is needed for the grid, the compressed air is expanded and converted back to electricity in an expander-driven generator. The motor, storage vessel, and second generator are all housed at ground level, below the turbine.[45]

> **Transition highlight**
> Another variation of compressed air storage is the use of excess renewable electricity to cool, then compress, air until it condenses as a liquid. The high-pressure, cold liquid is then stored in a container until electricity is needed. At that point, the air is warmed until it re-evaporates. The expanding air then drives a turbine to generate electricity. The process is made efficient by storing the heat released during condensation and using that heat to help re-evaporate the air. The process does not require special materials, such as lithium used in batteries. A 50-megawatt (million watts) storage facility based on this concept, which was developed by inventor Peter Dearman, is being built in the north of England.[46]

3.7 Gravitational Storage with Solid Masses

A form of electricity storage similar to pumped hydropower storage, except that it uses solid material instead of water, is **gravitational storage with solid masses**. In one version of this storage, excess electricity from the grid is used to power an electric motor in a crane to lift cement blocks against the force of gravity and stack them, one at a time, on a tower. When electricity is needed, the crane grabs each block and slowly drops the block toward the ground. The downward motion uncoils the hoist chain holding the block. The rotating motion during the uncoiling is translated to a rotating motion inside the same electric motor that lifted the block, turning the motor into an electric generator.[47] The electricity produced by the generator is sent to the grid.

Transition highlight
One company that is building a gravitational storage system is using a large crane with multiple arms so that multiple blocks can be lifted or dropped simultaneously.[48] The efficiencies for charging and discharging storage are stated both to be 90.5 percent, resulting in a round-trip efficiency (electricity consumed per unit electricity input) of 82 percent. The one-way efficiencies break down into a crane/pulley efficiency of 96 percent, a gear efficiency of 98 percent, a motor/generator efficiency of 98.2 percent, a variable frequency drive efficiency of 98.7 percent, and a transformer efficiency of 99.3 percent.[49] So long as the crane is holding a block, electricity can be produced almost instantaneously when needed. The company suggests that the cost should be less than that of batteries.[48]

In another version of gravitational storage, excess grid electricity is used in an electric motor to move a train carrying rocks or concrete blocks up a hill. When electricity is needed, the train rolls down the hill, almost instantly producing electricity from a generator (motor running in reverse). The round-trip efficiency is stated to be about 86 percent, similar to that from moving mass vertically.[50]

In a third version of this technology, excess grid electricity is run through a motor to lift sand or gravel in a container up a mountain slope via a cable. Containers of sand or gravel are stored at the top of the mountain. When electricity is needed, the containers are lowered down the slope via the cable. The rotating motion of the coil at the top of the mountain produces electricity in the generator. This type of storage is referred to as **mountain gravity energy storage**.[51] It can store electricity for use on the timescale of seconds, minutes, days, weeks, or months. It is most efficient in locations with tall, steep mountain slopes.

4 WWS SOLUTIONS FOR TRANSPORTATION

Transportation is the conveyance of people, animals, or goods from one place to another. Types of transportation include skateboards, bicycles, motorcycles, passenger vehicles, sport utility vehicles, small trucks, large trucks, semi-trucks, buses, forklifts, cranes, tractors, bulldozers, asphalt pavers, backhoe loaders, cold planers, compactors, excavators, harvesters, graders, off-road vehicles, trains, ferries, motorboats, yachts, ships, helicopters, aircraft, battle tanks, infantry fighting vehicles, armored personnel carriers, armored combat support vehicles, mine-protected vehicles, light armored vehicles, light utility vehicles, amphibious vehicles, and more. All of these, except skateboards and bicycles, currently run primarily by burning gasoline, diesel, biodiesel, methanol, ethanol, liquefied natural gas, bunker fuel, or jet fuel in an internal combustion engine. These fuels are all derivatives of fossil fuels or biofuels. The cleanest and most efficient method of replacing these fossil-fuel and biofuel vehicles is to convert them to battery-electric vehicles or hydrogen fuel-cell vehicles. This chapter discusses these two WWS solutions.

4.1   Battery-Electric Vehicles

Battery-electric vehicles store electricity in batteries then draw power from the batteries to run an electric motor that moves the vehicle. Battery-electric vehicles have zero tailpipe emissions. However, in many countries today, a good portion of electricity used for battery-electric vehicles comes from fossil-fuel power plants. In a 100 percent WWS world, though, all electricity for battery-electric vehicles will come from zero-emission sources.

Some argue that if electricity for a battery-electric vehicle comes from a polluting source, then the battery-electric vehicle is just as bad for health as a fossil-fuel vehicle, which emits pollution continuously from its tailpipe. However, that is not the case. The reason is that the intake fraction of pollution from street traffic is 15 to 30 times that of pollution from power plants.[52] The **intake fraction** is the ratio of the mass of pollutant inhaled by a population to the mass of pollutant emitted by a source. In other words, people breathe in a much greater fraction of vehicle exhaust than they do of power-plant exhaust. As a result, taking pollution off the street and moving it to a power plant reduces health impacts significantly. With 100 percent WWS, though, all power-plant emissions will also be eliminated.

In addition to producing tailpipe emissions, fossil-fuel vehicles produce emissions from the mining, transporting, and processing of oil to produce gasoline, diesel, jet fuel, and bunker fuel. In a WWS world, all such emissions are eliminated.

Fossil-fuel and biofuel vehicles also emit particles during the use of brake pads. In such vehicles, the car slows when a brake pedal is pushed. This pushes a brake against a turning tire disc, reducing the rotation rate of the tire and slowing the vehicle. The resulting friction creates heat and causes some particles to break off the brake pads and float into the air. Electric vehicles eliminate most brake pad emissions because they use mostly regenerative braking rather than brake pads, although drivers still occasionally apply brake pads.

Regenerative braking works as follows: to slow a car, the driver's foot is taken off the accelerator pedal. Doing so causes rotational energy from the rotating wheels to be fed into a generator (the motor running in reverse) to "regenerate" electricity, which is then stored in the vehicle's batteries. Thus, regenerative braking, which converts rotational energy into electricity, slows the car, avoiding the need for the use of brake pads and thus avoiding brake pad particle emissions.

Regenerative braking was invented in 1886 by the Sprague Electric Railway & Motor Company, founded in 1884 by **Frank Sprague** (1857 to 1934). Sprague had worked for Thomas Edison starting in 1883, but because Sprague wanted to develop motors, which Edison was less interested in, he left Edison's company in 1884 to start his own. Sprague's first application of regenerative braking was in electric streetcars that his company put in place in Richmond, Virginia, in 1888. Sprague's company also developed regenerative braking for elevators.

All vehicles require energy for mining materials for the vehicle, manufacturing the vehicle, and recycling the vehicle at the end of its life. In a WWS world, such energy will come from clean, renewable sources with zero emissions.

Also, all vehicles today emit particles due to tire wear. However, such particles are mostly larger and fewer in number than are fossil-fuel combustion particles. Because tire particles are generally large, most drop out of the air faster and do not penetrate so deep into people's lungs as combustion particles do.

In sum, transitioning to battery-electric vehicles substantially reduces air pollution and its health problems. Providing the electricity needed to manufacture the car with WWS reduces pollution even more.

4.1.1   Efficiency of Battery-Electric Vehicles

Another advantage of today's battery-electric vehicles compared with fossil-fuel vehicles is that battery-electric vehicles can travel 3 to 5 times further per unit energy input obtained from the plug outlet than a fossil-fuel vehicle can travel per unit energy in gasoline.

For example, only 17 to 20 percent of the energy in gasoline goes toward moving a gasoline passenger vehicle. This is the **tank-to-wheel efficiency** of the vehicle. The rest of the energy (80 to 83 percent) is lost as waste heat.

On the other hand, 64 to 89 percent of the electricity from a plug outlet moves a battery-electric passenger vehicle, and the rest is waste heat. This is the **plug-to-wheel efficiency** of a battery-electric vehicle. This efficiency accounts for the efficiency loss of charging the vehicle.

As such, a battery-electric vehicle requires much less energy input than does a fossil-fuel vehicle. Alternatively, a conversion from fossil-fuel vehicles to battery-electric vehicles reduces energy demand for transportation fuel by a factor of 3 to 5.

The plug-to-wheel efficiency of a battery-electric vehicle is determined as follows. First, a permanent magnet electric motor has an efficiency of 89 to 96 percent. An induction electric motor has an efficiency of 84 to 94 percent.[53] Thus, the range of efficiencies of electric car motors is 84 to 96 percent. Whereas the Tesla Model S and Model X use an induction motor, for example, the Tesla Model 3 uses a permanent magnet motor.

In addition, efficiency losses occur due to converting electricity from the grid to chemical energy in a battery. Such vehicle charging losses can range from 4 percent to 20 percent of the electricity going into the battery, depending on the current and voltage used to charge the vehicle. Another 1 to 2 percent of energy is lost during conversion of the battery's chemical energy to DC electricity. In addition, 2 to 3 percent is lost with the combination of converting DC to AC electricity in an inverter, adjusting the voltage for use in the motor, and using power electronic controls in the vehicle. Inverter losses are only 1 percent of energy when the inverter uses silicon carbide as the semiconductor material.[54]

Accounting for all losses gives the overall plug-to-wheel efficiency of a battery-electric passenger vehicle as 64 to 89 percent, with an average of 77 percent.

A battery-electric vehicle can increase its efficiency, thus range, by using regenerative braking. As discussed, regenerative braking converts rotational energy into electricity, slowing the vehicle.

Transition highlight

Regenerative braking can not only extend battery-electric vehicle range substantially, but it can also eliminate, in at least one special case, the need to recharge an electric vehicle entirely. The largest stand-alone electric vehicle in the world in 2019 was an enormous electric dump truck operating in a quarry in Biel, Switzerland. The empty truck used electricity to climb from the bottom to the top of the quarry. At the top, it was filled with ore, more than doubling the truck's weight compared with its uphill trip. The additional weight created so much kinetic energy that the electricity produced from regenerative braking during the downhill trip exceeded the electricity consumed during the uphill trip. As such, the truck did not need recharging during its daily operation.[55]

In a WWS world, battery electricity will be used not only to power light-duty vehicles, but also short- and medium-range semi-trucks, short-range aircraft, some military vehicles, and short-distance boats and ships. The history of battery-electric transportation is discussed next.

4.1.2 History of Battery-Electric Transportation

Today, battery-electric vehicles are being developed for ground, water, and air transport. Here, the evolution of battery-electric vehicles is explored for each of these transport types.

4.1.2.1 Ground Vehicles

Passenger Vehicles, Pickup Trucks, Semi-trucks, and Buses
Scotland's **Robert Anderson** built the world's first battery-electric vehicle, which was also the world's first horseless carriage, sometime between 1832 and 1839. To accomplish this feat, he affixed a battery and a motor to a carriage. At the time, batteries were not chargeable, so the carriage's battery needed to be replaced after each use. Thus, Anderson also developed the first swappable battery system. Rechargeable battery-electric vehicles were not possible until 1859 when the French physicist Gaston Planté (1834 to 1889) invented the first chargeable battery, which was made of lead.

Around 1884 in England, Thomas Parker invented several prototype electric cars with chargeable batteries. In 1888, William Morrison of Des Moines, Iowa, showed off an electric carriage in the city parade. The carriage had 24 batteries, a top speed of 32.2 kilometers per hour, and an overall range of 81 kilometers.[56]

In 1894, Pedro Salom and Henry G. Morris of Philadelphia patented several battery-electric streetcars and boats. One battery-electric carriage they developed, called the Electrobat, travelled 40 kilometers at a top speed of 32 kilometers per hour. Salom and Morris turned this idea into a startup, which they sold to Isaac L. Rice, who then started the Electric Vehicle Company in New Jersey. By the early 1900s, the company had produced more than 600 electric taxi cabs used in New York City, Boston, Baltimore, and other cities. In New York City, the cab batteries were replaced at a battery-swapping station. Conflicts among investors caused the company to fold in 1907.

In 1898, Ransom Olds, who had founded the Olds Motor Vehicle Company in 1897, decided to manufacture electric cars, which were less complicated and more reliable than were gasoline cars of the day that he was also producing. Olds retrofitted a part of his Detroit, Michigan, factory to mass produce battery-electric vehicles. However, on March 9, 1901, a worker's mistake caused a fire that destroyed the entire factory and Olds' electric car dream. Only one electric car survived the blaze.

Nevertheless, by 1900, about one-third of all vehicles on the road were battery-electric.[57] The public wanted electric vehicles because they were quiet, easy to drive, required few repairs, and did not produce pollution. Unfortunately, once Henry Ford started producing, in 1908, the Model T, which cost one-third as much as a battery-electric vehicle, the electric vehicle market collapsed. The expansion of gas stations and the greater range and top speed of gasoline vehicles through the 1920s sealed the fate of battery-electric vehicles for the next several decades.

It was not until 1973 that General Motors developed a new-generation tiny two-seat urban electric car prototype. The car was never sold commercially. Only following the Arab Oil Embargo of October 1973 to March 1974 did interest in funding the research and development of battery-electric vehicles revive. This resulted in the American Motor Company producing electric delivery jeeps for a U.S. Postal Service test program in 1975. These vehicles, though, were still limited in their range (64 kilometers) and top speed (72 kilometers per hour). More work was needed. In 1976, the U.S. Department of Energy began to research and develop electric and hybrid vehicles.

In the early 1990s, interest in battery-electric vehicles rose again owing to tougher emission standards for internal combustion engine vehicles enacted under the 1990 U.S. Clean Air Act Amendment and by the California Air Resources Board. As a consequence, General Motors developed the EV1 electric car, which had a range of 129 kilometers and good acceleration. Because the cost to produce it was high, the EV1 was never made available commercially, and General Motors scrapped its development in 2001.

In 1997, Toyota produced the first mass-produced hybrid gasoline-electric vehicle, the Prius. The vehicle was released worldwide in 2000. In 1999, Honda also released a gasoline–electric hybrid, the Insight. The commercial availability of both vehicles motivated the research and development of a new generation of pure battery-electric vehicles.

Arguably the most important event in the history of battery-electric transportation was the announcement in 2006 by a startup company in California's Silicon Valley, **Tesla Motors**, that it would produce an all-electric luxury sports car. The car would have a range of over 390 kilometers on a single charge.

Tesla Motors was founded on July 1, 2003, in San Carlos, California, by **Martin Eberhard** and **Marc Tarpenning**. **Elon Musk** was the major initial funder of Tesla Motors. **Ian Wright** and **J. B. Straubel** joined Tesla during the next few months. All five are considered cofounders of the company. Elon Musk took over as chief executive officer of the company during October 2008. Straubel was the chief technical officer. Tesla Motors was named after inventor **Nikola Tesla** (1856 to 1943), who designed and improved electric motors, electric generators, and the AC electricity grid.

Tesla Motors (later changed to Tesla, Inc.) focused on an all-electric car rather than on a hybrid car because of the simplicity and reliability of using only electricity rather than electricity plus gasoline. They built a sports car first to change the perception in the public's mind about electric cars. At the time, the public had a poor image of electric vehicles, associating them with short range, low speed, and clunkiness.

The sports car built was the Tesla Roadster. The first one was delivered during February 2008. Between 2008 and 2012, 2,450 Roadsters were sold in over 30 countries. The car had a peak range of 244 miles (393 kilometers) on a single charge and an acceleration of 0 to 60 miles per hour (96.6 kilometers per hour) in 3.7 or 3.9 seconds, depending on model. It had a Lotus body that looked sleek. The amazing thing about the Tesla Roadster was not the number of cars sold but how it changed people's view of electric vehicles. Instead of electric cars being stereotyped as transport for tree huggers over short distances, they were now considered sleek, beautiful, powerful, and capable of being used for at least 90 percent of travel. However, the cost of the Roadster was out of the price range of most people.

Tesla, though, had already made plans to capitalize on the popularity of the Roadster by developing a series of lower-priced electric cars. These included the Model S (five-door liftback sedan), first delivered in 2012; the Model X (three-row crossover sport utility vehicle), first delivered in 2015; the Model 3 (midsize four-door sedan), first delivered in 2017; and the Model Y (compact sport utility vehicle), first delivered in 2020.

After Tesla single-handedly created a market for battery-electric vehicles, many other car manufacturers began developing prototypes and then vehicles for sale. By 2021, 21 electric vehicles were available for sale in the United States. The longest-range vehicle was the Tesla Model S Long Range (range of 663 kilometers, or 412 miles). The least expensive cars per dollar of range were the Tesla Model 3 Long Range ($85 per kilometer of range) and the Chevrolet Bolt ($88 per kilometer). Only two, the Kandi K27 (at $19,999) and the Mini Cooper ($30,750), cost less than $31,000.[58]

By the end of 2021, about 12 million electric cars had been sold worldwide, including 5 million in 2021 alone. While this is progress, these cars represent only 0.86 percent of the 1.4 billion vehicles on the road worldwide (1.06 billion passenger cars and 363 million trucks and buses).

Tesla and several other companies have also moved beyond cars, developing electric pickup trucks, semi-trucks, and buses.

Electric pickup trucks under production or planning in 2022 include the Rivian R1T, the Tesla Cybertruck, the Ford F-150, the Lordstown Motors Endurance pickup, the Fisker Alaska, the Bollinger boxy B2, the Chevrolet EV truck, the GMC Hummer EV, and the Stellantis Ram pickup.

Transition highlight

The 2021 Rivian R1T, which is the first electric pickup truck to reach market, has an official range of 506 kilometers (314 miles) on a single charge. The battery pack storage capacity is 135 kilowatt-hours. So, if the truck is charging at a rate of 20 kilowatts (typical for a relatively fast home garage charger), the time to fully charge it will be 6.75 hours.

Electric semi-trucks under production or planning include the Tesla Semi and the Nikola Tre electric semi. The Tesla Semi has an estimated range of 805 kilometers (500 miles) and can travel up a 5 percent grade road at 97 kilometers per hour (60 miles per hour).[59] The Nikola Tre electric semi has a range of 564 kilometers (350 miles) and a battery capacity of 753 kilowatt-hours, which is about 9.4 times that of the Tesla Model S passenger car (80 kilowatt-hours).

With regard to buses, 98 percent of the 600,000 or so electric buses built in the world, as of the end of 2020, were on the road in China. The United States had about 2,800 and Europe, about 10,000. One major bus manufacturer is BYD. They are headquartered in China and have bus manufacturing plants in several countries. BYD manufactured its first bus in 2010. One of their most popular buses is the BYD

K9, also known as the BYD ebus. It uses a lithium iron phosphate battery and has a range of 250 kilometers (155 miles) on a single charge under typical urban street conditions.

Trains, Funiculars, Cable Cars, Trams, Streetcars, Trolleys, and Light Rail

Another area of progress is with electric trains. Electric trains eliminate the iconic belching smoke from trains that once ran on coal. Electric trains use an electric motor, which is very efficient, converting 84 to 96 percent of input electricity into motion. They also use regenerative braking, thereby generating electricity when they slow and reducing emissions of pollution particles that arise when brake pads or brake blocks are applied to a train's wheels. An electric train obtains its electricity from either overhead lines, a third rail, or onboard batteries.

Built in 1802, in Shropshire, United Kingdom, the first railway locomotive worldwide ran on coal. Only in 1837 did Scottish inventor **Robert Davidson** (1804 to 1894) built a prototype electric locomotive, which ran on non-rechargeable batteries. He then built an improved version in 1841, a train that could travel at a speed of 6.4 kilometers per hour for 2.4 kilometers while towing 5.4 tonnes. Unfortunately, coal-locomotive workers destroyed the train out of fear that they would lose their jobs to this new technology.

Possibly the first non-polluting (both at the vehicle and at the energy source) non-animal-powered, non-sailboat commercial public transportation vehicle in history was the funicular. A **funicular** is a railway on a steep slope with two parallel tracks and one car on each track. The two cars are connected by a cable that loops over a pulley at the top of the track. The cars move in concert. As one slides down, it pulls the other up. The word funicular derives from the Latin word *funis*, which means rope.

Transition highlight

The first known funicular in the world was the Prospect Park Incline Railway on the United States side of Niagara Falls. It was built in 1845 but closed in 1907 following a deadly accident that killed one person. Each car carried 15 to 20 passengers from the top to the bottom of the falls. The whole system was covered. Water was originally used to make one car heavier than the other. Enough water was added to a water tank under the floor of the car at the top of the falls to make the car heavy enough to roll down the incline and pull the bottom car up. When the first car reached the bottom, water was emptied from its tank, and water was added to the tank of the second car, which was now at the top. In this way, this railway system operated with zero emissions from either the vehicle or its source of power. The use of water to move the cars was replaced by the use of electricity once the Niagara Falls hydroelectric facility opened in 1895. Even then, the whole system was pollution-free. Several funiculars using water were built in the 1800s, including in Giessbach, Switzerland (1879), and Braga, Portugal (1882). Most other funiculars were powered by steam engines running on coal.

Another non-polluting (but only at the vehicle), non-animal-powered commercial public transportation vehicle developed was the **cable car**, invented in San Francisco, California. In 1869, **Andrew Smith Hallidie** (1836 to 1900) witnessed an accident in San Francisco in which horses pulling a horsecar up a steep, wet cobblestone road (Jackson Street, between Kearny and Stockton Streets) were killed when the horsecar slid backwards, pulling them down. He thought that there must be a safer way to transport people and goods up the steep hills of San Francisco.

Hallidie had a history, starting in 1856, of producing and improving wire ropes for mines, aerial trams, and bridges. He had also built bridges himself. Guided by his expertise, Hallidie invented the cable car, which was first tested on August 2, 1873. The first cable car line in San Francisco was on Clay Street. It began regular service on September 1, 1873 and was immediately successful. Between 1873 and 1890, 23 cable car lines were built in the city. However, the 1906 earthquake and fire in San Francisco destroyed most of them. Despite a rapid rebuild, by 1912 only eight lines remained. Owing to competition by electric streetcars and buses, only three cable car lines remain in San Francisco today, and they are used primarily for tourism. Two operate between Hallidie Plaza and Fisherman's Wharf, each via different paths, and one operates between Market Street and Van Ness Avenue.

Cable cars move on steel rails. A slot between the tracks opens to a cable attached to the car that extends inside the track all the way to a central powerhouse full of large wheels. Each wheel winds a cable for a different car, pulling the car. Cable cars often have grips on both ends so they do not need to be turned around. The cars have no way of moving themselves. They are pulled solely by the cables. As such, cable cars emit zero pollution at the car. Whereas the powerhouse wheels used to pull the cable cars were originally turned using steam from coal burning, this method was eventually replaced with the use of electric motors to pull the cars.

In 1879, 42 years after Davidson invented the electric locomotive and 6 years after Hallidie invented the cable car, the German inventor **Werner Siemens** (1816 to 1892) operated the first electric passenger train, in Berlin, Germany. It consisted of three cars pulled by a locomotive and traveled at a speed up to 13 kilometers per hour. Over a 4-month period, it carried 90,000 passengers around a 300-meter-long circular track. It obtained its electricity from a third rail between the tracks.

In 1881, Siemens opened the world's first electric tram line, in Lichterfelde, Germany, near Berlin. A **tram** generally has the same meaning as a **streetcar, trolley,** or **light rail** in different parts of the world. Trams are smaller and lighter than regular trains and run on a track through public urban streets. They are powered by electricity from either an overhead wire or a third rail. Trams differ from cable cars, which are pulled by underground cables. The first tram with an overhead line appeared near Vienna, Austria, in 1883.

Tram (streetcar) technology improved substantially following inventions by Frank Sprague and the Sprague Electric Railway & Motor Company. In 1886, Sprague's company invented both a constant-speed, non-sparking motor and regenerative braking. The company used these technologies in the world's first large electric streetcar system, completed in Richmond, Virginia, in 1888. The streetcars used overhead lines. In 1889, Boston, Massachusetts, added streetcars as did many other cities. In San Francisco, the first electric streetcars began operating in 1892. Following the 1906 earthquake and fire that destroyed most cable car lines, streetcars took over all but eight routes that the cable car ran.

The first subway in the world running on electricity was installed in London in 1890. Electric subways in Budapest, Hungary (1896), and Boston (1897) followed.

Aside from the first locomotives running on batteries in 1837 and 1841, only a few locomotives have run on batteries since then. In 1917, the Kennecott Copper Mine in Latouche, Alaska, used two battery-powered locomotives to haul dirt. From 1968 to 2009, the

Toronto Transit Commission operated a battery-electric locomotive on the Toronto subway. The London Underground also regularly uses battery-electric locomotives for maintenance work.

Electric (using a third rail or overhead wire) trains, subways, and streetcars are now ubiquitous worldwide. In China, for example, more than 100,000 kilometers of rail is electrified. In India, about 46,000 kilometers of rail is electrified, and 85 percent of freight and passenger traffic is carried on electric rails. The railways in Japan are almost completely electrified. By 2021, Uzbekistan had almost 1,000 kilometers of electrified rail between major cities, the bulk of which is high-speed rail. In 2021, Germany ordered 31 new electric trains and Turkey began building its first electric train. The trains in most countries, however, still run on fossil fuels. A challenge going forward is to electrify all rail for all distances everywhere worldwide and power that electricity with WWS.

Agricultural and Construction Machines

Agricultural and construction machines have only recently joined the transition to electricity. In 2017, Solectrac manufactured the first two commercially available electric tractors, the Compact Electric Tractor (with a 22-kilowatt-hour battery pack) and the eUtility (with a 28-kilowatt-hour battery pack). In 2021, Monarch Tractor developed the first autonomous (no driver needed) electric tractor. In 2019, the largest stand-alone electric vehicle in the world was completed, the electric dump truck operating in a quarry in Biel, Switzerland, discussed earlier. Today, many electric options for agricultural and construction machines are available. However, the main limitation in transitioning such machines is a sufficient charging infrastructure. An alternative to machines that use onboard batteries is machines that use electricity directly through a cable. Some such machines are autonomous as well.

4.1.2.2 Water Vessels

Batteries are not limited to land-based vehicles. Several battery-electric marine vessels have been developed for short-distance water transport.

In 1893, the Electric Launch Company began manufacturing battery-driven electric boats in the United States. It first built 55 boats, each 11 meters long, to ferry over a million passengers during the

1893 World's Columbian Exposition in Chicago. By 1900, the number of electric-powered pleasure boats worldwide exceeded the combined number of coal- and gasoline-powered pleasure boats. Around 1910, though, electric boat adoption plummeted as coal- and gasoline-powered boats improved their range and lowered their cost.

One of the electric boats built by the Electric Launch Company was a pleasure-craft for Anheuser-Busch cofounder Adolphus Busch, delivered in 1912. This 16.8-meter all-mahogany boat, *Chief Uncas*, was used by the Busch family during summers on Otsego Lake, near Cooperstown, New York. The boat was converted to gasoline in 1950 but then restored to battery electricity in 2012 with 16 batteries that power two propellers.

In a renaissance of electric marine transportation, Norway's Norled began operating the all-electric ferry *Ampere* in 2015. The 80-meter-long ferry moves cars and people 34 times per day between the villages of Lavik and Oppedal, Norway, which are 5.7 kilometers apart. The battery storage capacity is 1,000 kilowatt-hours.

In 2017, India commissioned a 50-kilowatt-hour, 75-passenger solar-powered ferry, *Aditya*, that operates between Vaikkom and Thavanakkadavu (2.5 kilometers apart), in the state of Kerala. It has 20 kilowatts of solar PV on its roof.

Also in 2017, Denmark retrofitted the 1,250-passenger, 240-car *Tycho Brahe* ferry to make it all-electric (4,160 kilowatt-hours of battery storage). This ferry operates between Denmark and Sweden.

In 2017, Finland retrofitted its oldest ferry (operating since 1904), *Elektra*, to be all-electric (1,060 kilowatt-hours of battery storage). The ferry operates between Nauvo and Parainen in the Turku archipelago.

In 2019, Denmark began operating another all-electric (4,300 kilowatt-hours of batteries) ferry, *Ellen*, between two Danish Baltic islands. Also in 2019, the U.S. state of Alabama began operating the all-electric Gee's Bend ferry (270 kilowatt-hours of battery storage), which carries 15 vehicles and 132 passengers.

In July 2020, Damen Shipyard delivered five 50-passenger electric ferries, each with 120 kilowatt-hours of battery storage, to Copenhagen, Denmark.

In October 2020, two tourist ferries (316 kilowatt-hours of battery storage each) began operating at the base of the Niagara Falls in New York.

In 2021, Norway launched a 4,300-kilowatt-hour, 139.2-meter-long ferry that operates between Moss and Horten, which are 10 kilometers apart. The battery pack sizes of these ferries pale in comparison with the proposed retrofit (50,000 kilowatt-hours) of the Stena Jutlandica ferry, which operates between Gothenburg, Sweden, and Frederikshavn, Denmark.

In July 2021, Hollands Shipyards Group provided Norway's Brevik Fergeeselskap company its first electric ferry, which has a battery capacity of 1,300 kilowatt-hours and carries 98 passengers and 16 cars.

Several recent developments have occurred with respect to nonferry water transport. In 2017, China developed an all-electric coal-carrying container ship with 2,400 kilowatt-hours of batteries and a range of 80 kilometers.

Transition highlight

In 2017, a 30-meter hybrid (battery-electric/hydrogen fuel-cell) catamaran, the *Energy Observer*, was built. It is powered by solar PV, which covers all available surface area of the vessel. During sunny days, the PV provides electricity for the two electric motors, desalination equipment, and appliances on the boat. The solar electricity also charges the batteries and produces hydrogen through an electrolyzer that splits water molecules. During the night and during cloudy days, either the batteries or fuel cells running on the stored hydrogen provide electricity for the motors. Sufficient hydrogen is available to run the vessel for 6 days if no other power is available.

In August 2020, Turkey launched the world's first all-electric tugboat, with a battery capacity of 2,900 kilowatt-hours, for use in the waters near Istanbul. In December 2020, Vietnam launched an all-electric tugboat, this one with a battery capacity of 2,800 kilowatt-hours, for use in New Zealand. In July 2021, the first United States electric tugboat, the *eWolf*, was announced for use in the Port of San Diego. This tug has a 6,200-kilowatt-hour battery pack and will eliminate the combustion of 30,000 gallons of diesel per year. It will begin operation in 2023.

In May 2020, Damen Shipyards Group unveiled the world's first all-electric dredger, a 550-tonne vessel.

The German AIDA Cruises added a 10,000-kilowatt-hour battery pack to a 300-meter-long cruise ship in 2021. This was the first such battery pack added to a cruise ship. In January 2022, China completed the first trial voyage of the world's largest electric-only cruise ship, the *Yangtze River Three Gorges 1*. The ship is 100 meters long, is designed to carry up to 1,300 passengers up and down the Yangtze River, and has a battery capacity of 7,500 kWh.

In 2021, Japan developed two electric propulsion tankers, each with a 4,000-kilowatt-hour battery pack, to operate in the bay of Tokyo.[60]

In August 2021, the Swedish boat maker Candela launched an electric hydrofoil powerboat, *C-8*, that seats eight people and can cruise at 22 knots for 93 kilometers. Its top speed is 30 knots. The boat was designed for mass production.

In November 2021, the *Yara Birkeland* became the first all-electric cargo ship to travel autonomously (with no crew and guided remotely). It is 80 meters long and has 7 megawatt-hours of battery storage. It will travel between ports off the coast of Norway.

Also in November 2021, Costa Rica began constructing the world's largest modern-day zero-emission cargo ship. It will run on sails and a backup battery-electric motor. The batteries will be supplied by electricity from solar PV and regenerative braking. The ship will carry up to 227 tonnes of coffee.

In December 2021, the world's first battery-electric bunker tanker (a small tanker ship used to load fuel oil onto ocean-going vessels) was launched in Tokyo Bay. It has a 3,480-kilowatt-hour battery system.

4.1.2.3 Aircraft

Aircraft emit about 2 to 3 percent of global carbon dioxide emissions, but mostly high in the air.[61] They also emit black and brown carbon particles and many gaseous pollutants. The particles and water vapor emitted by aircraft produce contrails, which also affect climate. Whereas aircraft may cause about 1.3 percent of near-surface global warming, they may also cause about 4 percent of upper-tropospheric global warming and 6 percent of Arctic warming.[62] As such, transitioning aircraft to battery-electric or hydrogen fuel-cell electric propulsion is critical for helping to address climate damage.

In 2009, the Swiss Company, Solar Impulse, built the first electric aircraft, *Solar Impulse 1*, which contained solar cells and batteries.

In 2010, a single pilot flew the plane for 26 hours, including for 9 hours at night, opening the world's eyes to the potential for electric flight. *Solar Impulse 2*, a second plane with more solar cells and batteries than the first, was built in 2014 and flown, from March 9, 2015, to July 26, 2016, around the world in stages.

> **Transition highlight**
>
> In 2015, the Israeli electric aircraft manufacturer Eviation began developing an electric plane, *Alice*. The plane is intended to fly either nine passengers or 1,200 kilograms of cargo. It has a range of 815 kilometers (507 miles) on a single charge and flies at a top speed of 400 kilometers per hour (250 miles per hour). The plane has 3.6 tonnes of lithium-ion batteries, which comprise 60 percent of the plane's takeoff weight, and two motors. *Alice* made its first test flight in June 2019. Later that year, the U.S. airline Cape Air ordered several of the planes. In 2021, the package carrier DHL Express ordered 12 planes. Whereas a jet-fuel plane costs between $1,200 and $2,000 per flight hour to operate, this plane will cost an estimated $200 per flight hour to operate, owing to the efficiency of electricity over combustion and the greater efficiency of the airplane design itself.[63]

In October 2019, China test-flew the *RX4E* electric airplane. It has four seats, 70 kilowatt-hours of battery storage, and a range of 300 kilometers (186 miles).

In 2019, Harbour Air, a North American seaplane airline, retrofitted one of its internal combustion engine planes with batteries installed by MagniX to form the *eBeaver*. The plane was tested on December 11, 2019. In April 2021, Harbour Air announced it would retrofit all 40 of its planes with batteries and propulsion systems with the goal of creating the world's first all-electric airline fleet.

In June 2019, the Slovenian company Pipistrel announced that it was building a vertical takeoff and landing (VTOL) electric aircraft, the *801*, that seats a pilot and four passengers. It has a range of 97 kilometers (60 miles) and a top speed of 282 kph (175 mph). VTOL electric aircraft will replace helicopters. In March 2020, Pipistrel also announced orders for 15 two-seat electric aircraft for the Australian flight-school company, Eyre to There.

In April 2020, Jaunt Air Mobility introduced its own VTOL electric aircraft, *Journey*, also with five seats and a top speed of

282 kph (175 mph). The company said that this aircraft is 63 percent quieter than a helicopter.

In May 2020, a nine-seat Cessna 208B Grand Caravan, retrofitted with a battery-electric propulsion system by MagniX and Aero-TEC, flew for 30 minutes above central Washington State. This test flight was of the largest electric plane at the time.

In June 2021, Ravn Alaska airlines reported it will purchase 50 nine-seat electric planes from California-based Airflow. The planes, which are expected to be available in 2025, have a range of 161 kilometers (100 miles).

In July 2021, United Airlines and its regional partner Mesa Airlines announced the purchase of 200 19-seat electric aircraft, the ES-19, from Sweden's Heart Aerospace. The aircraft has a range of 400 kilometers (248 miles) and will be available in 2026. In August 2021, New Zealand's Sounds Air announced it will order three of the same aircraft and that it is striving to be an all-electric airline by 2030.

In September 2021, the startup company Wright Electric announced that it was testing the most powerful (2 megawatt) motor to date for use in an aircraft. The motor will be used for battery-electric and hydrogen fuel-cell planes. Ten of the motors provide enough power for a 140- to 170-seat single-aisle plane. In November 2021, Wright Electric announced that it will have a 100-seat electric plane with a range of 740 kilometers (460 miles) commercially available by 2027.

In December 2021, a Boston company, Regent, announced that they will test-fly an electric boat-plane hybrid called a seaglider, over Tampa Bay. The seaglider rests at a dock and floats like a boat in areas with no waves. It then takes off from the water, hovering over waves using underwater wings, or hydrofoils. The first commercial version of the seaglider will carry 12 passengers.

4.1.3   Lithium and Neodymium Mining

Today's battery-electric vehicles use lithium-ion batteries. Lithium itself is not a toxic element. However, lithium-ion batteries also contain nickel, cobalt, aluminum, manganese, titanium, and/or phosphorus,[64] some of which are toxic if breathed in high enough concentrations. On the other hand, these chemicals are not generally breathed in, as they are confined to the batteries that contain them. Further, the quantities of those toxic chemicals are less than the quantities of lead, nickel, or cadmium in lead acid or nickel cadmium batteries. If lithium-ion batteries are thrown into a landfill, toxic chemicals from them will leak out over time, contaminating ground water. However, lithium-battery recycling, which is already occurring, avoids this problem.[65]

Nevertheless, most mining for lithium causes environmental degradation. Such damage, though, should be put in perspective. Batteries that now last 15 to 20 years require one-time mining at the beginning of their life. Battery recycling allows the lithium to be reused. Gasoline and diesel vehicles, on the other hand, require continuous mining of oil forever, so the mining damage is orders of magnitude greater than with lithium mining. For example, one study found that just 1 gigawatt of wind nameplate capacity replacing a coal-fired power plant on the Texas grid reduces total mining by 25 million tonnes over 20 years.[19] That accounts for changes in the quantity of ores for steel, copper, and rare earth elements needed and for all soil and waste moved in the process.

Further, traditional lithium mines will become cleaner as they begin to run on WWS energy. For example, in July 2020, a Texas company announced it will use solar PV to provide 100 percent of the electricity required in the annual average for its Round Top Mountain rare earth and lithium mining project in Hudspeth County, West Texas. Scientists have also developed a technique to extract lithium with no new mining.

> **Transition highlight**
> Environmental damage due to lithium mining can be averted almost entirely. Existing geothermal electricity production in the Salton Sea, California,[66] and the Upper Rhine River, Germany,[67] entails lithium-rich brine water (water rich in salt) being pulled from deep geothermal wells and recycled back into the wells after high-temperature heat is extracted for geothermal electricity generation. Efforts are currently underway to extract lithium from such brines using the electricity and heat from the geothermal plant itself. If the plants are binary geothermal plants, such efforts will result in no greenhouse gas emissions, no air pollution emissions, and no additional mining. Similarly, lithium can be extracted from the same brine that bromine is extracted from, as done in Arkansas.[68]

Powering mining operations with WWS is not limited to lithium mining. In April 2021, a gold mine in Mali began using solar PV plus batteries for its electricity. In August 2021, a company in Western Australia announced it would power one nickel mine with solar PV and another with solar PV plus batteries. A second company announced it would power an ilmenite mine in Madagascar with solar PV, wind, and batteries.[69] In September 2021, a potash mining company in Australia announced it would provide two-thirds of its power with a solar PV–wind–battery microgrid.[70]

A question related to lithium-ion batteries is whether sufficient resource exists worldwide to provide enough lithium to power all electric vehicles, equipment, and stationary batteries needed for a 100 percent WWS world. Based on current data, enough lithium is known to exist to power over 5 billion electric vehicles of substantial range. Currently the world has only 1.4 billion vehicles. But lithium batteries will also be used for stationary storage, cell phones, and other tools and appliances. It is possible that more lithium resources will be discovered during the next decade or two to cover all possible uses of lithium. In addition, lithium and other rare components are already being recycled, which will help to ensure that shortages do not occur.[65] Stresses on lithium supplies will also be alleviated by the fact that batteries without lithium, such as iron–air batteries, are being developed as well.

Another important element used in many electric vehicles is neodymium. Neodymium is used in permanent-magnet AC motors. However, the amount of neodymium required per vehicle is about an order of magnitude less than that required per wind turbine, and neodymium is not a limiting factor in wind turbine production. In addition, one common type of electric motor, an induction motor, does not use neodymium. Finally, an alternative to neodymium, which is just now being commercialized, is iron nitride. Iron nitride magnets are stronger than neodymium magnets and may be more environmentally benign to mass-produce since they contain common elements.[71]

4.2   Hydrogen Fuel-Cell Vehicles

Hydrogen gas (H_2), or just hydrogen, has been present in the Earth's atmosphere since the formation of the Earth 4.6 billion years ago. Today, it is a well-mixed gas in the lower atmosphere whose main natural source is bacteria living in the oceans and soils. Humans contribute to hydrogen's atmospheric burden too, as hydrogen is a byproduct of fossil-fuel combustion. Hydrogen is removed from the air primarily by soil and ocean bacteria.

Paracelsus (1493 to 1541), a Swiss physician and alchemist, born Theophrastus von Hohenheim, may have been the first to isolate hydrogen. He discovered that pouring sulfuric acid over the metals iron, zinc, or tin gave off a highly flammable vapor. In 1766, **Henry Cavendish** also made this determination and isolated hydrogen's properties.

Because hydrogen is much lighter than air, hydrogen has been used to lift passenger balloons and airships for flight as well as to lift party balloons. Hydrogen has also been burned to propel rockets and the Space Shuttles into space. Its use in the **Hindenburg** airship, which caught fire and crashed, killing 36 people on May 6, 1937, set back its use for transportation for many decades. Even in that case, though, the fire damage was due to the burning of the airship's shell rather than of the hydrogen. Owing to its lightness, a hydrogen flame shoots straight up, minimizing damage to a vehicle carrying it. On the other hand, jet fuel, diesel, and gasoline explode outward, destroying a vehicle and its contents.

4.2.1   Green Hydrogen Production

Hydrogen can be produced synthetically by steam reforming of methane, autothermal reforming of methane, coal gasification, electrolysis, and photoelectrochemical water splitting. In a WWS world, only electrolysis and photoelectrochemical water splitting should be used, where the energy in both cases comes from WWS (green hydrogen). Section 8.8 discusses why the other techniques should not be used.

The cleanest and simplest way to produce hydrogen is by **electrolysis**, where the electricity source is WWS. This type of hydrogen is referred to as **green hydrogen**. Electrolysis involves using electricity in an **electrolyzer** to split liquid water into hydrogen and oxygen. This process has no emissions if the electricity originates from a WWS source. The resulting hydrogen is usually compressed and stored in a tank for later use in a fuel cell. Electrolysis is the simplest way to produce hydrogen because it requires only an electrolyzer and a WWS electricity source. Electrolysis also eliminates the mining, transport, processing, pollution emissions, methane emissions, and carbon dioxide emissions associated with producing hydrogen from fossil fuels.

Three types of electrolyzers are the polymer electrolyte membrane electrolyzer, the alkaline water electrolyzer, and the solid oxide electrolyzer. The polymer electrolyte membrane electrolyzer is the most commonly used electrolyzer, but the solid oxide electrolyzer may be the most efficient.[72] Once hydrogen is produced, it is usually compressed or liquefied and stored before use.

Several green hydrogen plants based on electrolysis have been built or are under construction. In July 2021, the Nordic and U.S.-based steel company SSAB produced the world's first fossil-free steel, in Lulea, Sweden, using a green hydrogen process. An electrolyzer powered by WWS electricity (mostly wind) produced the hydrogen.

In August 2021, Plug Power Inc. broke ground on the development of a hydrogen electrolyzer plant in Camden County, Georgia, powered by 100 percent WWS, that will produce 5,500 tonnes of hydrogen per year. The plant is the first of five plants that they are building.

Also in August 2021, China approved the building of 1.85 gigawatts of solar PV and 370 megawatts of wind to produce almost 67,000 tonnes of green hydrogen per year by electrolysis in Mongolia.

Transition highlight

An emerging method of producing hydrogen from WWS electricity is **photoelectrochemical water splitting**. With this technology, sunlight is used to produce electricity inside of a solar PV cell, and the electricity is then used within the same cell to split water into hydrogen and oxygen. As such, the solar PV cell and electrolyzer are merged into one device. Because this method involves a WWS energy source and has no emissions, the resulting hydrogen is green. This process reduces electricity losses though transmission lines compared with solar PV producing electricity that is transmitted on the grid to an electrolyzer. However, the technology may not reach commercialization until about 2028.

4.2.2 Hydrogen Storage

In a 100 percent WWS world, hydrogen will be used primarily in fuel cells to provide electricity to run an electric motor in certain vehicles. Hydrogen will also be used for some industrial processes, such as steel production, for electricity and heat production in remote microgrids, and for ammonia production. The modes of transport that will benefit most from hydrogen fuel cells will be long-distance, heavy transport, namely mid- and long-distance ships and aircraft, long-distance trucks and trains, and some military vehicles. Electrolysis, with electricity from WWS, will produce the hydrogen.

Hydrogen for fuel-cell vehicles will either be used immediately or stored in a storage tank. An advantage of using hydrogen in a 100 percent WWS economy is that WWS electric power generators can produce hydrogen when more WWS electricity is available on the grid than is needed to meet electricity demand. Otherwise, extra electricity that is produced by WWS might be curtailed or shed (wasted), resulting in lost electricity and a higher cost of WWS electricity than necessary. If excess WWS electricity is used to produce hydrogen, and if the hydrogen is stored, the cost of WWS generation drops to less than if the WWS electricity is wasted. In a similar manner, excess WWS electricity can be used to power heat pumps that produce heat or cold that is stored in water or soil for later use. As such, electricity, heat, cold, and hydrogen can work together to help power a 100 percent WWS economy at low cost.

At sea-level air pressure, gasoline density is over 8,000 times that of hydrogen gas density. Since hydrogen gas has a low density, it requires a large storage volume. Alternatively, hydrogen can be compressed into a small volume or its temperature can be dropped until the hydrogen liquefies.

Compressed hydrogen can be used in hydrogen fuel-cell ground vehicles or stored in tanks for later use. Compressed hydrogen storage containers for vehicles are made of either low-alloy steel or a composite material or both. The steel can become brittle if it is exposed to hydrogen over a long time, which is why composites are often used. Stationary hydrogen storage containers are made of low-alloy steel, a composite material, or stainless steel and carbon steel encapsulated within prestressed concrete.[73]

Aircraft, rockets, and Space Shuttles require much more energy and thus much more hydrogen than cars. Given their limited volumes, the former then require even more dense, liquid (**cryogenic**) hydrogen. Liquid hydrogen was used in demonstration flights for the Soviet Union's commercial aircraft Tupolev Tu-154B, beginning on April 15, 1988. The aircraft was fitted with a thermally insulated fuel tank behind the passenger cabin that contained liquid hydrogen. Combusted hydrogen powered a third engine on the aircraft. Combusted liquid hydrogen was also used to power the Space Shuttles and most rockets into space.

Liquid hydrogen is obtained by reducing hydrogen's temperature to below its boiling point, which is near −253 degrees Celsius at sea-level air pressure. Liquid hydrogen's density is much greater than is compressed hydrogen's density. Liquid hydrogen is stored in cold tanks at a temperature below the evaporation temperature of hydrogen. In a closed tank, even a small increase in temperature will evaporate enough hydrogen to explode the pressure by a factor of 10,000. As such, liquid hydrogen tanks require substantial insulation to prevent hydrogen from evaporating (boiling).

About 13 percent of the energy stored in liquid hydrogen is needed to cool the hydrogen enough to liquefy it. On top of that, liquid hydrogen has only one-quarter the energy density of jet fuel. However, a given mass of hydrogen propels an airplane further than does the same mass of jet fuel. As such, the volume of hydrogen fuel in a hydrogen fuel-cell plane is only 1.8 to 4.8 times that of jet fuel in a jet-fueled aircraft for the two aircraft to go the same distance. A liquid hydrogen storage tank therefore needs to be larger than does a jet fuel storage tank.

4.2.3   Hydrogen Fuel Cells

A **fuel cell** is a device that converts chemical energy from a fuel into electricity and heat. A hydrogen fuel cell converts hydrogen and oxygen to electricity, heat, and water vapor.

Fuel cells are connected in series to form a **fuel-cell stack**. Stacks contain a few to hundreds of fuel cells. Fuel-cell stacks of 50- to 125-kilowatt nameplate capacity are used in most vehicles. Stacks of 1 to 200 megawatts are used in power plants.

In a 100 percent WWS world, hydrogen fuel cells will be used mostly for certain transport. In a hydrogen fuel-cell ground vehicle, the fuel cell combines hydrogen from an onboard storage tank with oxygen from the air to produce electricity, water vapor, and waste heat. The electricity runs an electric motor, which produces rotational motion to turn the vehicle's wheels. The water vapor and waste heat are expelled.

In a WWS world, some hydrogen may also be used in stationary fuel cells to provide electricity for buildings or industry. However, using electricity to produce hydrogen and then using the hydrogen to reproduce electricity is inefficient compared with storing electricity in a battery and then discharging the battery to reproduce electricity. The overall efficiency of using hydrogen for normal electric power can be improved if heat released during fuel-cell use is captured and used.

Transition highlight

One useful application of a hydrogen fuel cell, aside from powering certain vehicles, is for producing electricity and usable heat in a microgrid. As mentioned in Chapter 1, a **microgrid** is an isolated grid that provides power for a single home, a few homes, a few buildings, a hospital, a university campus, a remote community, or a remote military base. In 2016, for example, an apartment complex was built in Brutten, Switzerland, that included 127 kilowatts of solar PV cells, an electrolyzer, two hydrogen storage tanks, a fuel cell, and a ground-source heat pump. Excess electricity from the solar PV was used to electrolyze water into hydrogen, which was stored until needed. During winter, hydrogen was run through the fuel cell to generate electricity and heat. The heat was then used to heat water. Ground-source heat pumps that ran on electricity from either the fuel cell or the solar PV were used to heat and cool the air.

When evaluating battery-electric versus hydrogen fuel-cell vehicles, analysts often compare the plug-to-wheel efficiency of one versus the other.

The overall **plug-to-wheel efficiency of a hydrogen fuel-cell passenger car** ranges from 23 to 43 percent. That is the percent of electrical energy used to produce hydrogen that is converted to motion. The rest is waste heat. This efficiency is much lower than the plug-to-wheel efficiency of a battery-electric passenger car (64 to 89 percent), but higher than the tank-to-wheel efficiency of an internal combustion engine car (17 to 20 percent).

The plug-to-wheel efficiency of a hydrogen fuel-cell car depends on several factors. First, electricity is needed to produce the hydrogen by electrolysis, then to compress or liquefy the hydrogen for storage in a holding tank and, eventually, in a vehicle fuel tank. Second, some of the hydrogen leaks between production and the vehicle fuel tank. Third, energy in hydrogen's chemical bonds is lost as heat in the fuel cell. More energy is lost from the fuel cell when water vapor produced from hydrogen escapes to the air, carrying with it latent heat. Finally, some electricity produced by the fuel cell is lost in an inverter and wires between the fuel cell and the electric motor, and more electricity is lost as heat in the motor.

The **electrolyzer efficiency** is the efficiency of electricity conversion in an electrolyzer to energy stored in hydrogen chemical bonds. Efficiencies range from 73.8 percent for standard electrolyzers to 85.6 percent for new solid oxide electrolyzers.[72] The **compressor efficiency** is about 90.4 percent. In other words, of the total energy needed to produce and compress 1 kilogram of hydrogen, 14.4 to 26.2 percent is lost due to electrolysis and another 9.6 percent is lost due to compression. The remainder is stored in the bonds of hydrogen molecules.

Once hydrogen is produced, some of it may leak. Hydrogen is a tiny molecule, much smaller than methane, so it can leak easily from pipes unless the pipes are well sealed. In a 100 percent WWS system, hydrogen will not be produced far away and sent by pipeline or truck. Instead, most will be produced locally after electricity is transmitted long-distance. This will eliminate hydrogen pipeline losses. Leaks, however, will still occur. An estimated 0.3 to 1 percent of hydrogen produced will leak between production and consumption.

Of the energy in hydrogen entering a vehicle, 30 to 50 percent is lost as the fuel cell converts the hydrogen to DC electricity. Another 15.4 percent is lost as heat used to evaporate water produced in the fuel cell. This lost energy is stored as latent heat in the water vapor produced by the hydrogen reaction in the fuel cell. The water vapor is released back to the air and ultimately condenses to liquid cloud water in the atmosphere. Another 1 to 3 percent of energy is lost in the inverter that converts DC to AC electricity, wiring, and power electronics. Finally, another 4 to 16 percent of energy is lost converting electricity to motion in the motor

Thus, the efficiency of the fuel-cell system inside a hydrogen fuel-cell passenger vehicle itself is 34 to 56 percent. Accounting also for the electrolyzer, compressor, and hydrogen leaks, the overall plug-to-wheel efficiency of a hydrogen fuel-cell vehicle is about 23 to 43 percent.

Because hydrogen fuel-cell passenger vehicles are much less efficient than are battery-electric passenger vehicles, a hydrogen fuel-cell passenger vehicle requires the energy from many more wind turbines or solar panels to move the same distance as does an equivalent battery-electric vehicle. However, aside from the leaked hydrogen and the water vapor emissions from hydrogen fuel-cell vehicles, the tailpipe emissions from both are zero.

With larger vehicles, such as long-distance, heavy commercial trucks, trains, ships, airplanes, and military vehicles, however, the tables turn and hydrogen fuel-cell vehicles become more efficient than battery-electric vehicles. The reason is that heavy, long-distance battery-electric vehicles end up using more and more energy just to carry the batteries. Battery-electric vehicles also do not consume fuel so do not decrease their weight during a trip. They consume only weightless electricity.

Hydrogen fuel-cell vehicles must also carry more equipment and fuel as their size increases. They need large tanks to store H_2, even if the H_2 is liquefied. Also, the longer the range, the greater the fuel-cell stack size needed, which increases vehicle weight and volume. However, hydrogen fuel itself is light and is converted to oxygen and water vapor, which are both expelled to the air as the vehicle travels. As such, hydrogen fuel-cell vehicle overall weight decreases during a trip.

In sum, the heavier a vehicle and the further it must travel, the more likely a hydrogen fuel-cell vehicle is to overtake a battery-electric vehicle in terms of efficiency. Hydrogen fuel-cell trucks, in fact, become advantageous over battery-electric trucks for a vehicle range greater than about 800 kilometers. Most long-haul commercial truckers in the United States drive 950 to 1,050 kilometers per day with only a 30-minute break during the drive. Thus, U.S. commercial trucks ideally have a range of 1,100 kilometers or a refueling or recharging time of less than 30 minutes. In Europe, truck drivers can drive a maximum of 8 hours per day, so the driving range is limited to closer to 800 kilometers per day.

Transition highlight

In 2021, the Daimler hydrogen fuel-cell prototype semi-truck was stated to have a range of up to 1,000 kilometers (621 miles),[74] and the Tesla electric semi was stated to have a range of 805 kilometers (500 miles).[59] A hydrogen fuel-cell truck can be refueled fully in 20 minutes. An electric truck, even with a supercharger, requires a few hours. Thus, electric trucks can meet the 800-kilometer range threshold of most European long-distance truckers but not the 950 to 1,050-kilometer range threshold of U.S. truckers. Hydrogen fuel-cell trucks, on the other hand, are closer to meeting the range threshold of U.S. truckers and can already meet the refueling threshold. Given rapid battery improvements, it is likely that both electric and hydrogen fuel-cell trucks will be used eventually for long-distance trucking.

Aside from their efficiency advantage, hydrogen fuel-cell vehicles have other advantages over fossil-fuel vehicles. For example, hydrogen fuel-cell vehicles eliminate all tailpipe particle and gas emissions, aside from water vapor, which is a waste product of the hydrogen fuel cell. But, during combustion, fossil-fuel vehicles also produce and emit water vapor and more than do hydrogen fuel-cell vehicles. Hydrogen production, transport, and storage also result in hydrogen leaks. However, fossil-fuel vehicles also emit hydrogen, but as a combustion product, and they emit more hydrogen than do hydrogen fuel-cell vehicles. In sum, replacing fossil-fuel vehicles with hydrogen fuel-cell vehicles will reduce global hydrogen and water vapor emissions.[75]

For a battery-electric or hydrogen fuel-cell aircraft, ship or military vehicle to replace a fossil-fuel vehicle, the former would ideally have an overall mass and volume similar to or lower than those of the fossil-fuel vehicle, and a range and power-to-weight ratio or thrust-to-weight ratio similar to or higher than those of the fossil-fuel vehicle. A vehicle's mass and volume affect not only its capabilities, but also how it interacts with roads, bridges, tunnels, runways, locks, canals, and ports. The power-to-weight ratio defines how fast a vehicle can accelerate to a top speed. For an airplane, the thrust-to-weight ratio defines its ability to take off within a runway's length. Those considering adopting new battery-electric or hydrogen fuel-cell vehicles are more likely to do so if the vehicle performance and range are similar to or better than those of existing fossil-fuel vehicle options.

The major components in a fossil-fuel vehicle that can be removed when building a battery-electric or hydrogen fuel-cell vehicle include the vehicle's engine, transmission, oil, coolant, automatic transmission fluid, fuel tank, and fuel. A battery-electric vehicle replaces these fossil-fuel vehicle components with a battery pack, inverter, electric motor, gearbox (optional), and wiring. A hydrogen fuel-cell vehicle replaces these fossil-fuel vehicle components with a hydrogen storage tank, compressed or liquefied hydrogen fuel, a fuel-cell stack, an electric motor, a gearbox, and wiring.

Transitioning long-haul aircraft, while challenging, can be done with near-future hydrogen fuel-cell technology. A future hydrogen fuel-cell aircraft could have the same thrust-to-weight and range as today's combustion aircraft while simultaneously having about a 22 percent smaller mass, but with a 21 percent larger volume. The larger volume is not an issue for aircraft. Long-distance ships and trains can similarly be transitioned to hydrogen fuel-cell versions. Military tanks and wheeled vehicles can be transitioned to either battery-electric or hydrogen fuel-cell vehicles.[76]

The main advantage of a hydrogen fuel-cell aircraft over a jet fuel aircraft is that the former eliminates all aircraft exhaust emissions except for water vapor (any hydrogen leaks would occur before or during fueling of the aircraft). Thus, hydrogen fuel-cell aircraft eliminate exhaust emissions of particles containing black carbon, brown carbon, and sulfate and gases, such as carbon dioxide, carbon monoxide, nitrogen oxides, and sulfur oxides, along with a soup of organic gases. Owing to the water vapor emissions, contrails will still form behind hydrogen fuel-cell aircraft under the right temperature and humidity conditions in the background atmosphere. However, such contrails will form only on background particles rather than on exhaust plus background particles. As such, contrails from hydrogen fuel-cell aircraft will be about 70 percent thinner, and thus dissipate faster, than contrails from jet fuel aircraft. Pure battery-electric aircraft emit zero water vapor so will eliminate all contrails.

4.2.4   History of Hydrogen Fuel-Cell Vehicles

Since the late 1990s, several types of hydrogen fuel-cell vehicles have been developed for land, sea, and air transport.

4.2.4.1 Ground Vehicles

In 1998, six fossil-fuel buses were retrofitted with Ballard hydrogen fuel-cell propulsion systems and dispatched equally in Vancouver and Chicago. They carried over 200,000 passengers, collectively, over a 3-year period.

From 2003 to 2006, several hydrogen fuel-cell buses, manufactured by Hino Motors and Toyota Motors, were tested in Japan.

Oakland, California's AC Transit trialed three hydrogen fuel-cell buses from 2004 to 2006 and 12 buses in 2009.

In 2006, Daimler dispatched three hydrogen fuel-cell buses to China. In 2007, they dispatched 36 more across 11 cities in China.

In 2008, Honda offered the first hydrogen fuel-cell passenger car to retail customers, the FCX Clarity. It was available only for lease. Between 2008 and 2015, Honda leased 48 of the vehicles in the United States. The range was 386 kilometers (240 miles). Its distribution was limited by the paucity of fuel-cell charging stations.

In 2014, Toyota unveiled the hydrogen fuel-cell passenger car, Mirai, at the Los Angeles auto show. Through 2020, over 12,000 were sold worldwide, including 8,200 in the United States. The range of its 2021 model is 644 kilometers (400 miles). Refueling takes less than 5 minutes.

Transition highlight

In September 2018, the company Alstom began operating the first hydrogen fuel-cell train, *Coradia iLint*, in Germany. The train reaches 140 kph (87 mph) and creates little noise. By June 2021, 41 of the trains had been ordered in Germany. The train has also been tested and considered for use in Austria, the Netherlands, and Poland.

In July 2021, Hyzon Motors delivered a 55-tonne hydrogen fuel-cell milk truck with a range of 520 kilometers (323 miles) for operation in the Netherlands. Other trucks being developed by Hyzon are for hauling food, sewage, construction equipment, and concrete. Overall, Hyzon delivered 87 heavy-duty hydrogen fuel-cell trucks in Asia, Australia, Europe, and North America in 2021.

As of 2022, Nikola, Toyota, Hyundai, and Daimler were also manufacturing hydrogen fuel-cell commercial trucks. The Daimler truck, for example, will have a range of 1,000 kilometers (621 miles).

4.2.4.2 Marine Vessels

In March 2018, Yanmar and Toyota tested a 16.5-meter hydrogen fuel-cell boat, *Shimpo*, with a 60-kilowatt-hour battery pack.

In 2021, SWITCH Maritime began testing the world's first hydrogen fuel-cell ferry, *Sea Change*. It will seat 75 passengers. In December 2021, it entered operational sea trials in Bellingham, Washington State. The first of these ferries will be deployed in San Francisco Bay. The company's goal is to replace many of the nearly 1,000 ferries in the United States, all of which burn diesel, with hydrogen fuel-cell ferries.

Also in 2022, Norway's Statkraft was building a 88-meter-long hydrogen fuel-cell cargo ship to travel between locations in Norway.[77] In addition, a group of Norwegian companies began retrofitting a cruise ship to run along the Norwegian coast on batteries and hydrogen fuel cells by 2023.

4.2.4.3 Aircraft

The first flight of a hydrogen fuel-cell aircraft occurred on September 29, 2016, when the four-seat Hy4 propeller aircraft, developed by the German Aerospace Center (DLR) together with the companies H2Fly, Pipistrel, and Hydrogenics, took to the skies over Stuttgart, Germany. The Hy4 is a hybrid with both hydrogen fuel cells and batteries. The batteries are used for takeoff and landing; the fuel cells, for cruising. It has a maximum range of 1,500 kilometers (938 miles) and a top speed of 200 kph (124 mph). The empty weight of the aircraft is about 42 percent of the maximum weight. In December 2020, a version of the plane with Cummins fuel cells and motor system was test-flown over Stuttgart.

In September 2020, the British/American company, ZeroAvia, conducted a test of a six-seat hydrogen fuel-cell plane, Piper Malibu, north of London, United Kingdom. This plane is a precursor to a 10- to 20-seat hydrogen fuel-cell plane with a range of 805 kilometers (500 miles) that they expect to deploy by 2023. Their goal after that is to produce a 100-seat plane with a range of 1,610 kilometers (1,000 miles). In the meantime, ZeroAvia has committed to convert ten 72-seat airplanes owned by ASL Airlines in Ireland to those powered by hydrogen and to develop green hydrogen fueling plans for some European airports.

In September 2020, Airbus announced that it will develop a hydrogen turbofan (jet) plane that will combine hydrogen fuel cells to generate electricity with hydrogen combustion for propulsion. The plane will carry 200 passengers more than 3,700 kilometers (2,298 miles). A propeller version of the plane that will use only hydrogen fuel-cell electricity is also being developed. It will travel 1,850 kilometers (1,149 miles) and carry 100 passengers.

In December 2020, Airbus announced the development of a liquid hydrogen fuel-cell aircraft with detachable pods along the wings, each of which contains a hydrogen propulsion system and a storage tank.

In July 2021, Universal Hydrogen announced that it will retrofit more than 16 regional planes among three airlines in Spain, Alaska, and Iceland, to become hydrogen fuel-cell planes. In one case, a 56-seat plane will be reduced to 40 seats, and the range of the resulting hydrogen fuel-cell plane will be 740 kilometers (460 miles).

In December 2021, United Airlines announced that it is ordering up to 100 hydrogen fuel-cell electric motors from ZeroAvia to retrofit 50 regional twin-motor aircraft by 2028.

Also in December 2021, the California startup H2 Clipper announced it was developing a hydrogen fuel-cell cargo airship, which looks like a blimp, that will carry 8 to 10 times the payload of a cargo plane at one-quarter the cost. It will travel at a cruising speed of 282 kilometers per hour (175 miles per hour).

4.2.5 Platinum for Hydrogen Fuel-Cell Vehicles

One concern that arises with the large-scale production of hydrogen fuel-cell vehicles is the potential shortage of the scarce metal platinum, which is used in many types of fuel cells. However, because hydrogen fuel-cell vehicles will replace gasoline and diesel vehicles, which both use platinum in their catalytic converters, the increase in platinum demand for hydrogen fuel-cell vehicles will be partly offset by a decrease in platinum demand for use in catalytic converters. Neither hydrogen fuel-cell nor battery-electric vehicles require catalytic converters, which convert some harmful air pollutants into more benign chemicals in the exhaust stream of gasoline and diesel vehicles.

Further, because battery-electric passenger vehicles will comprise the largest share of vehicles in a WWS world owing to their greater efficiency, the number of hydrogen fuel-cell passenger vehicles required will be limited. If even 10 percent of the world's vehicle fleet in 2050 were hydrogen fuel-cell vehicles, this would translate to about 150 million 50-kilowatt hydrogen fuel-cell vehicles. Since each 50-kilowatt fuel-cell stack requires about 12.5 grams of platinum,[78] 150 million such vehicles require about 1.875 million kilograms of platinum. This is much less than the 17 million kilograms of platinum in known resources that can be extracted from ore worldwide.

In addition, platinum is already recycled, and additional platinum will be available owing to the reduced need for it in catalytic converters upon a transition to battery-electric and hydrogen fuel-cell vehicles. As such, platinum is not considered a limiting element in a WWS economy.

5 WWS SOLUTIONS FOR BUILDINGS

A significant portion of a building's energy is used to provide air heating, water heating, air conditioning, and refrigeration. Heat is also used for cooking, dishwashing, and clothes washing and drying. Away from the tropics, demand for air heating in buildings is usually greatest during winter, whereas demand for air conditioning is usually greatest during summer. In all locations, hot water and refrigeration are needed year-round, although energy demand for hot water peaks during cold months and demand for refrigeration peaks during warm months.

In a 100 percent WWS world, air and water heating and air conditioning in buildings will be provided either by district heating and cooling systems or by individual heating and cooling systems. In both cases, heating and cooling will be provided primarily by electric heat pumps, where the electricity comes from WWS sources and the heat or cold is extracted from either the air, the ground, water, or a waste stream of hot or cold air or water. The heat and cold may be stored or used immediately. Additional heat may come from geothermal and solar heat. The remaining energy used in buildings is for electric appliances and gadgets, such as lights, televisions, computers, and phone chargers. This chapter discusses how WWS will power district and individual-building heating and cooling systems, including hot and cold storage options for both. It also discusses electric appliances and machines that will replace natural gas ones in buildings and their gardens. The chapter also examines energy efficiency in buildings and techniques to reduce building energy use. Finally, it discusses a modern district heating and cooling system and an all-electric home.

5.1 District Heating and Cooling

District heating is the name given to a heating system in which hot water in a centralized boiler is distributed in a closed loop through insulated pipes to multiple buildings for either air heating (through radiators), potable water heating, or both. Once the heat is transferred from the water to the building, the resulting cooler water is returned to the centralized boilers for reheating. District cooling is the production of cold water at a centralized location (usually the same location as district heating) and the distribution of the cold water by pipes to buildings to provide cold air and/or refrigeration.

The first district heating system worldwide began operating on March 3, 1882. On that day, the New York Steam Company sent high-temperature (up to 200 degrees Celsius) steam through pipes to buildings in lower Manhattan, New York City, to heat them. Burning coal to boil water created the steam. Heat losses from coal combustion and through poorly insulated and leaky pipes were significant, but coal was plentiful. Steam district heating still exists today in Manhattan, with Consolidated Edison providing steam to over 1,700 buildings. From 1882 to 1930, steam-based district heating powered by coal combustion (first-generation district heating) was used in many U.S. and European cities.

From 1930 to 1970, second-generation district heating emerged. This system was based on burning coal and oil to heat water and sending the hot water through pressurized pipes to buildings. Input water temperatures exceeded 100 degrees Celsius, and pipe losses were less than with steam heating, partly due to the use of concrete ducts around the pipes for insulation. These systems were installed worldwide but particularly in Eastern Europe after World War II.

Third-generation district heating uses prefabricated, insulated water pipes buried underground to reduce heat loss. The insulation allows input water temperatures to be below 100 degrees Celsius, which in turn reduces the energy needed to heat the water. The heat for water in these systems was originally provided by a combination of coal, natural gas, and biomass combustion and by recycling the heat of combustion from electricity generated by these fuels. More recently, heat has also been recycled from data centers and created from excess electricity produced by wind, solar, and geothermal plants. In some district heating systems, the heat is stored underground instead of in aboveground boilers.

Third-generation district heating evolved from 1970 to the present. It was developed and used first in Scandinavian countries. Today, 50 to 65 percent of building air and potable water heating in Denmark, Iceland, Sweden, Finland, Estonia, Latvia, Lithuania, Poland, Russia, and Northern China is from district heating. About 12.5 percent of all heat delivered in the European Union and about 8 percent delivered worldwide is district heating.[79]

Fourth-generation district heating emerged in 2015 and is currently evolving. It combines district cooling with district heating. District cooling involves a centralized chiller that stores water that has been cooled. Like with district heating, the cold water is sent through an insulated piping network to buildings to cool air in the buildings. With fourth-generation district heating, heat pumps are used to heat water in boilers and cool water in chillers. The heat pumps, which run on electricity, extract heat or cold from the air, water, or the ground.

Waste hot water is obtained after cold water is sent to a building to cool it, and the water absorbs and carries away waste heat. Waste cold water is obtained after hot water is sent to a building to warm it, and the water absorbs and carries away waste cold. If waste hot water from buildings, data centers, or manufacturing processes is used as the source of heat for a heat pump, the heat pump requires less electricity and thus runs more efficiently than if room-temperature water is used. Ideally, in a fourth-generation system, all electricity for a heat pump is obtained from 100 percent WWS.

An example of a fourth-generation district heating system is the Stanford Energy System Innovations project.[80] The round-trip efficiency of the heating portion of this system (ratio of the energy returned as heating after storage to the energy in the electricity used to heat the water for storage) is around 83 percent. That of the cooling portion is around 84.7 percent.[81] Hot water temperatures in a fourth-generation system are below 70 degrees Celsius to minimize losses.

In a 100 percent WWS world, fourth-generation district heating and cooling will be used as much as possible in densely populated areas, such as cities, where centralized systems are most advantageous. District heating can also be used on college campuses, military bases, and remote communities. Rural and many suburban homes and buildings are less ideal for district heating and cooling owing to the length of trenches and pipes required and the resulting energy losses through the pipes. Homes and buildings that are not on district heating and cooling loops will have their own domestic hot water tanks and use heat pumps for air and water heating and air conditioning.

In a WWS system, electric heat pumps, solar heat, or geothermal heat will provide district heat. WWS sources will provide the electricity for the heat pumps. The source of heat for the heat pumps will either be the air, the ground, water, or recycled waste heat. For example, waste heat from industry or buildings will be captured and piped back to a district-heating center. The heat produced for district heating will either be used immediately, stored in water tanks, or stored underground for later use.

The cold for district cooling will be provided mostly by electric heat pumps, where the electricity is from WWS. The source of cold will either be the air, the ground, water, or recycled waste cold air or water. The cold produced for district cooling will either be used immediately or stored in water tanks, ice, or aquifers for later use. In some cases, the cold for district cooling will be drawn directly from cold lake water and passed through a heat exchanger, where the cold water will absorb heat from the air inside a building, thereby cooling the building.[82]

An advantage of district heating and cooling is that it avoids the need for individual buildings to have their own heat pumps. It also allows heat and cold to be stored for later use. Such storage helps to keep the electric grid stable in a WWS world. A disadvantage of district heating and cooling is that they require dedicated water piping to all buildings in a town or city. This can be costly if the buildings are far apart. As a result, district heating is most effective in densely populated cities.

District heating and cooling requires storage of the heat and cold, respectively. The main short-term storage for heat is in water tanks; the main short-term storages for cold are in water tanks and ice. Heat and/or cold can also be stored seasonally underground in boreholes, water pits, and aquifers. These types of thermal energy storage are discussed next.

5.1.1   Heat and Cold Storage in Water Tanks

Tank thermal energy storage is the most common type of heat and cold storage worldwide. It involves heating or chilling water as it sits in a storage tank or before it is piped to the storage tank. Water tanks are used primarily as part of district heating and/or cooling systems.

Domestic hot water tanks for individual homes or buildings are also a form of tank storage, but such tanks are much smaller than are those used for district heating.

District heating tanks are made of concrete, steel, or fiber-reinforced plastic. Concrete tanks usually contain an interior polymer or stainless steel liner to minimize water vapor and heat diffusion out of the tank. They are also insulated on the outside. Steel tanks are insulated as well.

Hot water tanks are called **boilers**. The hot water is used to heat the air or water in a building, where the hot water is used for showering, bathing, cooking, hand washing, and drinking. Cold-water tanks are called **chillers**. The cold water is used for air conditioning and, sometimes, refrigeration.

Although community-scale boilers and chillers can be large, they are generally used to store heat and cold for only days to weeks. Underground thermal energy storage is less costly per unit energy. As such, it is now the main type of seasonal (between summer and winter) heat and cold storage. However, an advantage of aboveground boilers and chillers is that they can be built almost anywhere. Underground borehole, water pit, and aquifer storage can be built in more limited locations.

Water stored in tanks can be heated directly with solar or geothermal heat or with heat from a heat pump powered by electricity. Similarly, water in a tank can be cooled with a heat pump. Ideally, electricity used to heat water is excess electricity that is not needed on the grid. Using excess electricity to produce heat is an ideal way of decreasing the cost of electricity because it reduces the **curtailment** (**shedding** or wasting) of excess WWS electricity.

Using a heat pump to heat water in a boiler reduces winter energy demand by 68 to 81 percent compared with burning natural gas to heat water because heat pumps are much more efficient than are natural gas heaters or even electric resistance heaters. Thus, boilers powered by heat pumps reduce energy requirements significantly.

Hot water stored in a water tank heats building air when the water passes through a radiator. **Radiators** are heat exchangers. **Heat exchangers** are devices that allow a heated liquid or gas to transfer its heat to another liquid or gas without the two fluids mixing together or otherwise coming into direct contact. The transfer occurs by conduction, where heat energy is passed from one atom to the next in the metal or other material that separates the two fluids.

In the case of a radiator, water flows through pipes, often with fins to increase the surface area exposed to the air. Heat from the water in a radiator conducts through the metal in the radiator to the air, then the energized air molecules disperse throughout the room. Radiators can be placed in or against a wall, under a floor (**radiant floor heating**), or in a ceiling. They may or may not be accompanied by a fan. After hot water passes through a radiator, it is piped to subsequent radiators until it returns back cold to the district heating and cooling center. Ideally, the waste cold is then extracted by a heat pump to cool down the water in the cold water tank. Since cold is extracted from the piped water, that water is now warm and is sent back to the hot water tank.

Similarly, cold water piped to a building is used to cool air in the building after the cold water passes through a heat exchanger. The heat exchanger moves heat from the warm air in the building to the cool water, which warms up. The warm water is then circulated back to the district heating and cooling center. The waste heat is then extracted by a heat pump to warm the water in the hot water tank. Since heat is extracted from the piped water, that water is now cold and is sent back to the cold water tank.

5.1.2   Cold Storage in Ice

A type of thermal energy storage that is effectively a type of electricity storage is cold storage in ice. **Ice storage** has been used for decades in universities, hospitals, stadiums, and other large facilities. It involves freezing liquid water to produce ice when excess electricity is available or when electricity prices are low, then running water through coils embedded in the ice when cooling is needed. Water passing through the coils is chilled, and the cold water is piped to a building. In the building, the cold water is run through a radiator to air condition the building. In this way, ice storage avoids the need for daytime electricity use for air conditioning. Since summer electricity demand peaks during the late afternoon in many places, ice storage is a method of reducing peak electricity demand in these places. The main advantage of ice storage over battery storage is the lower cost of ice storage. Ice storage costs $35 to $40 per kilowatt-hour of thermal storage versus $100 to $200 per kilowatt-hour of electricity storage for a battery. The round-trip efficiency (cooling energy output to electricity input) of ice storage is around 82.5 percent.[83]

5.1.3 Underground Heat and Cold Storage

Whereas boilers are useful for meeting district or home heat storage for days to a week, longer-term heat storage can be obtained from underground heat storage reservoirs filled primarily during the summer. Such seasonal heat storage is called **underground thermal energy storage**. Three main types of underground thermal energy storage are borehole, water pit, and aquifer thermal energy storage.

5.1.3.1 Borehole Thermal Energy Storage

A **borehole thermal energy storage** system is effectively a large, underground heat exchanger that stores solar heat collected during summer for later use during winter. It consists of an array of boreholes drilled into soil. A plastic pipe with a U-shaped bend at the bottom is inserted down each borehole. The space around each pipe in each borehole is then filled with highly conducting grouting material to increase the molecule-to-molecule conduction of heat from hot water that passes through the pipe to the soil surrounding the grouting material. After hot water in the pipe transfers its heat to the soil, the water is now cold and is sent back to a solar collector, heat pump, or heat exchanger to collect more heat. Most ground heating occurs during summer. When hot water for buildings is needed during winter, cold water is sent down the pipe; heat that is stored in the soil transfers, molecule-by-molecule, through the grouting material and pipe wall to the water in the pipe, and the hot water flows to a boiler for distribution to the buildings.

A good example of a borehole storage system is the **Drake Landing Solar Community** in Okotoks, Alberta, Canada. Starting in 2004, 52 homes were constructed. On the garage roof of each home, solar collectors were installed. The collectors contain a glycol solution (mix of water and nontoxic glycol) that absorbs solar heat, particularly during long summer days. The heated solution is transferred from the roofs, through insulated pipes, underground to a building, where the heat from the solution is transferred, through a heat exchanger, to water stored in a short-term hot water storage tank (boiler). The water temperature in the boiler is maintained above 40 degrees Celsius.

Hot water in excess of what can be stored in the boiler is piped to the borehole field, which is about 35 meters in diameter. In this field, 144 holes were drilled to 35 meters depth, and each was filled with a pipe with a U-shaped bend at the bottom. Insulation was placed on top of the pipes, and a grassy field, used as a community park, was grown on top of the insulation. The boiler delivers hot water through a separate pipe to each hole in the borehole field. Each pipe extends to the bottom of its respective borehole. The heat from the water in each pipe is transferred by conduction to the surrounding soil, raising the soil temperature up to 80 degrees Celsius by the end of summer. Cold water in each pipe returns upward and is sent back to the boiler to be heated again.

During summer and autumn, all home heating needs in this community are met by solar heat supplied to the boiler, and excess heat is sent to the borehole field. During winter and other times of the year that the boiler alone cannot satisfy all building heat demands, cold water is piped through the boreholes to collect heat and bring it back to the boiler. The hot water from the boiler is then distributed to the 52 homes for air and domestic water heating. With this system, up to 100 percent of winter heat demand can be satisfied by summer heat collection.

The Drake Landing system has been operational since 2007. The efficiency of the underground system, which is defined as the fraction of heated fluid entering underground storage that is ultimately returned during the year for air or water heat, is about 56 percent.[84]

In sum, Drake Landing is a district heating system with a community hot water tank but where much of the wintertime hot water is stored underground between summer and winter. The technology is repeatable for most any district heating system and requires little land. The borehole field is not visible and can be used simultaneously as grazing land, parkland, or open space, or for the solar collectors needed to provide heat stored in the field.

Transition highlight
Another borehole thermal energy storage facility serving a district heating system resides in Braedstrup, Denmark. The Braedstrup district heating system services 1,500 customers. The borehole field, installed in 2012, is honeycomb shaped and has 48 boreholes, each 45 meters deep and 3 meters apart. U-shaped pipes are used, just as in Drake Landing. The storage field, which has a volume of 19,000 cubic meters for heating and a storage capacity of 400 megawatt-hours of heat, is heated to 55 to 60 degrees Celsius during summer and cooled down to 12 to 15 degrees Celsius during winter, after heat is drawn from the field. The boreholes, along with two additional steel water tanks totaling 7,700 cubic

meters, are fed by 18,600 square meters of solar collectors. The round-trip efficiency is about 63 percent, slightly higher than that of Drake Landing. In other words, of all the heat that enters storage, 63 percent is used by consumers and the rest is lost to soil or air. The investment cost is $0.77 per kilowatt-hour.[85]

5.1.3.2 Pit Thermal Energy Storage

Pit thermal energy storage consists of a lined pit dug into the ground and filled with a storage material, such as water or a mix of gravel and water, then covered with insulation and soil. The pit can be small or large. The material in the pit is supplied with heat primarily during the summer. The heat is extracted mostly during the winter and sent to a boiler, where it is distributed via district heating to buildings for air and water heating.

As of 2021, the largest pit storage worldwide was the water pit storage facility in Vojens, Denmark, completed in 2015. It is 13 meters deep, 610 meters in circumference, and filled with 200,000 cubic meters of water. The pit is lined underneath and on the sides with welded plastic to ensure water does not leak through it. Soil, a few meters thick outside the lining, becomes warm, insulating the storage pit. The surface of the water is also lined with a floating plastic cover topped with 60 centimeters of insulating expanding clay and a draining system to remove rainwater. The water temperature during summer can reach 95 degrees Celsius but is maintained at a maximum of 80 degrees Celsius to extend the life of the liner.[86] Almost 5,500 solar thermal panels heat the water. The pit is connected to the district heating system of Vojens, which has 2,000 customers. The solar collectors plus pit storage facility provide 45 to 50 percent of the city's annual heat demand.[87] The cost of wintertime heat from the solar heating combined with this pit storage system in Vojens is competitive, even without subsidy, with the cost of heat from gas boilers, owing to economies of scale. In fact, heating bills declined by 10 to 15 percent with the implementation of this seasonal heat storage system.[86]

Transition highlight

A modified version of the Vojens water pit storage plant was built for the district heating system in Gram, Denmark, in 2015. Gram is a town with 2,500 residents, 99 percent of whom are connected to the district heating system. The storage pit in this case is 15 meters deep with 10 meters of it below ground and 5 meters above ground. A sloped dam was built to raise the height of the pit. The water volume in the pit is 122,000 cubic meters. The facility also has 44,800 square meters of solar collectors, which directly and through pit storage, provide 61 percent of the town's heat demand.[88]

Three other pit storage systems coupled with solar collectors and district heating reside in Marstal, Dronninglund, and Toftlund, Denmark.

The Marstal system, completed in 2012, serves 2,200 residents. It consists of 33,365 square meters of solar collectors, tank water storage, and 75,000 cubic meters of water pit storage with a thermal capacity of 6 gigawatt-hours of heat and maximum charge and discharge rates of 10 megawatts. It has a maximum measured temperature of 88 degrees Celsius, an efficiency of 52 percent, and an investment cost of $0.52 per kilowatt-hour.[85]

The Dronninglund system, which serves 3,300 residents, consists of 37,573 square meters of solar collectors and 62,000 cubic meters of water pit storage. It began operating during March of 2014. The storage has a capacity of 5.4 gigawatt-hours of heat, a maximum charge and discharge rate of 27 megawatts, a maximum measured temperature of 89 degrees Celsius, an efficiency of 81 percent, and an investment cost of $0.51 per kilowatt-hour.[85]

The Toftlund system, which serves 3,250 residents, consists of 27,000 square meters of solar collectors, and 85,000 cubic meters of water pit storage. Heat is supplied from the solar collectors or from a heat pump, where the heat can be obtained from an electric boiler. The storage system began operating during June of 2017. The storage has a maximum capacity of 6.9 gigawatt-hours of heat, a maximum charge rate of 18 megawatts, a maximum discharge rate of 8 megawatts, and an investment cost of $0.70 per kilowatt-hour.[89] Its average efficiency is 72 percent.

5.1.3.3 Aquifer Thermal Energy Storage

Aquifer thermal energy storage is similar to pit storage, except that the water used for storing heat with aquifer storage is naturally occurring water in underground layers of groundwater, called aquifers. Aquifers can also be used to store cold water. Aquifers contain permeable sand, gravel, sandstone, or limestone. Aquifers can be used for underground thermal energy storage if they are encapsulated by impervious layers above and below them and if natural groundwater flow is slow.

With aquifer storage, one pair or several pairs of wells are drilled into an aquifer. During summer, cold water is extracted from one well in each pair, heated with building heat (thereby cooling the building), then returned to the other well, where it mixes with and heats the water in that well. During winter, warm water from the warm well is extracted, the heat is removed to heat the building, and the cold water is returned to the cold well.

Because aquifers cannot be insulated, heat storage at temperatures greater than 50 degrees Celsius is efficient only for large, deep aquifers with volumes greater than 50,000 cubic meters and a low surface-to-volume ratio.[90] As such, the maximum temperature in a shallow aquifer for heating is generally limited to 20 degrees Celsius. For cooling, shallow aquifers are generally used.

Some aquifer systems built to date include the following.

The **Eindhoven aquifer system** is a 1.7-million-cubic-meter heat and cold storage reservoir under Eindhoven University of Technology, the Netherlands. Completed in 2001, it supplies both direct cooling during summer and low-temperature heating during winter to heat pumps to maintain comfort in 20 university buildings. Sixteen wells for cooling and 16 wells for heating were constructed. The aquifer lies between 28 and 80 meters below the surface and has a natural temperature of 11.8 degrees Celsius. The temperature in the aquifer varies from 6 to 16 degrees Celsius during the year owing to the extraction or addition of heat. The maximum heating and cooling energy delivered is 15 to 30 gigawatt-hours per year. The maximum charge and discharge rate of the reservoir for both heating and cooling is 17 megawatts.[90]

The **Stockholm airport aquifer system**, completed in 2011, consists of a 1-million-cubic-meter storage reservoir under the Arlanda Airport in Stockholm. It is used for both heating and cooling. During winter, heat stored in the aquifer from the summer is used to preheat air used for ventilation in the terminal buildings and to melt snow at the gates. The extraction of heat cools the aquifer during winter. As such, by summer, the aquifer water is cold. The cold water is extracted during summer and used to provide air conditioning for the terminals. The aquifer well lies 15 to 30 meters below the surface. The maximum heating and cooling energy delivered is 20 gigawatt-hours per year. The maximum charge and discharge rates of the reservoir for both heating and cooling are both 10 megawatts.[90] The temperature in the aquifer varies from 2 to 25 degrees Celsius during the year owing to the extraction or addition of heat.

> **Transition highlight**
> The **Riverlight Project aquifer system** consists of a 180,000-cubic-meter storage reservoir under the Riverlight Project in London, along the River Thames. Completed in 2014, it serves 806 residential apartments in addition to several commercial businesses. The water pipes are linked to heat pumps, which provide cooling during the summer and heating during the winter. The aquifer system consists of four warm wells and four cold wells drilled into the London chalk aquifer. The well depths are about 100 meters each. The thickness of the aquifer in this layer is about 25 meters. The maximum heating and cooling energy delivered is 1.4 gigawatt-hours per year. The maximum charge and discharge rate is 3.7 megawatts for cooling and 1.6 megawatts for heating.[90] The temperature in the aquifer varies from 8 to 24 degrees Celsius during the year owing to the extraction or addition of heat.

5.1.4 Stanford University 100 Percent Renewable Electricity, Heat, and Cold Energy System

Stanford University, in California, developed a district heating and cooling system powered by solar PV. The PV also powers other electrical needs in the university. In fact, sufficient PV was purchased to provide 100 percent of the campus annual electricity demand and almost 100 percent of the heating and cooling demands with WWS as of 2022. In 2015, the electricity, heat, and cold demands were about 208, 158, and 225 gigawatt-hours per year, respectively. For an average California home that uses 7 megawatt-hours per year of electricity, this corresponds to the demand from roughly 32,000 homes.

The **Stanford Energy System Innovations** project is a fourth-generation district heating and cooling system. It was first conceived in 2009 and became operational in March 2015, replacing a natural gas cogeneration plant that supplied 80 percent of the electricity and heat for the university campus.

The system consists of a 2.3-million-gallon insulated steel hot water tank, two 4.75-million-gallon insulated steel cold water tanks, three waste-heat-driven 2,270-tonne heat pumps (heat recovery chillers) that raise the temperature of the hot water tank and expel cold to the cold water tanks, three hot water generators (boilers) to provide additional heat to the hot water tank, four 2,720-tonne cold water generators (chillers) to provide additional cold to the cold water tanks, cooling towers (13,200 tonnes total) to cool water further for the cold water tanks and provide additional waste heat for the heat pumps, an elaborate 35.4-kilometer hot water pipeline system, a 42.2-kilometer cold water pipeline system, a heat recovery and distribution system that takes advantage of the fact that different parts of the university have both heating and cooling needs at the same time, a planning model for optimizing the distribution of heating and cooling around campus every 15 minutes, and a high voltage substation that receives electricity from the grid when necessary.[80]

The heat recovery system collects waste heat from buildings and moves it to a hot water loop, where it is fed to the heat pumps. The recycling of waste heat eliminates the discharge of waste heat to the air.

The project requires electricity, as do all buildings on campus. To provide this electricity, the university first built 5 megawatts of solar PV on university rooftops, then signed a power purchase agreement over 25 years to generate 67 megawatts from a new solar PV farm in Rosamond, California, dedicated to the campus. The farm came online during December 2016. The university then signed another power purchase agreement for dedicated power from an 88-megawatt solar PV farm near Lemoore California. However, a 2019 wildfire damaged the Rosamond farm. Repairs were completed in early 2022. The combination of the district heating and cooling project and the 160 megawatts of solar PV mean that 100 percent of the Stanford University campus annual electricity demand and nearly 100 percent of the university heating and cooling demand are supplied by clean, renewable energy.

The 100 percent renewable electricity portfolio of the campus reduces its overall greenhouse gas footprint by 80 percent compared with 1990 levels and its water use by 18 percent compared with 2014. The water savings are due mostly to eliminating the water needed to cool the natural gas turbines, which were removed. The remaining greenhouse gas emissions are from university vehicles, natural gas steam heating units in some buildings that have not yet been tied to the district heating and cooling system, and emergency diesel generators. Plans exist to transition these remaining sources by 2025.

5.2   Individual Heating and Cooling Units in Buildings

Buildings not connected to district heating and cooling systems need their own source of heating and cooling. In this section, rooftop solar water heaters, heat pump air and water heaters, and passive heating and cooling technologies are discussed.

5.2.1   Rooftop Solar Water Heaters

Rooftop solar water heaters (or domestic hot water systems) have been used for over 130 years. In 1891, inventor **Clarence Kemp** of Baltimore, Maryland, patented the first commercial solar water heater. The technology was adopted regionally right away. By 1897, solar water heaters had been installed on the rooftops of 30 percent of the homes in Pasadena, California.[91]

In 1909, **William Bailey** redesigned the rooftop solar water heater by separating the solar collector from the storage tank. Cold water from the storage tank in the house or garage would now flow through a copper pipe that was coiled as it passed through a shallow, insulated box under glass that was placed on the roof. The coils maximized the area of the pipe exposed to the sun, which heated the water in the pipe. The hot water then flowed to the top of the tank. The name given for this new device was the Day and Night Solar Heater.[91] Whereas 1,000 of the units were sold in 1920 in Los Angeles alone, the number declined to only 40 in 1930, not only owing to the Great Depression but also because natural gas was now being piped into people's homes, and solar heating could not compete in price. However, in southern Florida, 100,000 solar water heaters were sold prior to World War II. In the mid-1950s, Frank Bridgers designed the Bridgers & Paxton Solar Building, which is the world's first commercial office building to use solar water heating plus passive heating.

Today, all solar hot water systems include a solar collector and a storage tank, but there are two primary types of systems, active and passive. Active systems use a pump to circulate water and use either direct or indirect circulation.

An **active direct-circulation system** has a pump that circulates water through a rooftop solar collector. The water heated by the collector goes to a storage tank, where it sits or is used immediately. Direct systems work well so long as the temperature of water in the collector does not drop below the freezing point of water.

An **active indirect-circulation system** has a pump that circulates a nonfreezing heat-transfer fluid through the rooftop solar collector. The heat from the fluid is then transferred to water with a heat exchanger. The water is then stored in the storage tank or used directly.

Passive systems generally use city water pressure to push water directly from a city water source through a rooftop solar collector that holds a fair amount of water. Sunlight heats the water in the collector. As water is used in the building, the water in the collector moves to the household storage tank, where a heat pump water heater warms the water further. Thus, the solar collector serves to preheat the water, reducing the energy required by the heat pump.

Although solar hot water systems are useful, they are often more costly and less efficient than is using a heat pump water heater powered exclusively by rooftop solar PV. Having only rooftop PV, but with more PV panels to provide electricity for a heat pump to heat water is more cost effective than having both PV (but less of it) and a solar water heater on a roof. The reason is that hiring two different contractors is almost always more expensive than hiring one to do a larger job.

5.2.2 Heat Pumps

Heat pumps are efficient devices that extract heat or cold from the air, the ground, or water, or from a hot or cold waste stream, and move the heat or cold to where it is needed. A heat pump that extracts heat or cold from the ground is a ground-source heat pump, whereas one that extracts heat or cold from the ambient air or from a waste stream of air is an air-source heat pump. One that extracts heat or cold from water, such as from a swimming pool or lake, is a water-source heat pump.

The advantage of a ground- or water-source heat pump over an air-source heat pump is that temperatures under the ground or in the water are relatively stable, even when the air outside is very cold or very hot; thus, ground- and water-source heat pumps are more efficient under extremely cold or hot air conditions than are air-source heat pumps. The advantage of an air-source heat pump is that it is easier and less expensive to install and maintain because none of its parts is buried underground or in the water.

Traditionally, ground-source heat pumps extracted heat and cold from the ground through coils buried in a shallow backyard pit spread over a large area and covered with soil. More recently, the development of a technology to drill two narrow but deep holes vertically has reduced the cost of installing ground-source heat pumps significantly.[92]

When heat pumps are used for cooling, they operate just like air conditioners or refrigerators, which move heat from a room or refrigerator, respectively, to the outside of the house or outside of the refrigerator, respectively. The difference between a heat pump and an air conditioner or refrigerator is that the heat pump provides both heating and cooling. An air conditioner and refrigerator provide only cooling.

One type of heat pump is the **ductless mini-split heat pump.** As its name implies, it does not require ducts to move heat or cold air around a building. Instead, an indoor air handling unit is placed in each of several rooms or zones of the building. The indoor units are connected to an outdoor compressor/condenser unit by thin pipes containing a power cable, suction tubing, refrigerant, and a condensate drain line.

For air conditioning (cooling of warm indoor air), a fan in the indoor unit blows the warm air from the room over evaporator coils. Simultaneously, a thin pipe brings liquid refrigerant to the evaporator. Inside the evaporator, the liquid pressure is dropped, cooling the refrigerant. Heat is transferred by conduction from the room air to the cold refrigerant, bringing the refrigerant to its boiling point and evaporating the refrigerant. The room heat is therefore converted to latent heat stored in the refrigerant molecules. The room air that passed over the coils is now cold and flows back to the room, cooling the room. The gaseous refrigerant containing the heat then travels by tubing to the outside unit, where it is heated further by compression. In a condenser, the hot gas then transfers some of its heat by conduction to the cooler (relative to the hot refrigerant) outside air, causing the refrigerant's temperature to drop below its boiling point and condense back to a liquid. The condensation releases the stored latent heat, which also passes to the outside air. In sum, a ductless heat pump moves heat from a room to the outside air, cooling the room, using only electricity as a power source.

For heating cold indoor air, the heat pump runs in reverse. Gaseous refrigerant in the outdoor unit is compressed and lique-fied, with both processes heating the refrigerant. The hot liquid is then piped to the evaporator coils. Simultaneously, cold air from the room is blown over the coils. At the coils, heat from the refriger-ant transfers by conduction to the cold air blowing over the coils, warming the air, which then flows back to the room. The refriger-ant is now cold and transfers back to the outside unit. Dropping the pressure cools the liquid refrigerant even more. At the outside unit, heat from the outside air is transferred by conduction to the cold liquid refrigerant, raising its temperature above its boiling point and evaporating the refrigerant while cooling the outside air. The cycle then starts again. In sum, heat from outside air is transferred to the room to warm it.

Air-source heat pumps become inefficient for heating a build-ing when the outdoor temperature is very low and inefficient for cooling when the outdoor temperature is very high. For very low tem-peratures, sometimes a resistance-heating element is used as a backup to ensure a room stays warm. Alternatively, ground- and water-source heat pumps largely solve this problem because ground and water tem-peratures rarely become extremely low or high. Thus, ground- and water-source heat pumps are recommended for locations that experi-ence extreme seasonal heat or cold or both. Air-source heat pumps are recommended for milder climates.

Whereas heat pump air heaters and air conditioners move heat between a building and the outside, **heat pump water heaters** often extract heat from air within a room they sit in. This can reduce the tem-perature of the air in the room by 1 to 3 degrees Celsius. In the summer, such air cooling is advantageous, because the door to the room can be opened, providing additional cooling to the house and reducing energy requirements for air heating. In the winter, the extra cooling slightly increases the air heating requirement.

An air-source heat pump water heater works as follows. First, liquid refrigerant inside the water heater is brought to an evaporator. At the evaporator, the refrigerant's pressure is dropped, cooling the refrigerant. A fan inhales warm (relative to the cool refrigerant) room air and blows it over the evaporator's coils. Heat from the room air transfers by conduction to the refrigerant, raising the refrigerant's temperature to its boiling point and evaporating the refrigerant. During evaporation heat from the room is stored as latent heat in the gas molecules of the refrigerant. As such, the inhaled room air is now cooler and is expelled back to the room.

The gaseous coolant is heated further by compression in a compressor. The gas then goes to a condenser, where it condenses back to a liquid, releasing its latent heat and compressional heat to water in the tank, heating the water. The net result is hot water and a slightly cooler mechanical room.

Transition highlight

In 1938, J. Ross Moore of North Dakota commercialized the first electric clothes dryer after tinkering with the technology since 1915. The first electric heat pump clothes dryer was commercialized in Europe in 1997. By 2007, heat pump dryers were being manufactured in Japan as well. By 2009, about 4 percent of European dryers were heat pump dryers.[93] Today, heat pump dryers are available worldwide. Heat pump dryers require about half the electricity of conventional electric dryers. However, heat pump dryers themselves cost more upfront. This upfront cost difference can be erased for new construction homes if a ventless heat pump dryer is purchased because the construction cost of vents is usually more than the cost of the dryer itself. As such, a ventless heat pump dryer in a new construction home will almost always have a lower lifecycle cost than a conventional electric dryer due to both the energy cost savings and the construction cost savings, despite the higher dryer cost. A heat pump dryer will also have a lower lifecycle cost if it is installed in an existing building and used a lot.[93]

Because heat pumps move, rather than create, heat and cold, they reduce the energy required to heat or cool air or water in a building by a factor of 3 to 5 compared with conventional gas heaters or electric resistance heaters. The efficiency of a heat pump is defined by its coefficient of performance, which is the ratio of useful heating or cooling required to work (electricity input) required. The coefficient of performance of a heat pump exceeds 1, which means that the heat pump moves more than one unit of heat or cold for every unit of electricity required to run the heat pump.

Air-source heat pumps have a coefficient of performance of 3.2 to 4.5, whereas ground-source heat pumps have a coefficient of performance of 4.2 to 5.2.[94] This compares with **electric resistance heaters,** which have a coefficient of performance of about 0.97, and fossil-fuel-powered boilers, which have a typical coefficient of performance of about 0.8. Since only 1 joule (unit of energy) of electricity is needed to move 3.2 to 5.2 joules of hot or cold air with a heat pump, heat pumps reduce power demand compared with natural gas boilers by 75 to 85 percent. The use of heat pumps for all air and water heating worldwide is estimated to reduce world all-purpose end-use power demand compared with a business-as-usual case by about 13.3 percent.[5]

5.2.3   Passive Heating and Cooling in Buildings

The energy required to heat and cool buildings can be reduced significantly with **passive heating and cooling techniques.** These techniques include installing thermal mass, ventilated facades, window blinds, window films, and night ventilation.[95] Each is discussed, in turn.

5.2.3.1 Thermal Mass

Thermal mass is a building material used to absorb, store and release heat in a building in such a way as to keep building temperature relatively constant during day and night. The two main types of thermal mass materials are sensible heat storage materials and latent heat storage materials.

Sensible heat storage materials are materials that do not change temperature much during the day or night, thus helping to keep buildings near constant temperature. Good sensible heat storage materials are materials with a combination of a high heat capacity and a moderate thermal conductivity.

Heat capacity is the energy required to increase the temperature of 1 cubic meter of a substance by 1 degree Celsius. Substances with high heat capacities can absorb lots of sunlight during the day without their temperature increasing much and release lots of heat radiation at night without their temperature decreasing much. On the other hand, substances with low heat capacities warm rapidly during the day upon absorbing sunlight and cool rapidly at night upon releasing heat radiation. The heat capacity of liquid water is 3.5 times that of sand. As such, the same amount of sunlight heats up a given volume of sand 3.5 times as much as it does water. This is why ocean beach water feels relatively cool during the day whereas sand feels hot. Conversely, the loss of the same amount of heat radiation at night cools sand 3.5 times as much as it cools water.

Conduction is the transfer of energy from molecule to molecule. The thermal conductivity of a substance is a measure of its ability to conduct heat in the presence of a temperature difference between one end of the material and the other. Steel is extremely conductive. Pinewood, on the other hand, is hardly conductive.

A good thermal mass material should have a moderate thermal conductivity. If the thermal conductivity is too high, such as with steel, the material will transfer its heat (or cold) too fast to the air or other materials, even if the material has a high heat capacity. If the conductivity is too low, such as with wood, the material will not conduct its heat (or cold) to the air when the heat or cold is needed.

Effective thermal mass materials are materials with high heat capacities and moderate thermal conductivities. Such materials include water, concrete, marble, granite, or brick in walls or floors. These materials modulate the temperature of the building in comparison with using wood or steel. Another way to think of it is that these materials store heat during the day without raising the temperature of the building and slowly release this heat to the building air at night, keeping the building warm at night. Water can be used as a thermal mass material when it is placed in tubes in a wall or under a floor. A disadvantage of sensible heat storage is that it requires a large volume of storage material to have an impact on building temperature.

Latent heat storage materials are materials that modulate building temperature by changing phase near room temperature. As such, they are also called phase-change materials. Phase-change materials used in buildings include paraffin wax, fatty acids, and salt hydrates.

Much less mass of a phase-change material is needed than of a high heat capacity material to modulate building temperature. When a phase-change material absorbs heat, its temperature quickly reaches its melting point, at which point the material absorbs more heat in order to melt without raising its temperature further.

For example, if paraffin wax is distributed within south-facing walls of a Northern Hemisphere building, the wax absorbs heat, increasing its temperature until the melting point is reached. Melting points can be 37 to 65 degrees Celsius, depending on which wax is used. During melting, the wax still absorbs heat, but that heat is used to melt the wax, so the wax temperature does not rise further until all the wax is melted. By absorbing heat without increasing its temperature, the wax keeps the building cool. After the sun goes down, the wax releases heat to the air, warming the air and cooling the wax. Once it cools to its freezing point (similar to its melting point), the melted wax freezes (solidifies), releasing latent heat to the air, warming the air more without dropping the wax temperature further. Thus, phase-change materials keep the temperature of a wall relatively constant during the day and night. A phase-change material can be encapsulated within a wallboard, mixed in concrete, or mixed with insulation material.

5.2.3.2 Ventilated Façades

A **ventilated façade** is a weather-protective wall, separated from the main wall of a building by air. Air in the space between the façade and the main wall can flow and mix with ambient air, providing ventilation. The air space also provides acoustic benefits to the building. The façade itself can be made of a thermal mass material to modulate temperatures.

5.2.3.3 Window Blinds

Window blinds are used to control sunlight into buildings. Lowering window blinds reduces sunlight, including damaging ultraviolet sunlight, into a building during hot, sunny days. Raising the blinds increases such sunlight penetrating into the building during cloudy days.

5.2.3.4 Window Films

Similarly, putting a transparent or translucent **window film** on a window reduces the penetration of sunlight into a building by increasing the window's solar reflectivity. Window film also reduces the outgoing heat radiation from the building. An advantage of window film is that it can significantly reduce the penetration of ultraviolet radiation into a building, reducing damage to furniture and people's skin and eyes. During the winter, though, the reduction in sunlight penetrating into a room due to the window film may be greater than the reduced heat loss from the room, resulting in a cooler room than desired. However, in the annual average, a window film is usually beneficial in terms of heat and always beneficial in terms of reducing ultraviolet radiation.

5.2.3.5 Night Ventilation

Night ventilation is the use of natural ventilation to remove heat from a building during the night. During the day, thermal mass in a building heats up owing to its absorption of solar and heat radiation. During the night, thermal mass releases its heat to the building, keeping the building warm. However, in hot climates, building heat can accumulate during the day and night over periods of days to weeks, even with thermal mass. To shed some of this heat, night ventilation is useful. Night ventilation takes advantage of either nighttime wind or heat-generated pressure to push heat added to the air inside the building out of a stack on the roof to the outside air. In all cases, cool outside air flushes the warm inside air at night. Since the inside air is now cooler, the thermal mass in the building cools down more as well. The next day, the thermal mass does not warm so much as it does with no nighttime ventilation.

5.3   WWS Electric Appliances and Machines

A key step in the electrification of all energy sectors is to ensure electric appliances and technologies that replace fossil-fuel ones perform at least as well and at a similar or lower cost. Here, some electric replacement technologies are described.

5.3.1   Electric Induction Cookers

In a 100 percent WWS world, **electric induction cookers** (induction burners) will replace cookers that run on natural gas, propane, wood, waste, dung, or coal. Induction cookers are also an improvement over electric resistance cookers. An induction cooker consists of a ceramic plate with a coiled wire beneath it. When the cooker is turned on, an electric current runs through the coil, generating a fluctuating magnetic field, but not heat, in the cooker itself. This is why the cooker does not feel hot when it is touched.

Induction cookers work only with pots and pans that stick to a magnet. The base of such cookware must be made of a high-resistance material, such as iron or stainless steel, not a highly conducting material, such as aluminum or copper. Once an iron or stainless steel pot or pan is placed on the cooker, the magnetic field in the cooker induces many smaller electric currents in the cookware's metal base. Because iron does not conduct well, much of the electricity in the small currents turns to heat by resistance, heating the metal. The heat is produced in the metal base of the pot or pan, but not in the cooker. As such, the cooker itself feels only warm owing to conduction of heat from the base of the pot or pan to it.

Water boils with an induction cooker in about half the time that it does with a natural gas cooker. Induction cookers also result in less scorching of food because they have fewer hot spots within a pot or pan than do natural gas cookers.

Electric induction cookers have the potential to eliminate enormous loss of life. This is because millions of children and adults, primarily in developing countries, die each year as a result of breathing air pollutants from the indoor burning of wood, waste, dung, and coal for home cooking and heating. Induction cookers can eliminate the need for such burning. Induction cookers can also save time for the hundreds of millions of people who collect their own wood or waste daily. Single induction cookers cost $40 to $100. All that is needed aside from the burner is an electricity source and an iron or stainless steel pot or pan. While bringing electricity to remote villages in developing countries is a challenge, the advent of low-cost solar PV and batteries now permits the widespread adoption in remote communities of clean, renewable electricity and electric appliances, such as induction cookers.

5.3.2 Electric Fireplaces

Wood and natural gas fires in fireplaces are fire risks and sources of carbon monoxide, nitrogen oxide, organic gas, and particle air pollution inside and outside of a home. Such fires are cozy and warm and give a home character, but they also produce indoor and outdoor air pollution. Wood fires can also cause smoke damage and unpredictable impacts from their embers that burst from a fireplace into a living room. Fortunately, warm and visually appealing electric fireplaces are now available to replace wood and gas fireplaces. These eliminate the air pollution and smell of wood or gas burning while providing the same warmth and coziness.

5.3.3   Electric Leaf Blowers

Possibly the most annoying fossil-fuel machine today is the gasoline **leaf blower**. It not only smells and creates noxious air pollution, but it is also noisy, since it has no muffler or other means of noise suppression. Both corded and battery-powered electric leaf blowers are available. They perform the same task as a gasoline leaf blower but do not smell and are quiet. Some battery-powered leaf blowers run for 1 to 3 hours at low speed or 10 to 30 minutes at full speed and take 30 to 100 minutes to charge, depending on battery size. Most models allow the battery to be swapped, so an electric leaf blower can be used continuously if multiple charged batteries are available for swapping.

5.3.4   Electric Lawnmowers

Another noisy machine that creates a smell and unhealthy air pollution is the gasoline **lawnmower**. Cost-competitive (over the lifetime of the lawnmower) battery-powered lawnmowers are readily available to replace gasoline lawnmowers. Not only do electric lawnmowers eliminate the exhaust-related air pollution (unburned hydrocarbons, oxides of nitrogen, carbon monoxide, and particulate matter) that a gasoline lawnmower emits, but they also start instantly, are quiet, and require less maintenance (since they have no engine and fewer parts) than does a fossil-fuel mower. Although the upfront cost of an electric lawnmower may be more than that of a gasoline mower, electric mowers have no gasoline cost and need less equivalent electricity than do gasoline mowers because the work output to energy input ratio of electricity powering a motor is about 4 times that of gasoline powering an engine. As such, the energy and overall cost to run an electric mower over 10 years is less than that to run a gasoline mower even if the upfront cost of the electric mower is higher.[96] Runtimes of electric lawnmowers are currently 40 to 80 minutes before recharging or battery swapping is needed.

5.3.5   Other Appliances and Technologies

All other home, business, and industrial technologies that run on fossil fuels or biofuels have an electric counterpart that is needed as part of a 100 percent clean, renewable energy economy. One example technology is the chainsaw.

The first **electric chainsaw** for woodcutting was invented in 1926 in Germany by Andreas Stihl. He invented a gasoline-powered chainsaw 3 years later. Both electric and gasoline chainsaws are available commercially today. Advantages of electric over gasoline chainsaws are that they have minimal noise, emit no fumes, require no mixing of oil and gasoline, require minimal maintenance, are lighter, and cost less. Cordless versions have traditionally lasted about an hour before needing a recharge and have been used for smaller jobs than gasoline chainsaws. However, with the improved storage capacity of batteries and with battery swapping capabilities, both limits are no longer an issue.

5.4   Increasing Energy Efficiency and Reducing Energy Use

Two additional necessary components of a 100 percent WWS world are **reducing energy use** and **increasing the energy efficiency** of buildings and electric appliances, machines, and tools. Both of these methods are referred to as **demand-side energy conservation** measures. Such measures reduce the need for energy. Demand-side measures also reduce the cost of electricity.

Reducing energy use involves primarily changing behavior. Some examples of reducing energy use include

- Using public transit or telecommuting instead of driving;
- Carpooling and minimizing the number of driving trips required;
- Teleconferencing instead of flying on an airplane to a conference;
- Eating foods that require less energy to produce;
- Eating locally sourced food instead of food transported over a long distance;
- Ensuring electronic equipment is shut off when it is not in use.

Increasing energy efficiency involves relatively low-cost investments that result in large energy savings. Some example energy efficiency measures are

- Replacing incandescent and compact fluorescent light bulbs with light-emitting diode (LED) light bulbs;
- Replacing electric resistance air and water heaters with heat pump heaters;
- **Weatherizing** (sealing) windows, doors, and fireplaces;
- Improving wall, floor, ceiling, and pipe insulation;

- Performing building energy audits to identify where energy is wasted;
- Implementing **green building standards** in new-building construction to minimize building energy use;
- Adding advanced lighting controls;
- Using ductless heating and air conditioning to eliminate heat and cold losses from pipes;
- Using triple-paned windows to reduce heat loss from windows;
- Installing thermal mass in walls and floors to modulate temperature changes in a building;
- Installing a ventilated façade to shield a building from extreme heat or cold;
- Installing window blinds to control sunlight into buildings;
- Applying window films to reduce heat loss and sunlight intake;
- Installing night ventilation cooling in a building;
- Improving airflow management in a building;
- Installing passive solar heating in a home;
- Using more energy-efficient appliances;
- Improving data center design;
- Improving the efficiencies of solar cells, wind turbines, batteries, and electric cars;
- Designing cities to facilitate the use of public transit and bicycles and to improve traffic flow; and
- Constructing high-rise apartments instead of single family homes to reduce construction materials and heat loss.

For example, weatherizing a home by using caulk or another type of sealant to stop leaks around the edges of windows, doors, and fireplaces costs only a few to tens of dollars yet can save hundreds to thousands of dollars over several years by reducing continuous leaks of either hot or cold air during winter or summer, respectively. Similarly, purchasing an energy-efficient appliance when a new appliance is needed can save hundreds to thousands of dollars over the lifetime of the appliance.

Increasing energy efficiency and reducing energy use avoid the use of fossil fuels, thus avoiding pollution. As such, they have the same effect as but are usually less expensive than replacing fossil fuels with WWS. Because increasing efficiency and reducing energy use are low cost, they should be prioritized highly in any strategy to address global warming, air pollution, and energy security. However, increasing efficiency and reducing energy use cannot solve the problems on their own. Strategies to solve these problems must include not only increasing efficiency and reducing energy use, but also transitioning to 100 percent clean, renewable energy and storage and eliminating non-energy emissions.

5.5   All-Electric Home

Part of the process of transitioning society to 100 percent clean, renewable energy and storage is to transition individual buildings, including residences. To illustrate the components and benefits of transitioning a home, I provide an example of a 100 percent WWS new-construction home, my own home.

The home, completed in 2017, has two floors with approximately 278 square meters (3,000 square feet) of home floor space and 46.5 square meters (500 square feet) of garage space. The structure itself was made of prefabricated steel, 80 percent of it recycled.[97] The advantages of a prefabricated steel structure are that it eliminates wood waste during assembly of the structure, it eliminates mold and termites associated with the structure itself (the exterior of the house and floors are still wood), and it produces walls that are perfectly flat, with corners that are exactly at 90-degree angles. The precise construction reduces the risk of air leaks and mistakes in constructing the rest of the house.

The home uses double-glazed windows that include a krypton-gas-filled cavity in lieu of a third glazing to suppress conduction and convection between glazes and to reduce 99.5 percent of incoming ultraviolet radiation. The low-conductivity fiberglass window frames have insulating foam within them and triple weather stripping around them. The windows are insulated to a similar degree as triple-paned windows.

No natural gas pipes are connected to the property. Instead, the house generates 20 percent more electricity in the annual average than it uses, including for electric vehicles, from forty-three 320-watt roof-mounted solar PV panels (total nameplate capacity of 13.76 kilowatts). Since the expected output from each panel is 298.7 watts, the real expected peak output is 12.84 kilowatts. The panels are mounted at a 15-degree tilt, facing south-by-southwest, in four rows, with at least 0.914 meters (3 feet) between rows. The house is also connected to the electricity grid. Although the home's rooftop solar panels produce more annual electricity than the building and its electric vehicles consume, enough solar PV electricity is not always available at the exact times it is needed to meet instantaneous electricity demand.

To improve the matching of power demand with supply over time, the house has four wall-mounted batteries. Each battery can hold 6.4 kilowatt-hours of energy for a total of 25.6 kilowatt-hours among all batteries. The peak charge and discharge rate of each battery is 3.3 kilowatts, for a total of 13.2 kilowatts. In other words, if all batteries were filled to the brim with stored electricity, they could discharge at a rate of 13.2 kilowatts for 1.94 hours. Alternatively, they can discharge at a rate of 1 kilowatt for 25.6 hours before depleting.

Two batteries each are connected to each of two 7.6-kilowatt inverters (15.2 kilowatts for both inverters) that have a 97.5 percent efficiency. The solar PV system is also connected to the inverters. Although four batteries are available, the utility permitted only two batteries to operate. As such, the other two batteries, while still physically connected to an inverter, are idle, and the battery system really stores only 12.8 kilowatt-hours of energy and charges and discharges at a maximum rate of 6.6 kilowatts. During the year, up to three electric cars are charged with one of two electric car chargers.

Because the property has no natural gas, all heating, cooling, and cooking must be done with electric appliances. For air heating and air conditioning, ductless mini-split heat pump air heaters and air conditioners are used. Nine indoor units are placed in different rooms or zones of the house, and two compressor/condensers are placed outside either to release heat to or extract heat from the outside air. The indoor units are connected to the outdoor units by pipes containing a refrigerant, thus the system is ductless.

For water heating, an air-source heat pump water heater is used. The source of the heat is the air in the mechanical room in which the water heater sits. When the heat pump water heater is used, heat from the air is transferred to the water heater, and the room slightly cools. Thus, in summer, the mechanical room door can be opened to provide additional cooling to the rest of the home. With a timer, hot water is circulated around the house through pipes with a circulation pump. The timer is set so that the circulation pump runs only during times of low electricity price during the day or night.

For cooking, an electric induction cooktop is used. Induction cookers boil water in half the time of natural gas cookers, yet do not feel hot when they are touched because they induce resistance heating in the pot rather than create heat themselves. As such, the pot gets hot, but the cooktop stays warm, heated only by conduction from the pot.

Other major electric appliances in the house include a washer, dryer, refrigerator, microwave oven, convection oven, toaster, garbage disposal, LED lights, televisions, computers, garage door, and an air filtration system. Energy-efficient versions of these were purchased.

The overall strategy for powering the home with rooftop solar PV and wall-mounted batteries is as follows. During the night, the batteries supply electricity. Thus, by morning, the battery level is usually at its low limit. As the day progresses, the PV begins to generate electricity, which is consumed first by household appliances. Excess electricity after that is used to charge the batteries. Excess electricity after that is sold to the grid. After the batteries fill up completely, usually by mid-morning, they sit idle until near sunset. At that point, the batteries are used to provide household power until the batteries are drained. When the batteries are fully drained, grid electricity is used. If the batteries do not drain during the night, then the house runs 24 hours per day on solar PV and batteries.

Electric cars at the home are charged at night for cost reasons, as discussed shortly. The two electric cars charged during the week both draw power quickly (one draws up to 20 kilowatts and the other, up to 8 kilowatts) and draw a large amount of energy (up to a maximum 85 kilowatt-hours and 53 kilowatt-hours, respectively). The two batteries alone cannot normally supply either the full charge rate or the total energy requirement of either car. However, the peak charge rate of both cars can be limited to the maximum discharge rate of the batteries. Even then, though, grid electricity is still needed to provide the total energy drawn during the night for the cars. Thus, more batteries would be needed to eliminate the use of grid electricity.

For the first 4 years of household electricity use, including during winter, the PV system produced 120 percent of its household electricity use. In other words, there were no electric bills (aside from a $10 per month hookup fee), no natural gas heating bills (because no natural gas was connected to the property), and no gasoline bills (because the cars were electric). In fact, excess electricity was sent back to the grid, and the utility paid an average of $800 per year for it.

In California and several other states, many **Community Choice Aggregation (CCA)** utilities have sprung up. These utilities procure clean, renewable electricity by signing power purchase agreements with wind, solar, geothermal, and hydroelectric energy farms. The community choice utilities provide the electricity at competitive prices to customers who sign up with them. They also procure electricity from homeowners who have solar PV on their roof or a small wind turbine in their backyard. In some cases, the community choice utility pays for the solar PV electricity at the same rate a ratepayer would pay for electricity at the same time of day. Thus, if a ratepayer pays 25 cents per kilowatt-hour for electricity at 4 p.m., a homeowner who sends electricity back to the grid at that time will also receive a 25 cents per kilowatt-hour payment for that electricity. The transmission and distribution portion of a ratepayer's bill is handled by a different utility, but transmission and distribution costs are often applied only when more electricity is drawn from the grid than sent to the grid.

Paying different electricity rates at different times of the day is called **time-of-use pricing**. Often, there is a period of the weekday (peak period), such as from 2 to 9 p.m., during which the electricity price is at its highest, another period (off-peak period), such as from 11 p.m. to 7 a.m., during which the price is at its lowest, and periods of intermediate-price (**partial-peak**) electricity price (all other times). Time-of-use pricing may differ between weekends and weekdays. For example, one weekend distribution may be to charge peak prices between 3 to 7 p.m. and off-peak prices for all other hours.

An alternative to time-of-use pricing is to use constant pricing all day and to charge increasingly higher rates per kilowatt-hour of energy used the more electricity that is used. The advantage of time-of-use pricing over this method is that the former is a built-in method of **demand response** management of the grid. In other words, time-of-use pricing helps utility operators balance supply of electricity with its demand.

With demand response, utility operators give individuals and businesses financial incentives not to use electricity at certain times of the day when electricity demand is high. This is done by increasing electricity prices or by paying customers not to use electricity during those times of day. Higher prices between 2 and 9 p.m., for example, shift electricity use from that period to another period, such as between

11 p.m. and 7 a.m., when electricity prices are low. Thus, nighttime car charging is encouraged by low nighttime electricity prices.

The building of a super-efficient all-electric home with heat pumps, electric appliances, and electric cars and powered by solar PV and batteries not only eliminates bills for electricity, natural gas, and gasoline and results in payments for the excess electricity provided, but it also eliminates installation costs related to natural gas. For example, it eliminates the cost of a gas hookup fee and the cost of gas pipes.

Transition highlight

Eliminating natural gas and electrifying a new home can save $6,000 to $23,000 in upfront costs and $3,000 to $10,000 per year in electricity and fuel bills. Upfront costs avoided include those for a natural gas hookup fee ($3,000 to $8,000) and natural gas pipes ($3,000 to $15,000). In addition, homeowners can receive $0 to $1,000 per year by selling excess electricity to their utility. These savings offset part of the capital cost of installing solar and battery systems that power all household and vehicle needs. Such systems have a total upfront capital cost ranging from $30,000 to $115,000. The capital and maintenance cost of heat pumps for air and water heating and air conditioning and other electric appliances and vehicles roughly cancel those of fossil-fuel equivalent appliances and vehicles used. As such, a solar plus battery system powering the home can pay itself off in 4.5 to 4.7 years with government incentives (which have been about 35 percent of the capital cost in the United States) and in 8 to 9.2 years without subsidies This payback time is decreasing yearly, as solar and battery costs are declining yearly.

6 WWS SOLUTIONS FOR INDUSTRY

The industrial sector creates products made of metal, plastic, rubber, concrete, glass, and ceramics, among other materials. Energy is needed in industry for heating, cooling, drying, curing, melting, and electricity. In the United States, energy for creating industrial heat comprises about 70 percent of industrial energy demand. Energy for cooling and refrigeration comprises about 3.4 percent; energy for machines, about 19 percent; energy for electrochemical processes, 2.6 percent; and energy for all other processes, 5 percent.[98] Industrial heat ranges from low to high temperature heat. About half of industrial heat is high-temperature heat (above 400 degrees Celsius) and the other half, low-temperature (30 to 200 degrees Celsius) and medium-temperature (200 to 400 degrees Celsius) heat.[99] High-temperature heat is used for plastics and rubber manufacturing, casting, steel production, other metal production, glass production, lime calcining in cement manufacturing, metal heat-treating and reheating, ironmaking, and silicon extraction from sand. Low- and medium-temperature heat are used for drying and washing during food production, chemical manufacturing, distilling, cracking, pulp and paper manufacturing, and petroleum refining, among other processes. This chapter first discusses the current sources of energy used in industry, then discusses WWS replacements for these sources. The chapter also includes methods of eliminating chemical emissions from steel, concrete, and silicon manufacturing.

6.1 Current Energy Sources for Industry

Energy for industrial heat is currently obtained from fuels, electricity, and steam.

Fuels producing industrial heat include natural gas, coal, fuel oils, liquefied gases, and biomass. These fuels are burned in ovens, fired heaters, kilns, and melters. The resulting heat is transferred either directly or indirectly to the material being melted. With **direct (convection) heating**, combusted gases come in direct contact with the material. With **indirect (radiant) heating**, the hot gases pass through radiant burner tubes or panels separated from the material, and the heat is transferred radiantly.

Electricity-based heating technologies include electric arc furnaces, induction furnaces, resistance furnaces, dielectric heaters (including radio frequency dryers and microwave processors), and electron beam heaters. The first three produce high-temperature heat (greater than 400 degrees Celsius). The remaining technologies produce primarily medium- and low-temperature heat. All technologies use either direct or indirect electric heating. With direct electric heating, an electric current is sent directly to a material, heating the material by resistance heating. With indirect electric heating, high-frequency energy is inductively coupled with a specific material to heat it.

Steam-based heating technologies heat materials directly with steam or indirectly through a heat-transfer mechanism. Almost all steam heating is for low-temperature (less than 200 degrees Celsius) processes, such as pulp and paper manufacturing, chemical manufacturing, and petroleum refining. Currently, most of the steam for these processes is generated simultaneously (cogenerated) with electricity produced by the burning of a fossil fuel or biomass.

In the United States in 2010, fuels comprised 48.5 percent of energy supplied to industry; electricity supplied 22.7 percent, and steam provided 28.8 percent.[98] Of all the energy consumed, only 58 percent was applied usefully, and the rest was lost as waste heat.

Transitioning industrial-sector energy to 100 percent clean, renewable WWS energy and storage requires moving high-, medium- and low-temperature fuel-based combustion to electric alternatives. Such alternatives include electric arc furnaces, induction furnaces, resistance furnaces, dielectric heaters, electron beam heaters, heat pump steam, and CSP steam.

In addition, it is necessary to eliminate non-energy carbon dioxide emissions (called **process emissions**) produced chemically during steel manufacturing, concrete production, and silicon extraction from sand. This requires alternative methods of producing steel, concrete, and silicon. Industrial-sector electrification and alternatives to steel, concrete, and silicon manufacturing are discussed next.

6.2 Arc Furnaces

An **electric arc furnace** is a spherical-shaped furnace for melting metal. It has a retractable roof and contains three graphite electric rods. The floor of the furnace is coated with a heat-resistant material that is used for collecting the molten metal. Scrap steel or iron is fed into the furnace, the roof is closed, and the electric rods (cathodes) are lowered onto the metal. An electric current that passes between the rods and an anode mounted at the bottom of the furnace creates an arc (extremely hot, bright light). The current that passes from the negatively charged cathode rods to the positively charged anode base melts the metal, as does the radiant heat emitted by the arc. When the metal is melted, the metal alloy, slag (stony waste matter separated from metals during smelting), and oxygen formed by the process are removed through side doors.

The arc furnace derives from the **carbon arc lamp.** In 1800, Sir Humphrey Davy invented the carbon arc lamp, which consists of two carbon rods with an electric current running through them. When the rods first contact each other, a spark is ignited. The rods are then pulled apart slowly. The current forms an extremely bright arc of light across the air gap. The bright arc forms because the hot (3,600 to 6,300 degrees Celsius) carbon rod tips vaporize and ionize, creating a plasma that contains positive carbon ions and free electrons at high temperature. The electrons turn the gas into a good conductor that can be maintained at the high temperature. When the current strikes the ionized carbon vapor, the result is a bright light. The carbon rods slowly burn away as they gasify, requiring the distance between them to be adjusted as well. Carbon arc lamps were the only form of light used for street lighting and industrial indoor lighting from 1801 to 1901. Their disadvantages are that they produce short ultraviolet wavelengths of radiation, which are dangerous to humans; they create a buzzing sound; and they produce flickering light and sparks, which can set fires.

In 1878, **Carl W. Siemens** extended the idea of the arc lamp to build and patent an electric arc furnace. James B. Readman subsequently invented (1888) and patented (1889) an arc furnace in Edinburgh, Scotland and applied it to produce phosphorus. Paul Heroult of France developed a commercial arc furnace for steel production in 1900. In 1905, he went to the United States to work with several steel companies, including the Sanderson Brothers Steel Company in Syracuse, New York, which installed the first commercial arc furnace worldwide in 1906.

Arc furnaces are used in foundries (workshops in which metals are melted and cast into different shapes), steel mills, and silicon extraction facilities. In steel mills, they are used to produce steel from recycled scrap metal, reducing the energy needed versus making steel from raw ores. Because arc furnaces require a lot of energy, they are often used when electricity prices are low, thus their use responds well to electricity pricing, which is a central feature of demand response management of the grid.

In a 100 percent clean, renewable energy world, arc furnaces are one technology that will replace fossil fuels and biomass to produce heat for making steel and for casting metals.

6.3   Induction Furnaces

Another method of melting metals, such as iron, steel, copper, aluminum, and precious metals, is with an electric **induction furnace**. Induction furnaces can melt from a kilogram to 100 tonnes of metal at a time.

An induction furnace consists of a nonconductive crucible surrounded by a large coil of copper wire. Metal is placed inside the crucible to be melted. A strong alternating electric current is sent through the wire coil, creating a rapidly reversing magnetic field that induces circular electric currents (called eddy currents) inside the metal. The metal heats by resistance heating as the eddy currents pass through it. Because the heating of the targeted material results from electric currents that are induced by a magnetic field created from electricity passing through a coiled wire, the heating is by electromagnetic induction.

Because the metal is heated by induction, the temperature of the metal rises to no higher than the temperature required to melt the metal. This prevents the loss of some alloying elements. Thus, an induction furnace differs from an arc furnace, where the temperature rises above that required to melt the metal.

In a 100 percent WWS world, induction furnaces are another electric technology that will replace fossil fuels and biomass for producing high-temperature heat.

6.4   Resistance Furnaces

A third method of obtaining high-temperature heat is with an electric resistance furnace. With this technology, DC electricity passes from a negative electrode (cathode), through a conducting material, to a positive electrode (anode). The conducting material heats up due to resistance within the material. In **direct-resistance heating**, the material targeted for heating or melting is itself the conductor of electricity. In **indirect-resistance heating**, a separate heating element is heated and transfers its heat by a combination of conduction, convection, and radiation to the material targeted for heating or melting.

The most common applications of direct-resistance heating are the heating of long rods, the heating of iron-containing metals prior to forging, and the continuous annealing (heating followed by slow cooling) of wire.[100]

Indirect-resistance heating is used to heat solids, liquids, and gases in many industries. It is used in the heating and metals industries for melting, hot working, plasma-heating processes, stress relieving, and preheating. It is also used in the food, paper, print, textile, rubber, plastic, glass, and ceramic industries.

6.5   Dielectric Heaters

A fourth method of electric heating is **dielectric heating**, which is what microwave ovens use to cook food. Dielectric heating is used for low-temperature applications.

Dielectric heating, also referred to as electronic heating, uses electromagnetic radiation in the frequency range covering either radio frequencies or microwave frequencies. The former type of dielectric heating is referred to as **radio frequency heating** and the latter, **microwave heating**.

In a dielectric heater, the radio wave or microwave is used to heat a **dielectric material**, which is a poor conductor of electric current. Radio frequency heating is often applied to heat large materials because the heating is more uniform than with microwave heating. As such, radio frequency heating is used for most dielectric industrial heating applications, including gluing, welding, plastic production, preheating, bread baking, textile drying, adhesive and paper drying, microwave preheating, and vulcanizing rubber. Microwave heating is used primarily for the tempering of meat and other food processing applications.[101]

6.6 Electron Beam Heaters

Electron beam heating is a method of melting metals or modifying materials with a fine beam of electrons. When the high concentration of electrons hits a solid material, the kinetic energy of the electrons is converted to heat, which melts the material. The electrons are produced with an **electron gun**, which ejects a narrow stream of electrons from a heated cathode and accelerates them using high voltage. The beam is obtained by using electric and magnetic fields to force free electrons in a vacuum into a straight line. The power per unit area of electron beam heating is 1,000 times that of peak sunlight reaching the surface of the Earth. Around 50 to 80 percent of the energy in the electron beam is transferred to the material.

Electron beam heating takes place in a large vacuum furnace and is applied to mass-produce steel and purify metals, such as titanium, vanadium, tantalum, molybdenum, tungsten, zirconium, niobium, and hafnium. The electronics industry uses many of these metals. A **vacuum furnace** is a furnace in which the material being operated on is surrounded by a vacuum, or absence of air. Temperatures in a vacuum furnace reach up to 3,000 degrees Celsius. Electron beam heating is also used in vacuum chambers to weld and precisely cut materials to make machines, evaporate and deposit thin layers of metal on solar cells, and modify surface layers of metals.

6.7 Steam Production from Heat Pumps and CSP

Steam is needed for many low- and some medium-temperature processes in a 100 percent WWS world. Such steam is currently obtained as a co-product of burning fossil fuels or biomass for electricity production. These sources will be replaced by steam from an electric heat pump or steam co-generated with electricity at a CSP plant. A heat pump can produce heat up to 160 degrees Celsius.[102] This heat can be used to boil water to produce steam of a similar temperature.

Parabolic trough CSP plants produce temperatures of 60 to 350 degrees Celsius. Parabolic dish CSP plants produce temperatures of 100 to 500 degrees Celsius. Central tower receiver CSP plants produce temperatures of up to 600 degrees Celsius.[103] The heat in all cases, if not used to produce electricity, can be used to produce steam directly. However, a manufacturing plant that uses such steam needs to be nearby, otherwise heat losses from pipes transporting the steam long-distance will be significant.

6.8 Steel Manufacturing

Steel manufacturing is a significant user of energy and source of air pollution and carbon dioxide. With steel manufacturing, carbon dioxide emissions arise not only from burning a fossil fuel or biomass for high-temperature heat, but also from chemical reaction during the extraction of pure iron from iron ore. Carbon dioxide emissions arising from chemical reaction are called **process emissions**.

Steel can be produced from raw iron ore or recycled metal. Steel produced from iron ore is produced in two stages. The first is called ironmaking, and the second, steelmaking.

In the **ironmaking** step, molten pure iron metal is extracted from solid iron ore (iron oxide). A **blast furnace** is filled with iron ore, some relatively pure solid carbon in the form of coke (coal heated in the absence of air), and **limestone** (calcium carbonate). Hot air containing oxygen is then forced through the bottom of the blast furnace. It reacts with the coke to form carbon monoxide gas and heat. The carbon monoxide then reacts with the iron ore and the coke to produce molten pure iron metal and carbon dioxide. Owing to the heat, the limestone simultaneously decomposes to calcium oxide and carbon dioxide. The calcium oxide then reacts with and removes sandy remnants of the iron ore to form a waste product, called slag. Slag is less dense than is molten iron, so it floats above the pure iron. The slag is then cooled and removed for use in roads, leaving pure iron behind. Thus, traditional steelmaking releases carbon dioxide, not only through fossil-fuel and biofuel burning to produce high temperatures, but also from chemical reactions during the production of pure iron.

Steelmaking is the second step in steel production. In this step, impurities are removed from the raw iron. Then, carbon and other alloying elements are added to make crude steel. Impurities removed include phosphorus, sulfur, and excess carbon. Alloying elements added include chromium, nickel, vanadium, and manganese.

Steelmaking is performed in one of two ways. **Primary steelmaking** involves the use of new iron from ironmaking. **Secondary steelmaking** involves the melting of recycled scrap steel in an electric arc furnace to produce new steel.

The main method of primary steelmaking is the **basic oxygen steelmaking** method. With this method, the molten iron and impurities from the blast furnace are mixed with scrap steel and placed in a **basic oxygen furnace**. Oxygen is then blown through the furnace and reacts with carbon in the molten mix to form carbon dioxide, which is released to the air. Calcium oxide in the molten mix also reacts with phosphorus and sulfur, the products of which rise to the top as slag and are removed. Finally, alloys are mixed in, and the molten steel is poured into preshaped molds, where it cools and hardens.

With secondary steelmaking, scrap metal is melted in an arc furnace. Oxygen is blown through the metal to help remove the carbon and speed the meltdown of the metal by increasing combustion. Calcium, phosphorus, and sulfur are removed and alloys are mixed in, in a manner similar to the process in the basic oxygen furnace.

In sum, the sources of carbon dioxide during the two-step steel formation process are (1) its emissions during fossil-fuel combustion to produce heat in the blast furnace and in the basic oxygen furnace, (2) its chemical release during reaction of carbon with iron ore, (3) its release upon the thermal decomposition of limestone during ironmaking, and (4) its release during the chemical reaction of carbon with oxygen during steelmaking. In addition, in an arc furnace, there is a small amount of carbon dioxide released because of the vaporization of graphite and its reaction with oxygen. The overall carbon emissions during the ironmaking plus steelmaking process using a blast furnace and basic oxygen furnace are about 1,870 kilograms of carbon dioxide per tonne of steel.[104] Of this, ironmaking produces about 70 to 80 percent of the carbon dioxide emissions.

6.8.1 Reducing Carbon Emissions with Hydrogen Direct Reduction

An alternative to extracting molten iron from iron oxide with coke during ironmaking is to extract the pure iron with hydrogen gas (H_2), where the hydrogen is produced with 100 percent WWS.[104] This process is called the **hydrogen direct reduction** process. The main extraction reaction involves mixing iron oxide with hydrogen gas to produce pure molten iron plus steam. This reaction occurs optimally at a temperature of around 800 degrees Celsius, which is lower than the temperature needed in a blast furnace.[104] The reaction eliminates chemical carbon dioxide produced from the purification of iron oxide to iron during ironmaking. However, an injection of carbon into the molten iron during the steelmaking process is still needed to create an iron–carbon alloy (0.002 to 2.14 percent carbon) to strengthen the steel. In addition, some carbon is still emitted from the thermal decomposition of limestone and subsequent emission of carbon dioxide.

If the heat required for the hydrogen direct reduction ironmaking process is obtained with an electric resistance furnace instead of with fossil fuels, if the hydrogen for the process is produced by electrolysis (passing of electricity through water), if an electric arc furnace is used for the steelmaking process, and if all electricity is provided with 100 percent WWS, the hydrogen direct reduction process emits only 53 kilograms of carbon dioxide per tonne of steel, or only 2.8 percent of the emissions of the blast furnace/basic oxygen furnace process (1,870 kilograms of carbon dioxide per tonne of steel).[104] The only carbon dioxide emissions during the hydrogen direct reduction process are from oxidation of injected carbon in the arc furnace, the thermal decomposition of limestone, and oxidation of the vaporized carbon in the arc furnace electrodes.

The system just described is a 100 percent clean, renewable energy system, but still results in a residual of 53 kilograms of carbon dioxide per tonne of steel (2.8 percent of the original emissions). The remainder will likely be released to the air. Capturing the remaining carbon dioxide requires electricity, and even if the electricity is from WWS, it is better to use that WWS electricity to displace a fossil-fuel electricity source than to capture carbon dioxide, since displacing a fossil-fuel source eliminates not only carbon dioxide from the source but also air pollution and upstream mining and emissions. Carbon capture does not reduce mining or air pollution.

Transition highlight

During June 2021, a steel-manufacturing plant in Lulea, Sweden, created pure metallic iron (sponge iron) from hydrogen for the first time. The hydrogen was produced by electrolysis, where the electricity came from WWS sources; thus the hydrogen was **green hydrogen**. The plant subsequently produced nearly carbon-free steel commercially from the purified iron in July 2021.

6.8.2   Reducing Carbon Emissions with Molten Oxide Electrolysis

A second alternative method for extracting molten iron from iron ore during ironmaking is with **molten oxide electrolysis**.[105,106] With this technique, iron ore is first heated in a molten electrolyte soup above 1,961 degrees Celsius, where it decomposes to produce a different form of iron oxide, called magnetite, and oxygen. The molten electrolyte soup contains silicon dioxide, aluminum oxide, and magnesium oxide and helps electricity flow. The soup is heated further, past the melting point of pure iron, which is 2,084 degrees Celsius. Above this temperature, electricity is passed through the soup, turning magnetite into pure molten iron. The pure liquid iron sinks to the bottom of the cauldron, where it is drained. As such, the molten oxide electrolysis process produces pure iron without emitting chemically produced carbon dioxide.

6.9   Concrete Manufacturing

Concrete is a mixture of **aggregate** (sand, gravel, and crushed stone) and paste (water and **Portland cement**). The paste binds the aggregate together, making a hard surface. Concrete is used for roads, foundations, buildings, runways, sidewalks, driveways, and a variety of other purposes.

Joseph Aspdin (1778 to 1855) of Leeds, England, invented Portland cement in the early nineteenth century. He formed it by burning powdered limestone and clay on his kitchen stove. Today, cement contains limestone, shells, or chalk, all of which contain calcium carbonate mixed with clay, shale, slate, blast furnace slag, silica sand, or iron ore. These ingredients are heated to 1,500 degrees Celsius to form a hard substance that is ground into a fine, powdery cement.

The concrete industry produces about 8 percent of the world's fossil-fuel carbon dioxide emissions,[107] or about 6.4 percent of all anthropogenic (fossil-fuel plus permanent-deforestation) carbon dioxide emissions. These emissions are equivalent to about 1.18 tonnes of carbon dioxide per tonne of cement produced.[108] Of the total, 1.6 percent is from the quarrying of raw materials, 42.4 percent is from electricity and heat production during the cement manufacturing process, 46.3 percent is from chemical reaction (process emissions) during cement manufacturing, and 9.7 percent is from the production of concrete from cement and the transport of concrete.[108] As such, almost half of carbon dioxide emissions from concrete production are process emissions, and the rest are emissions related to energy (electricity production, heat production, and transport).

The process emissions during cement manufacturing arise owing to chemical reaction of calcium carbonate with clay at a high temperature to produce clinker (a mix of oxides of silicon, iron, aluminum, and calcium) and carbon dioxide. The clinker is then mixed with gypsum (plaster of Paris) to form cement. The cement is subsequently mixed with water to form a paste, which is combined with the aggregate to form concrete.

Four ways of reducing both process emissions and energy emissions from concrete manufacturing are to (1) use geopolymer concrete instead of concrete derived from Portland cement, (2) use a material called Ferrock instead of concrete, (3) recycle concrete, and (4) make concrete that traps carbon dioxide. In all cases, energy must be supplied by WWS to maximize carbon dioxide reductions.

6.9.1 Geopolymer Concrete

Geopolymer concrete was named and developed in the 1970s by Joseph Davidovits, a French material scientist. It is a hardened mixture of geopolymer cement, aggregate, and water. Geopolymer cement consists of any natural or industrial waste material containing aluminosilicate minerals mixed with an alkali solution.[109] Waste materials include fly ash, granulated blast-furnace slag, rice husk ash, or metakaolin. Fly ash is abundant in most countries since it is a waste product of 100 years of coal burning for electricity generation. Slag is a waste product of steel production. Rice husk ash is an abundant waste product of rice milling. Metakaolin is a form of the clay mineral kaolinite.

Alkali solution options include sodium hydroxide, potassium hydroxide, and/or sodium silicate mixed with water. The cement is cured at a temperature of 100 degrees Celsius to provide strength. The cement is then mixed with aggregate and water to form concrete.

Slag-based geopolymer cement consists of fly ash, ground-granulated blast-furnace slag, and an alkali solution. Rock-based geopolymer cement consists of volcanic rock, fly ash, slag, and an alkali solution.

Because geopolymer concrete does not use calcium carbonate and because it does not need energy for high-temperature kilns, producing it results in about 80 percent lower carbon dioxide emissions than producing concrete from Portland cement.[109] Most of the carbon dioxide reduction is due to eliminating the offgassing of carbon dioxide from calcium carbonate chemical reaction. The remaining reduction is due to the fact that geopolymer concrete does not require the use of extreme high-temperature kilns, and thus it reduces energy consumption by about 50 percent versus concrete derived from Portland cement. If the remaining energy is provided by WWS, geopolymer concrete emissions of carbon dioxide are almost eliminated. Other benefits of geopolymer concrete are that it is more resistant to freezing and thawing cycles and to corrosion by acid rain than is concrete from Portland cement. The costs of the two types of concrete are similar.

Transition highlight
Geopolymer concrete has been used in many projects to date. For example, 70,000 tonnes of it was used to construct the Toowoomba Wellcamp airport in Queensland, Australia.

6.9.2   Ferrock

Another commercialized alternative to concrete is **Ferrock**, or iron carbonate ($FeCO_3$).[110,111] Ferrock is derived by first mixing waste steel dust containing iron oxide with crushed glass containing silicon dioxide, limestone, kaolinite or another clay, stabilizers, promoters, and a catalyst into a mixer at room temperature. The mixture is then poured into a mold containing seawater. The filled mold is put into a curing chamber, where carbon dioxide from a furnace is injected. The iron, carbon dioxide, and salt water react together to form Ferrock and hydrogen gas. When the final product dries, it is about 5 times harder and more flexible than Portland cement. The production of Ferrock not only avoids the chemical carbon dioxide emissions and most energy emissions from concrete production, but it also traps carbon dioxide and produces hydrogen, which can be used for other applications.

6.9.3 Concrete Recycling

Another method of reducing carbon dioxide emissions from concrete manufacturing is with **concrete recycling**. Concrete structures or roads are often demolished. Historically, such concrete has been sent to a landfill. However, if the concrete is uncontaminated (free of trash, wood, and paper), it can be recycled. Rebar (steel reinforcement) in concrete can also be recycled, as magnets can remove it. The rebar can then be melted and used for other purposes. The broken concrete is crushed. Crushed concrete is often used as gravel in new construction projects or as aggregate in new concrete.

6.9.4 Sequestering Carbon Dioxide in Concrete

Trapping carbon dioxide in a material, as done in Ferrock, is a method of offsetting emissions of chemically produced carbon dioxide from the concrete production process. Trapping carbon dioxide within concrete itself is another option.[112] The clinker in Portland cement contains calcium oxide. If carbon dioxide from any source is mixed with the clinker, it will react with the calcium oxide to form calcium carbonate within the cement. Upon drying, the solid calcium carbonate strengthens the cement. Even if the cement breaks, the carbon dioxide, trapped in the calcium carbonate, will not break free because calcium carbonate is a solid bound to the cement.

Remaining carbon dioxide emitted chemically during the cement formation process can be captured upon emission. However, capturing carbon dioxide requires equipment and electricity. Even if the electricity is from WWS, it is far better to use that WWS electricity and the money for the equipment to displace coal or natural gas electricity than to capture carbon dioxide. This is because displacing a fossil-fuel electricity source eliminates not only carbon dioxide from the source but also air pollution, upstream mining and emissions, and fossil-fuel infrastructure. Capturing carbon dioxide eliminates only some carbon dioxide.

6.10   Silicon Purification

Pure silicon is a common chemical in alloys and semiconductors. Alloys, including aluminum–silicon and iron–silicon alloys, are the major application of pure silicon. Such alloys are used to make transformer plates, engine blocks, cylinder heads, and machine tools. Semiconductors are used in computers, microelectronics, and solar photovoltaic cells.

Pure silicon (Si) is extracted from silicon dioxide (SiO_2), also called silica and quartz. The source of silicon dioxide is either sand or a mine. The most common process of silicon extraction is **carbothermic reduction of silica**. It involves heating silicon dioxide together with pure graphitic carbon up to a temperature of 2,200 degrees Celsius in an electric arc furnace. The reaction produces pure silicon and carbon dioxide. The silicon is extracted, and the carbon dioxide is released to the air. If electricity for the arc furnace is derived from a fossil-fuel energy source, additional carbon dioxide is emitted. For some applications, such as photovoltaics, the silicon is still not pure enough and more purification is needed.

Several alternative methods of producing pure silicon from silica exist that do not involve releasing carbon dioxide. These methods generally involve either dissolving silica in a mixture of organic alcohol and a base; reacting silica with aluminum or magnesium at moderate temperature (450 to 650 degrees Celsius); or placing silica in a molten salt electrolyte solution and running electricity through the solution.[113]

Alternatively, pure silicon can be obtained from calcium silicate instead of from silica. The calcium silicate is dissolved in a mixture of three salts – calcium chloride, magnesium chloride, and sodium chloride – at a warm temperature. In solution, the calcium silicate dissociates into pure silicon.[114] All materials for this process are relatively inexpensive.

Developing one of these alternative methods and using WWS electricity to provide the heat needed eliminates energy and process carbon dioxide emissions from silicon production.

7 SOLUTIONS FOR NONENERGY EMISSIONS

WWS technologies eliminate energy-related emissions. However, some emissions that affect human health and climate are not from energy sources but must still be reduced or eliminated in order to help solve the air pollution and climate problems we face. Such non-energy emissions include gases and particles from open biomass burning; methane from agriculture and landfill waste; halogens from leaks and their reckless disposal; and nitrous oxide from fertilizers, industry, and wastewater treatment. These sources of emissions and methods of controlling them are discussed in this chapter.

7.1   Open Biomass Burning and Waste Burning

Open biomass burning is the burning of evergreen forests, deciduous forests, woodland, grassland, and agricultural land, either to clear land for other use, to stimulate grass growth, to manage forest growth, or to satisfy a ritual. Biomass burning can also be triggered accidentally by a campfire, debris burning, an electrical spark, or a cigarette, or intentionally by arson. About 17 percent of all global carbon dioxide and air pollution emissions worldwide are from open biomass burning. Humans cause around 93 percent of biomass burning today. Nature causes the rest.[6]

Agricultural fields are often set on fire after a harvest to remove straw, clearing the way for a new crop the next spring. Sugarcane fields are usually burned before harvest to remove the outer leaves around the stalks to facilitate sugarcane extraction.

Waste burning is the burning of trash, such as in a landfill, open pit, garbage can, or backyard incinerator. Such waste burning is illegal in many countries but still occurs in others.

Biomass burning produces not only gases that warm the climate (carbon dioxide, methane, ozone precursors, carbon monoxide, and water vapor), but also climate-warming particles (black and brown carbon) and direct heat. Black and brown carbon, along with other particles emitted during biomass burning (ash, organic carbon aside from brown carbon, and sulfate) cause substantial health impacts to people and animals who breathe them in. In addition, the oxides of nitrogen and organic gases from biomass burning result in elevated levels of ozone, formaldehyde, and other gases that affect human health. Waste burning emits the same chemicals as biomass burning but also emits toxic chemicals from the burning of plastics, paints, varnishes, pesticides, medical waste, and chemical byproducts.

While some argue that biomass burning followed by regrowth of vegetation results in no net increase in carbon dioxide to the air, that contention is incorrect. Although the carbon dioxide released upon burning is offset by carbon dioxide taken from the air during photosynthesis to regrow the vegetation burned, the time lag between burning and regrowth (from 1 to 10 years for savannah and 80 years for a forest, for example) always increases carbon dioxide in the air.[115] The contention is also misleading because biomass burning emits black carbon, brown carbon, water vapor, methane, carbon monoxide, ozone precursors, and heat, all of which increase global temperatures. These are not recycled like carbon dioxide is. As such, biomass burning always causes net global warming.[6]

The only solution to biomass burning is to stop it. No technology exists to control its emissions. An alternative to burning agricultural waste straw is to till it into the soil. An alternative to sugarcane burning is to cut away the leafy parts of the sugarcane before harvest and mix it into the soil. Since humans cause more than 90 percent of all open biomass fires worldwide, biomass burning is largely (but not completely) preventable through government policies restricting burning and discouraging the conversion of forest land to agricultural land or another land use type.

Similarly, the only method of reducing the impacts of landfill waste burning is to stop it, because its emissions cannot be trapped. If waste is burned in an incinerator, many of its emissions can theoretically be controlled with emission control technologies; however, no technology eliminates all emissions, including of carbon dioxide. Thus, even incinerators with emission controls produce substantial pollution. The best control of waste burning is to instead recycle the waste or put it in a landfill. Waste should not be dumped into the oceans since plastics do not degrade for centuries. The accumulation of plastics in the oceans has caused an environmental catastrophe, resulting in the **Great Pacific Garbage Patch**. This is a plastic wasteland that consists of two distinct patches in the North Pacific Ocean, one near Japan (Western Garbage Patch) and the other between California and Hawaii (Eastern Garbage Patch). Collectively, the two are three times the size of France.

7.2   Methane from Agriculture and Waste

Methane is a long-lived greenhouse gas that selectively absorbs certain long wavelengths of heat radiation, trapping some of that radiation near the surface of the Earth. It not only increases global warming but also increases background levels of ozone, another greenhouse gas.

Anthropogenic sources of methane include open biomass burning, natural gas leaks, fossil-fuel combustion, and biological sources enabled by human activity. Biological sources include bacteria in landfills, rice paddies, the stomachs and manure of farm animals, and sewage treatment plants. A 100 percent WWS energy system will eliminate methane from energy. Policies controlling biomass burning and the remaining sources are needed to reduce methane from those sources as well. This section discusses controlling methane from human-enabled biological sources.

The root biological source of most methane is **methanogenic (methane-producing) bacteria**, which live in environments lacking oxygen. Methanogenic bacteria consume organic material and excrete methane. Ripe oxygen-depleted environments include the digestive tracts of cattle, sheep, and termites; manure from cows, sheep, pigs, and chickens; rice paddies; landfills; and wetlands. All of these sources, aside from termites and wetlands, are human-enabled.

The main method of reducing methane from bacteria in the digestive tracks and manure of animals is to reduce human consumption of meat and poultry. This requires people changing their diets, which can have additional health benefits.

Methane released from manure can also be captured with a methane digester. A **methane digester**, or manure digester, is an airtight tank in which manure is processed after water is separated from it. The manure is heated and stirred to simulate the inside of a cow's stomach. Methane rises from the manure to the top of the tank, where it is captured in a bag or piped out of the digester, into a storage tank.

In a 100 percent WWS world, the captured methane should be used only for steam reforming of methane to produce hydrogen for use in a hydrogen fuel cell, not for methane combustion. Whereas carbon dioxide emissions from steam reforming are similar to those of methane combustion, emissions of all other chemicals from steam reforming are small[116] in comparison with those from natural gas combustion. Although using methane from a landfill or rice paddy to produce hydrogen is better than releasing the methane to the air, using methane from natural gas to produce hydrogen is not good for either the climate or the environment.[117]

Methane gas from landfills can also be captured. **Landfill gas** is often a mixture of nearly 50 percent methane, nearly 50 percent carbon dioxide, and trace amounts of nonmethane organic gases. Landfill gas is extracted by drilling multiple half-meter-wide boreholes up to 30 meters deep into a section of the landfill. Trash is then removed so that a well, which consists of a perforated or slotted siding with a cap on the bottom, can be installed. Most of the top of the well is sealed to prevent gas escape. A wellhead is installed at the top through which gas is piped to its end destination. In some cases, a network of multiple vertical and horizontal pipes can capture the landfill gas.

Once captured, the landfill gas is usually filtered to separate out the methane from the other gases. Once methane is relatively isolated, it can be used to produce hydrogen by steam reforming.

Transition highlight
Rice paddies release a significant amount of methane globally, both through the leaves of the rice plants themselves and in the oxygen-depleted environment of the flooded soil in which rice plants usually grow. On average, rice paddy soil is flooded for 4 months of the year. Direct seeding of rice plants, instead of transplanting them into already-flooded paddies, can reduce the time needed for flooding down to 1 month. This reduces

methane emissions from rice paddies by 15 to 90 percent. A system of pipes can also be used to capture rice paddy methane just as it captures methane from landfill gas.

7.3 Halogens

Halogens are responsible for about 9 percent of global warming. Halogens are still used today as refrigerants, solvents, blowing agents, fire extinguishants, and fumigants. They enter the atmosphere primarily upon evaporation when they leak or when the appliances containing them are drained. Their persistence in the atmosphere and strong warming per molecule make them potent greenhouse gases.

The main methods of reducing halogen emissions and their impacts on climate are (1) substituting halogens that have low global warming potentials for halogens that have high warming potentials, (2) requiring more stringent standards for sealing halogens in the equipment or appliance that they are used in to reduce leakage, and (3) requiring tougher standards for disposing of halogens at the end of the life of the equipment or appliance they are contained in. For these suggestions to be effective, they need to be implemented and enforced worldwide.

7.4 Nitrous Oxide and Ammonia

Nitrous oxide is a potent greenhouse gas with a long lifetime. It is produced largely by bacteria and contributes about 4.3 percent of global warming. Sixty-seven to 80 percent of anthropogenic nitrous oxide originates from agriculture.[118] In particular, nitrogen-containing fertilizers emit substantial nitrous oxide. In addition, the cultivation of legumes (plants in the pea family) results in the conversion of atmospheric nitrogen gas to nitrous oxide, which is released to the air. A third agricultural source of nitrous oxide is the solid waste of domesticated animals.

Evaporation of ammonia from fertilizers also results in the emission of the **ammonia** to the air. Ammonia is a gas that dissolves in liquid aerosol particles to form the ammonium ion, which reacts with other chemicals inside of the particles. Ammonium ions inside of liquid particles cause the particles to swell by encouraging water vapor to condense. Swollen particles reduce visibility and absorb more toxic gases from the air, increasing health problems for those who breathe in the particles.

Some methods of reducing nitrous oxide and ammonia emissions from fertilizers are (1) using less nitrogen-based fertilizer, (2) cultivating leguminous crops that do not require fertilizer in the crop rotation, and (3) reducing tillage to reduce the breakdown of organic fertilizer, thereby reducing reaction and release of chemicals.

The remaining anthropogenic sources of nitrous oxide are fossil-fuel combustion, open biomass burning, industrial processes, and treatment of wastewater. Transitioning fossil energy to 100 percent WWS will eliminate fossil combustion sources of nitrous oxide.

The two main industrial sources of nitrous oxide are the production of nitric acid for use in fertilizers and the production of adipic acid for use in the production of nylon fibers and plastics. To date, nitrous oxide emissions from adipic acid production have been reduced effectively with emission control technologies in several plants, so the expansion of such technologies to adipic acid plants worldwide will help reduce nitrous oxide emissions. Similarly, nitrous oxide emission control technologies for nitric acid production plants are available[119] and can be implemented worldwide with stringent policies.

The source of nitrous oxide in wastewater is organic material from human or animal waste. Modulating the dissolved oxygen content in the wastewater treatment process can control the nitrous oxide content of wastewater.[120,121]

8 WHAT DOESN'T WORK

Although biomass, coal, natural gas, and oil have provided energy that has helped societies until today, such energy has come at a cost, namely increased air pollution, ground and water contamination, land despoilment, global warming, and energy insecurity. While such a cost was accepted in the past, scientists and policymakers no longer believe societies can function properly while allowing the emissions associated with these fuels to continue. As such, efforts are underway to reduce the emissions.

Aside from implementing WWS and storage technologies, the main suggestions for reducing or eliminating energy-related emissions have included using natural gas for electricity instead of coal, using natural gas or coal with carbon capture, using nuclear power instead of fossil fuels for electricity, using biomass with or without carbon capture for electricity, using liquid biofuels instead of gasoline or diesel for transportation, and using blue instead of green hydrogen.

Non-energy-producing methods have also been proposed to remediate global warming. These include primarily synthetic direct air carbon capture and geoengineering. Synthetic direct air capture involves removing carbon dioxide directly from the air by chemical reaction and storing the carbon dioxide or using it for an industrial application. Geoengineering consists of a variety of measures to increase the reflectivity (albedo) of the Earth to cool the Earth's surface.

Policies that include these technologies along with WWS technologies are referred to as "all-of-the-above" policies, since they involve promoting most all technologies, regardless of their side effects, cost, effectiveness, or length of time between planning and operation. The justification for using these technologies is that they are a bridge between current carbon-intense technologies and WWS.

The main issue with all of these non-WWS technologies going forward is that they result in greater global warming, air pollution, land degradation, or all three compared with WWS resources. Second, some of these technologies increase energy insecurity and/or take up to 20 years between planning and operation while costing substantially more than WWS technologies. As such, they are opportunity costs compared with investing in WWS technologies, so are not recommended.

This chapter discusses these non-WWS technologies and delineates the reasons why they are not needed or helpful for solving global warming, air pollution, and energy security problems.

8.1 Brief History of Fossil Fuels

Prior to the Industrial Revolution of the mid-1700s, the world relied primarily on biomass but also on some coal for its energy. Biomass and coal were used mostly for heat. Since the 1650s, however, discoveries and inventions resulted in these fuels, together with natural gas and oil, being used to produce mechanical energy for industry and transportation and to provide electricity.

8.1.1 Wood and Coal Burning Before the Industrial Revolution

In ancient Greece and Rome, wood was burned for heating, cooking, and industrial processes. One major industry was the smelting of silver, which contained lead, to produce silver coins. Smelting of lead itself arose during the Roman Empire because the Romans used lead in cookware, pipes, face powders, rouges, and paints. The lead emissions and transport resulting from smelting was so great that, from 500 BC to 300 AD, lead deposited on Greenland ice and buried over the centuries has measured ice-core concentrations 4 times those in earlier samples.[122] The Romans also burned wood to smelt ores containing copper, zinc, and mercury.[123] Ice-core data suggest that the smelting of copper to produce coins, not only by the Romans, but also in China during the Song Dynasty (960 to 1279 AD), increased atmospheric copper concentrations appreciably.[124]

As noted by the poet Horace, thousands of wood-burning fires in Rome blackened buildings.[125] Air pollution events in Rome were called "*heavy heavens*".[126]

During the fall of the Roman Empire, wood burning for smelters declined significantly. However, c. 1000 AD following the discovery of lead and silver mines in Germany, Austria, Hungary, and other parts of central and eastern Europe, smelting increased again.[122] Wood burning for heating and cooking continued throughout the decline of the Roman Empire in all population centers of the world.

Sea coal was introduced to London by 1228 and gradually replaced the use of wood as a fuel in lime kilns and forges. Wood shortages may have led to a surge in sea coal use by the mid-1200s. Lime kilns were used to produce quicklime, a building material. The pollution in London due to the burning of sea coal in lime kilns became sufficiently severe that a commission was ordered by King Edward I in 1285 to study and remedy the situation. The commission met for several years. Finally, in 1306, the king banned the use of coal in lime kilns. The punishment was "*grievous ransom,*" which may have meant fines and furnace confiscation.[127] However, by 1329, the ban had either been lifted or lost its effect.

Between the thirteenth and eighteenth centuries, the use of sea coal and charcoal increased in England. Coal was used not only in lime kilns and forges, but also in glass furnaces, brick furnaces, breweries, and home heating. One of the early writers on air pollution was **John Evelyn** (1620 to 1706), who wrote *Fumifugium or The Inconveniencie of the Aer and the Smoake of London Dissipated* in 1661. He explained how smoke in London was responsible for the fouling of churches, palaces, clothes, furnishings, paintings, rain, dew, water, and plants. He blamed "*Brewers, Diers, Limeburners, Salt and Sopeboylers*" for the problems.

8.1.2 The Industrial Revolution and the Growth of Coal

In 1679 the French-born English physicist **Denis Papin** (1647 to 1712) invented the pressure cooker while working with the English chemist and physicist **Robert Boyle** (1627 to 1691). In this device, water was boiled under a closed lid. The addition of steam to the air in the cooker increased the total air pressure exerted on the cooker's lid. Papin noticed that the high pressure pushed the lid up. The phenomenon gave him the idea that steam could be used to push a piston up in a cylinder, and the movement of the cylinder could be used to do work. Although he designed a model of such a cylinder-and-piston steam engine in 1690, Papin never built one. Nevertheless, the pressure cooker is the first device used to create mechanical energy from heat produced from either biomass or coal.

In 1698, **Thomas Savery** (1650 to 1715), an English engineer, used the concept of the pressure cooker to patent the first practical steam engine, which is a machine that converts heat from coal or biomass to produce mechanical energy. The first steam engine was used to pump water out of coal mines, replacing the use of horses for this task. The engine worked when water was boiled to produce water vapor that was transferred to a steam chamber. A closed pipe from the steam chamber to the liquid water source in the mine was then opened. Liquid water drops were sprayed on the hot water vapor in the steam chamber to recondense the vapor, creating a vacuum that sucked the liquid water from the mine into the pipe and to the steam chamber. The pipe from the chamber to the mine was then closed, and another pipe from the chamber to outside the mine was opened. Finally, the boiler was fired up again to produce more water vapor to force the liquid water from the steam chamber out of the mine through the second pipe.

In 1712, **Thomas Newcomen** (1663 to 1729), also an English engineer, developed a more efficient steam engine than Savery by introducing a piston. Water evaporated into the steam chamber under the piston, creating pressure that pushed the piston and an attached lever up, producing mechanical energy. Liquid water was then squirted into the steam chamber to recondense the water vapor, creating a vacuum that pulled the piston and lever down. Newcomen's engine was used to pump water out of mines and to power waterwheels. Steam engines in the early eighteenth century were inefficient, capturing only 1 percent of their input energy.

In 1763, Scottish engineer and inventor **James Watt** (1736 to 1819) was given a Newcomen steam engine to repair. He found an inefficiency with the original engine in that it allowed evaporation and condensation to occur in the same chamber. The squirting of cold water to condense vapor and create a vacuum for pulling the piston down resulted in more heat required to produce steam to push the piston back up if both operations occurred in the same chamber rather than in different chambers. Watt overcame this shortcoming by developing two chambers: one in which condensation occurred and the second in which evaporation occurred. The condensation chamber stayed cold because of the squirting of cold water into it and the evaporation chamber stayed warm because of the boiling of water under it. In 1769, Watt patented this revised steam engine, which had twice the efficiency of the previous one.

Watt made further modifications until 1800, including an engine in which the steam was supplied to both sides of the piston and an engine in which motions were circular instead of up and down. Watt's engines were used not only to pump water out of mines, but also to provide energy for paper, iron, flour, cotton, and steel mills; distilleries; canals; waterworks; and locomotives. For many of these uses, steam engines were located in urban areas, thus increasing urban air pollution. Pollution became particularly severe because, although Watt had improved the steam engine, it still captured only 5 percent of the energy it used by 1800.[128] Because the steam engine was a large, centralized source of energy, it was responsible for the shift from the artisan shop to the factory system of industrial production during the Industrial Revolution of 1760 to 1880.[129]

The first utility-scale electricity generation from coal in the world occurred on September 4, 1882, with the opening of the Pearl Street Power Station in New York City. After that, coal use for heat, mechanical energy, and electricity soared, reaching a peak of 8 billion tonnes in 2013. Owing to a slow transition away from coal, consumption declined to 7.4 billion tonnes by 2021, with China consuming 52.1 percent of the total. Given that there are still 1.05 trillion tonnes of known coal reserves remaining in the world, it would take 142 years to deplete all coal at the 2021 rate of consumption.

8.1.3   The History of Natural Gas

Natural gas is a colorless, flammable gas containing about 88.5 percent methane by mass plus smaller amounts of ethane, propane, butane, pentane, hexane, nitrogen, carbon dioxide, and oxygen.[130] It is often found near petroleum deposits. Its first recorded discovery was in the Greek Temple of Apollo in Delphi on Mount Parnassus, Greece. Starting between 1,400 BC and 800 BC and ending in 392 AD, when the Roman Emperor Theodosius forbade it, Greek and Roman rulers travelled to Delphi to listen to the high priestess of the temple, called the Pythia, or the **Oracle of Delphi**. The Pythia would enter into a trance inside a small chamber before channeling prophecies from the god Apollo. The Greek author, Plutarch, wrote that before entering the trance, the Pythia would inhale sweet-smelling, perfume-like vapors from crevices in the floor that arose from a fissure or spring below the temple. Recent studies indicate that underlying the Temple of Apollo are two earthquake faults that cross each other. Seismic activity is thought to have released natural gas, which contains methane and ethene, a sweet-smelling gas that can cause hallucinations. This theory is supported by the fact that water from a nearby spring also contains methane and ethene.[131]

Around 500 BC in Szechuan, China, natural gas was found seeping from the ground. Bamboo was used to pipe the gas to a nearby village, where it was burned to evaporate water from salty brine water that had been mined from deep underground in order to isolate the salt.

In 1609, the scientist **Jan Baptist van Helmont** (1580 to 1644), born in modern-day Belgium, introduced the term **gas** into the chemical vocabulary. He produced what he called *gas silvestre* (*"gas that is wild and dwells in out-of-the-way places"*) by fermenting alcoholic liquor, burning charcoal, and acidifying marble and chalk. The gas he discovered in all three cases, but did not know at the time, was **carbon dioxide**. It wasn't until 1756 that **Joseph Black** (1728 to 1799), a Scottish physician and chemist, isolated carbon dioxide and identified its properties. Black, though, called carbon dioxide "fixed air" because of its ability to attach or "fix" to compounds exposed to it. Fixed air was renamed to carbon dioxide in 1781 by the French chemist **Antoine Laurent de Lavoisier** (1743 to 1794). During his experiments, van Helmont also discovered that heating coal or wood produced a vapor, which is known now to be natural gas.

In 1626, French explorers observed Native Americans igniting gases that seeped from the ground near Lake Erie in North America. The gases were burned to produce light and heat.

In November 1776, the Italian physicist and chemist **Alessandro Volta** (1745 to 1827) found natural gas at Lake Maggiore, Italy. In 1778, he isolated methane from the gas, thereby discovering methane.

The first recorded use of natural gas for lighting occurred in 1783, when Professor **Jan Pieter Minckeleers** of the University of Louvain, in modern-day Belgium, produced flammable gases by heating coal and other materials. He then used the gas to light his lecture hall at the university. In 1794, the French civil engineer **Philippe LeBron** patented a technique to obtain natural gas by distilling wood or coal. The first commercial plant producing natural gas was built by the London and Westminster Gas Light and Coke Company in 1812. They installed wooden pipes to the Westminster Bridge and lit the bridge using gas on New Year's Eve in 1813.

In the United States, the first commercial gas plant was the Gas Light Company of Baltimore, established in 1816. However, natural gas was not consumed on a large scale anywhere in the world until after **William Hart** dug the first natural gas well in Fredonia, New York in 1821. Hart saw bubbles percolating up through creek water and dug 8.2 meters down with a shovel to create a well that allowed more gas to flow. He then used hollowed-out logs connected by tar and rags to create a pipeline for the gas. The gas was ultimately distributed by America's first natural gas company, the Fredonia Gas Light Company, established in 1858.

Natural gas in the 1800s was used primarily for lighting. In 1885, **Robert Bunsen** (1811 to 1899) invented the Bunsen burner at the University of Heidelberg, Germany. This device burns natural gas to produce a continuous flame for heating, sterilization, or combustion. After this invention, the use of natural gas exploded. Today, it is used for home air heating and cooking, water heating, dryers, high-temperature industrial processes, and electricity production. The first electricity production with natural gas did not occur until 1937, when Sun Oil used a gas turbine to generate electricity at a chemical plant in Philadelphia. The first public electricity generation with natural gas occurred in 1939, in a plant in Neuchatel, Switzerland, where a 4-megawatt turbine was powered by natural gas.

8.1.4 Discovery of Oil

On August 27, 1859, **Colonel Edwin Drake** (1819 to 1880) discovered oil 21 meters below ground near Titusville, Pennsylvania. The well also produced natural gas. Oil and gas were piped almost 9 kilometers to Titusville, cementing the ability of oil and natural gas to be piped long distance. Oil was subsequently used primarily to produce transportation fuels but also to produce heat, chemicals, asphalt, lubricants, waxes, plastics, clothing, cosmetics, and gum. Today, in the clothing industry, oil is used to produce acrylic, rayon, vegan leather, polyester, nylon, and spandex. In the sporting goods industry, it is used in basketballs, golf balls, football helmets, surfboards, fishing rods, skis, and tennis rackets. Oil products are also used in perfume, cosmetics, hand lotion, toothpaste, soap, shaving cream, combs, shampoo, contact lenses, and eyeglasses. Even in medicine, oil has deep roots. Its products are used in artificial limbs, hearing aids, heart valves, antihistamines, aspirin, and IV bags. In buildings, oil is used in roofing material, insulation, linoleum flooring, furniture, rugs, house paint, dishes, nonstick pans, detergents, and pillows.

Now that the origins of the use of coal, oil, and natural gas for energy have been explored, it is time to look at why they should not be used and are unnecessary in a clean, renewable energy future.

8.2 Comparison of Energy Technologies

When evaluating different proposed solutions to air pollution, global warming, and energy insecurity, agnostic metrics are needed to compare the damage caused by each technology performing the same task.

One such metric, which compares the potential climate impact of one energy technology over another, is **CO_2-equivalent emissions**. These are the emissions of carbon dioxide from a source plus the emissions of all other gas and particle pollutants from the source, each multiplied by the pollutant's ability to warm the planet over 20 or 100 years relative to carbon dioxide's ability to warm the planet over the same period. Different time frames are considered because carbon dioxide has a long life in the atmosphere. So, if another chemical, such as methane, which has a much shorter life but a more powerful warming potential per molecule during that short life, is emitted in sufficient quantities, the CO_2-equivalent emissions of carbon dioxide plus methane will be much higher over 20 years than over 100 years. Given that damaging climate impacts are occurring today over periods of only a few years, and short-lived powerful global warming agents such as black carbon and methane cause much more damage over the short term than over the long term, it is important to compare lifecycle emissions of different technologies over short periods, such as 20 years.[132] However, impacts over 100 years are also important because long-lived warming chemicals emitted today will impact the climate of future generations in addition to that of the current generation.

Energy technologies emit several categories of CO_2-equivalent emissions. These include lifecycle emissions, opportunity cost emissions, anthropogenic heat emissions, anthropogenic water vapor emissions, emissions risk due to the leakage of captured carbon dioxide, and emissions due to the covering or clearing land for energy development, among others. These emissions are discussed, in turn.

8.2.1 Lifecycle Emissions

Lifecycle emissions are all CO_2-equivalent emissions that arise during the construction, operation, and decommissioning of an electricity-generating plant per unit electricity generated, averaged over 20 or 100 years. For a fossil-fuel or nuclear plant, operating a plant includes mining, transporting, and processing the plant's fuel; running the plant equipment; repairing the plant; and disposing of waste during the plant's life. WWS requires no fuel mining and has no waste. Figure 8.1 provides lifecycle emissions of several electricity-generating technologies.

8.2.2 Opportunity Cost Emissions

Opportunity cost emissions are emissions from the background electric power grid due to the longer time lag between planning and operation of one energy technology relative to another and the longer downtime needed to refurbish one technology at the end of its life relative to the other.[133]

For example, if Plant A takes 4 years and Plant B takes 10 years between planning and operation, the background grid will emit pollution for 6 more years with Plant B than with Plant A. The emissions during those additional 6 years, when averaged over the lifetime energy production of the technology, are opportunity cost emissions.

Similarly, if Plant A has a useful life of 20 years and requires 2 years of refurbishing to last another 20 years, and Plant B has a useful life of 30 years but takes only 1 year of refurbishing, then Plant A is down an additional 5.9 years out of every 100 years for refurbishing compared with Plant B. During those additional years that Plant A is down, the background grid emits pollution that must be accounted for.

The time lag between planning and operation of a plant includes planning time and construction time. The planning time includes the

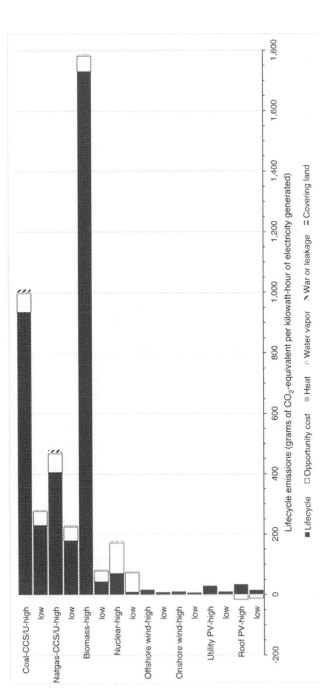

Figure 8.1 Low and high estimates of total 100-year-averaged CO_2-equivalent emissions from several electricity-producing technologies. The total for each bar is the sum of lifecycle emissions ("Lifecycle"), opportunity cost emissions ("Opportunity cost"), anthropogenic heat emissions, ("Heat"), anthropogenic water vapor emissions ("Water vapor"), emissions from nuclear weapons proliferation or carbon capture leakage ("War or leakage"), and emissions from loss of carbon storage by covering land and vegetation with building materials ("covering land"). CCS/U is carbon capture and storage or use. Opportunity cost emissions are relative to onshore wind, so negative opportunity cost emissions (for rooftop PV) are due to the shorter time lag between planning and operation of roof PV versus onshore wind. All wind and solar have negative heat emissions; wind has negative water vapor emissions (see text). Source: Jacobson, *100% Clean, Renewable Energy.*[8]

time required to identify a site, obtain a site permit, purchase or lease the land, obtain a construction permit, obtain financing and insurance for construction, install transmission lines, negotiate a power purchase agreement, and obtain an operating permit. The construction time includes the time required to build the plant, connect it to the transmission infrastructure, and obtain a final operating license.

The planning-to-operation time of a utility-scale wind or solar farm is generally 2 to 5 years, which includes a planning time of 1 to 3 years and a construction period of 1 to 2 years.[133] This time applies to both onshore and offshore wind. For example, the 407-megawatt (49-turbine) Horns Rev 3 offshore wind farm, located in the North Sea off of the west coast of Denmark, required 1 year and 10 months to build.[134] Wind turbines often last 30 years before refurbishing, and the refurbishing time is 3 months to 1 year. For rooftop PV, the time between planning and operation can be as little as 6 months. For geothermal electricity, the planning-to-operation time is 3 to 6 years; for hydroelectric, it is 8 to 16 years.

The planning phase of a coal-fired power plant without carbon capture equipment is generally 1 to 3 years, and the construction phase is another 5 to 8 years, for a total of 6 to 11 years between planning and operation. The planning-to-operation time of a coal plant with carbon capture is similar to one without since the carbon capture equipment can be installed during the long period of coal-plant construction. The planning-to-operation time of a natural gas power plant without carbon capture is less than that of a coal plant. But adding carbon capture is likely to extend the planning-to-operation time of a natural gas plant to that of a coal plant (6 to 11 years). The planning-to-operation time of most all nuclear plants worldwide is 10 to 19 years.

For technologies that take a long time before operation, such as nuclear and hydroelectric power, the opportunity cost emissions can be greater than the lifecycle emissions (Figure 8.1).

8.2.3 Anthropogenic Heat Emissions

In a coal, oil, natural gas, and biomass electric power plant, fuel is burned to generate electricity. For coal and oil, about 65 to 67 percent of the energy in the fuel is released as waste heat, and the rest is converted to electricity. For natural gas, 40 to 60 percent of the energy is released as waste heat. For biomass, about 74 percent is released. For nuclear, about 65 percent of the energy in the uranium is released as waste heat.[6]

In all cases, most of the waste heat is absorbed by liquid water and released to a river, a lake, or the ocean. The rest is released directly to the air. Some of the heat absorbed by the liquid water is released to the air as well.

The heat released to the air due to combustion or nuclear reaction is one source of **anthropogenic heat**. Other sources are the dissipation of electricity to heat, the burning of fuels to move vehicles, and the burning of biomass.

For example, when electricity is used to run an appliance, the electricity is converted to heat. This is most obvious in an electric resistance heater or an electric resistance cooktop, where the electricity is converted directly to heat to warm a room or to cook food, respectively. However any electrical device or appliance, such as a light bulb or a dishwasher, converts all electricity consumed to heat. Energy must be conserved, so even if electricity is first used to produce light in a light bulb, the wavelengths of visible light emitted by the bulb are absorbed by building materials, causing the temperature of the building to rise.

Similarly, when gasoline is used to move a car, only 17 to 20 percent of the energy in the gasoline is used to move the car. The rest is waste heat released to the air. Even the energy used to move the car is converted to heat due to the resistance caused by friction between the car's tires and the road and between the car and air molecules in its path. If no friction occurred, the car would be a perpetual-motion machine, running for an infinite distance with no heat production once moving. Even with an electric vehicle, where 80 to 90 percent of the electricity is converted to energy used to move the car, that electricity is also dissipated to heat by friction.

Whereas fossil-fuel and nuclear electricity generators produce anthropogenic heat, solar and wind generators decrease anthropogenic heat.

Solar PV and CSP convert sunlight to electricity, thereby reducing the amount of sunlight reaching the surface below the PV panel or CSP mirrors. Because less sunlight reaches the ground or rooftop, those surfaces cool. Neither solar PV nor CSP involves the burning of a fuel, so they do not have the same heat release problem as do fossil-fuel plants.

Wind turbines also cool the air by extracting kinetic energy from the air and converting it to electricity. **Kinetic energy** is the energy embodied in air due to its motion. For each kilowatt-hour of electricity produced by a wind turbine, 1 kilowatt-hour of kinetic energy is extracted from the wind. Since energy in the wind is ultimately converted to heat when the wind blows against objects or the ground, taking energy out of the wind reduces air heating.

As mentioned, all electricity produced for the grid will eventually dissipate back to heat as well. However, for purposes of comparing the heat released by different electricity-generating technologies, the heat from electricity consumption is treated separately and is the same regardless of the electricity-generating technology.

Anthropogenic heat fluxes are converted to positive or negative CO_2-equivalent emissions in order to account fully for the impacts of electricity-generating technologies on climate. Such a conversion reveals that the anthropogenic heat fluxes are relatively small (but are not zero) compared with other sources of heating (Figure 8.1).

8.2.4   Anthropogenic Water Vapor Emissions

Burning fossil fuels, bioenergy fuels, and vegetation also releases water vapor (H_2O) due to the chemical reaction between the hydrogen in the fuel and oxygen in the air. Electricity-generating plants running on fossil fuels, bioenergy fuels, and nuclear energy use liquid water for cooling. Much of this water evaporates from a cooling tower or from the surface of the water once the heated water is sent back to the river, lake, or ocean that it came from. Many CSP plants also use water-cooling although some use air-cooling. Similarly, whereas nonbinary geothermal plants and some binary plants use water-cooling, and thus emit water vapor, binary plants that use air-cooling do not emit any water vapor. Water also evaporates from reservoirs behind hydroelectric power plant dams. Water vapor from all anthropogenic sources worldwide may cause about 0.23 percent of gross global warming.[6]

On the other hand, wind turbines reduce water evaporation, since they reduce wind speeds behind them, and lower wind speeds mean less evaporation from ground water and surface water.[135,136]

In sum, while electricity-generating plants that run on fuel combustion add water vapor to the air, wind turbines reduce water vapor to the air. Since water vapor is a greenhouse gas, electricity generators that burn fuels slightly increase global warming, whereas wind turbines slightly decrease global warming due to this process. The small changes in warming and cooling due to changes in water vapor for each generator are reflected in changes in CO_2-equivalent emissions in Figure 8.1.

8.2.5 Leaks of Carbon Dioxide Sequestered Underground

The sequestration of carbon dioxide underground, in carbon capture and storage, runs the risk of carbon dioxide leaking back to the air through existing fractured rock or overly porous soil or through new fractures in rock or soil resulting from an earthquake. The ability of a geological formation to sequester carbon dioxide for decades to centuries varies with location and tectonic activity. Estimated carbon dioxide leakage rates have ranged from 0.1 to 10 percent per 1,000 years.[137] However, there is no certainty of this result since tectonic activity or natural leakage over 1,000 years is not possible to predict. Also, because liquefied carbon dioxide injected underground is under high pressure, it may leak more readily than expected through horizontal and vertical fractures in rocks. Because carbon dioxide is an acid, it will also erode rocks over time. If a leak from an underground formation to the atmosphere occurs, it may or may not be detected. If a leak is detected, it may or may not be sealed, particularly if it occurs over a large area. Figure 8.1 provides an estimate of the CO_2-equivalent emissions arising from leaks of carbon dioxide stored underground.

8.2.6 Emissions from Covering Land or Clearing Vegetation

Another source of carbon dioxide emissions associated with electricity generation is from covering land or clearing vegetation. Covering land with impervious construction material, such as asphalt or concrete, eliminates vegetation and its carbon above the soil and eliminates the ability of soil to accumulate carbon from the decay of vegetation. Similarly, clearing land of vegetation reduces the carbon stored in the vegetation and its roots.

Normally, when grass dies, the dead grass contributes to the soil organic carbon. The grass then regrows, removing carbon from the air by photosynthesis. If the soil is instead covered with concrete, the grass no longer exists, so it can no longer remove carbon from the air or store carbon in the soil. However, existing carbon stored underground remains.

The carbon emissions due to developing land for an energy facility can be estimated simplistically by first summing the land areas covered by the facility; the mine (if any) where the fuel is extracted from; the roads, railways, or pipelines needed to transport the fuel; and the waste disposal site associated with the facility. This summed area is then multiplied by the organic carbon contents stored in vegetation and in the soil under the vegetation per unit area that is lost. The loss of carbon is then converted to a loss of carbon per unit electricity produced by the energy facility over a period, such as 100 years. Figure 8.1 shows the resulting CO_2-equivalent emissions due to this process for each energy facility.

8.2.7   Total CO_2-Equivalent Emissions

A summation of the individual emission types just discussed indicates that WWS technologies (onshore and offshore wind, solar PV, CSP, geothermal, tidal, wave, and hydroelectricity) have lower overall CO_2-equivalent emissions than do nuclear, biomass, natural gas with carbon capture, or coal with carbon capture (Figure 8.1). This result accounts not only for lifecycle emissions, but also for several types of emissions not included in standard comparisons. The lower emissions of WWS are one reason WWS technologies are proposed. Other reasons include the fact that WWS also eliminates air pollution, whereas fossil fuels and biomass, even with carbon capture, do not. WWS also eliminates continuous fuel mining, thus land and ecosystem damage, associated with fossil fuels, biomass, and nuclear power. WWS further increases energy security compared with all other technologies. Finally, WWS reduces meltdown risk, weapons proliferation risk, uranium mining lung cancer risk, and waste storage risks associated with nuclear power. These issues are discussed next.

8.3   Why Not Natural Gas as a Bridge Fuel?

Most natural gas is used for electricity generation, heat for buildings, or industrial heat. Because natural gas is not very dense, it can be stored only in a large container or an underground storage facility. As such, natural gas is often compressed or liquefied for transport and storage. **Compressed natural gas (CNG)** is natural gas compressed to less than 1 percent of its gas volume at room temperature. **Liquefied natural gas (LNG)** is natural gas that has been cooled to minus 162 degrees Celsius, the temperature at which it condenses to a liquid at ambient pressure. LNG has a volume that is 1/600 the volume of the original gas.

Both compressed and liquefied natural gas can be sent through pipelines, although different pipelines are needed for each. Both can also be stored and used in automobiles that are designed to run on them. Both can further be transported by truck or bus with a special fuel tank and stored at a power plant when pipeline gas is not available.

For overseas transport, compressed natural gas in a pipeline is often converted to liquefied natural gas at a marine export terminal, put on a tanker ship with super-cooled storage tanks, then shipped overseas. At the import terminal, it is regasified and piped to its final destination – either a power plant, industrial company, or company that transmits and distributes the natural gas to buildings for heating or other purposes.

Natural gas is obtained from underground conventional or unconventional wells. Conventional wells are shallow and small in area and produce oil, natural gas, or both. Unconventional wells are deep and large in area and include shale gas and coal-bed methane wells. **Hydraulic fracturing (fracking)** is used in both well types, but mostly in unconventional shale gas wells. **Shale** is sedimentary rock composed of a muddy mix of clay mineral flakes and small fragments of quartz and calcite. Large shale formations containing natural gas can be found in many areas of North America and worldwide, some close to population centers. In the United States, about 67 percent of natural gas in 2015 was extracted from shale rock.[138] Extraction of natural gas from shale requires large volumes of water, laced with chemicals, forced under pressure to fracture and re-fracture the rock to increase the flow of natural gas. As the water returns to the surface over days to weeks, it is accompanied by methane that escapes to the air. As such, more methane leaks during the fracking of unconventional shale wells than from conventional gas and oil wells. Methane also leaks during the transmission, distribution, and processing of natural gas.[139,140,141,142]

For electricity production, natural gas is usually used in either an **open cycle gas turbine** or a **combined cycle gas turbine**. In an open cycle turbine, air is sent to a compressor, and the compressed air and natural gas are both sent to a combustion chamber, where the mixture is burned. The hot gas expands quickly, flowing through a turbine to perform work by spinning the turbine's blades. The rotating blades turn a shaft connected to a generator, which converts a portion of the rotating mechanical energy into electricity.

The main disadvantage of an open cycle turbine is that the exhaust contains a lot of waste heat that could otherwise be used to generate more electricity. A combined cycle turbine routes that heat to a heat recovery steam generator, which boils water with the heat to create steam. The steam is then sent to a steam turbine connected to the generator to generate 50 percent more electricity than with the open cycle turbine alone. Thus, a combined cycle turbine produces about 150 percent of the electricity of an open cycle turbine with the same input amount of natural gas and the same resulting carbon dioxide emissions.

On the other hand, the rate at which an open cycle turbine can produce electricity from a cold start is 20 percent of its maximum output per minute, which is 2 to 4 times that of a combined cycle turbine (5 to 10 percent per minute). In other words, the less efficient open cycle turbine, which emits 50 percent more carbon dioxide per unit electricity generated, is more useful for filling in short-term gaps in electricity supply on the grid than is a combined cycle turbine.

It has long been suggested that natural gas could be used as a **bridge fuel** between coal and renewables.[143] The two main arguments for this suggestion are (1) natural gas emits less CO_2-equivalent emissions per unit energy produced than coal and (2) natural gas electric power plants are better suited to be used with intermittent renewables than coal.

However, these justifications for using gas as a bridge fuel are incorrect and insufficient. Natural gas is not recommended for use together with WWS technologies for multiple reasons, as discussed in the following sections.

8.3.1 Climate Impacts of Natural Gas versus Coal

When used in an electric power plant, natural gas substantially increases, rather than decreases, global warming compared with coal over a 20-year time frame. Over a 100-year time frame, coal causes more warming than natural gas, but the difference is not large. Also, whereas coal causes more air pollution damage than does natural gas, both cause far more air pollution damage than does WWS. As such, spending money on either natural gas or coal represents an opportunity cost relative to spending the same money on WWS.

Considering only lifecycle emissions, electricity from natural gas using a combined cycle gas turbine or an open cycle gas turbine may cause 2.3 and 2.8 times, respectively, as much warming as coal over a 20-year time frame. Over a 100-year time frame, natural gas with those turbines may cause only 36 and 19 percent less warming, respectively, than coal.[8] The fact that natural gas causes far more global warming than coal over a 20-year time frame is a concern because of the severe damage global warming is already causing. More natural gas emissions in the short term may trigger difficult-to-reverse impacts, such as the complete melting of the Arctic sea ice.

The reasons that natural gas causes more warming than coal over a 20-year timeframe and only slightly less warming than coal over a 100-year timeframe are as follows.

First, although natural gas burned in an open cycle turbine or combined cycle turbine produces less carbon dioxide per unit electricity than does burning coal, more natural gas generally leaks to the air during the mining, transport, and processing of natural gas than it does during coal mining. Because methane is a potent greenhouse gas, causing substantially more warming per unit mass than carbon dioxide over both short and long timescales, the leakage of more methane with natural gas than with coal contributes to the greater short-term warming of natural gas over coal.

Second, and more important, coal combustion emits more nitrogen oxides and sulfur dioxide per unit electricity produced than does natural gas combustion, and nitrogen oxides and sulfur dioxide both produce cooling aerosol particles, which offset or mask much of coal's global warming. The cooling impacts of these particles are through their reflection of sunlight back to space and their enhancement of cloud thickness. Thicker clouds reflect more sunlight, cooling the ground. As such, coal's emission of nitrogen oxides and sulfur dioxide, which are both short-lived chemicals, causes substantial short-term cooling, offsetting much of the short-term warming caused by carbon dioxide from coal.

Regardless, neither natural gas nor coal is recommended in a 100 percent WWS world. Considering lifecycle emissions alone and before considering carbon capture, natural gas causes about 56 to 72 times the warming, per unit electricity generated, of wind electricity, averaged over 100 years.[8] Coal causes about 88 times the warming of wind. With carbon capture and considering all emissions, natural gas and coal also cause much more climate damage than does wind (Figure 8.1).

8.3.2   Air Pollution Impacts of Natural Gas versus Coal and Renewables

Whereas natural gas for electricity causes more warming than coal over 20 years and slightly less warming than coal over 100 years, coal emits more health-damaging air pollutants than does natural gas. Nevertheless, both natural gas and coal are much worse for human health than are WWS technologies, which emit zero air pollutants during their operation. WWS emits pollutants only during the manufacture and decommissioning of WWS devices. Such WWS-related emissions will disappear to zero as all energy is transitioned to 100 percent WWS, since at that point, even mining and manufacturing will be powered with WWS.

More carbon monoxide, volatile organic carbon, methane, and ammonia are emitted during natural gas production and use than during coal production and use in the United States. Coal production and use emit more nitrogen oxides, sulfur dioxide, and particulate matter.[144] Emissions from the mining, transport, processing, and consumption of natural gas may cause 5,000 to 10,000 deaths each year in the United States from air pollution.[145] Coal-related emissions may cause 20,000 to 50,000 deaths. Thus, both fuels cause air pollution deaths, but coal causes more. The reason is that nitrogen oxides and sulfur dioxide from coal convert to particles in the air, and those particles plus the particles emitted directly by coal cause severe health damage.

In sum, coal causes more deaths than does natural gas, but both coal and natural gas cause far more deaths than do WWS technologies. The combination of the much higher CO_2-equivalent emissions and air pollution deaths due to natural gas than to WWS technologies renders natural gas not an option as a bridge fuel.

8.3.3   Natural Gas for Peaking and Load Following Is Not Needed

Another argument for using natural gas as a bridge fuel is that it can be used for load following or peaking, thereby preventing black-outs on the electricity grid. WWS technologies, on the other hand, are intermittent and need load-following or peaking plants that use natural gas to back them up when wind and solar outputs are low.

Although natural gas plants do help with peaking and load fol-lowing, they are not needed. Other types of WWS electric power stor-age options available include CSP with storage, hydroelectric power reservoir storage, pumped hydropower storage, stationary batteries, flywheels, compressed air energy storage, and gravitational storage with solid masses.

Transition highlight
By 2021, the cost of a system consisting of wind, solar, and batteries was already less than that consisting of natural gas. For example, even in 2019, a Florida utility replaced two natu-ral gas plants with a combined solar–battery system because of the lower cost of the latter.[146]

8.3.4   Land Required for Natural Gas Infrastructure

The continuous use of natural gas for electricity and heat results in the continuous and cumulative degradation of land for as long as natural gas use continues. Wells must be dug, and pipes must be laid every year to supply a world thirsty for natural gas. When gas wells become depleted after 20 to 30 years, new wells must be drilled. Fifty thousand new natural gas wells are drilled each year in North America alone to satisfy natural gas demand.[147] The land area required for the well pads, roads, and storage facilities for these new wells amounts to 2,500 square kilometers of additional land consumed per year. Once a gas well is depleted, it is sealed and abandoned, and a portion of the abandoned land cannot be used for any other purpose. The United States alone has 1.3 million active oil and gas wells and 3.2 million abandoned ones. Worldwide, an estimated 29 million wells are aban-doned. Two-thirds of these are estimated to leak methane and other hydrocarbons.[148] The natural gas infrastructure also requires land for underground and aboveground pipes, power plants, fueling stations, and underground storage facilities.

The flammability of natural gas further results in explosions with fatal consequences in homes and urban areas. For example, on September 9, 2010, a natural gas explosion in a San Bruno, California, neighborhood destroyed 38 homes, killed 8 people, and injured 58 others.

Natural gas blowouts are also a danger to the local and regional community. For example, from October 23, 2015, to February 11, 2016, 97,100 tonnes of methane, 7,300 tonnes of ethane, and a host of other hydrocarbons blasted into the air from a major breach in the Aliso Canyon, California, underground natural gas storage facility. The result was not only damage to the health of nearby residents but also the death from air pollution of up to a few dozen people throughout California.[149]

Fossil-fuel infrastructure currently takes up about 1.3 percent of the United States land area.[8] This is due not only to active and abandoned oil and gas wells, but also to coal mines, oil refineries, millions of kilometers of pipelines and the accompanying deforestation to make room for them, power plants, fueling stations, and storage facilities. Whereas all fossil fuels contribute to this land area degradation, natural gas's share is growing because of the phase out of coal and the growth of natural gas, particularly of hydraulically fracked gas. The damage due to fracking includes damage not only to the landscape but also to groundwater, in which fracked natural gas often leaks. Additional damage occurs to roads, which must carry heavy trucks associated with natural gas development. Gas flaring is another form of damage, as flaring emits soot (particles containing black carbon), which causes health damage, warms the air, evaporates clouds, and melts snow.

8.4   Why Not Carbon Capture with Natural Gas or Coal?

A proposal to help solve the climate problem, but that also has the side effect of keeping the fossil-fuel and bioenergy industries in business, is to capture the carbon dioxide emitted from fossil-fuel or bioenergy power plants before the carbon dioxide escapes their exhaust stacks. The carbon dioxide is then either stored underground or used by industry. The carbon dioxide is captured with equipment added to the plant.

This solution is poor for five reasons: it increases emissions and the resulting health problems of all gases and particles aside from carbon dioxide compared with no capture; it only marginally reduces carbon dioxide; it increases the land degradation from the mining of fossil fuels compared with no capture; it increases fossil-fuel infrastructure; and it diverts funding from lower-cost renewables that reduce climate and air pollution problems more effectively than does carbon capture. These issues are discussed in detail next.

Carbon capture and storage (CCS) is the separation of carbon dioxide from other exhaust gases following fossil-fuel or biofuel combustion or chemical release, such as during cement or steel manufacturing. The captured carbon dioxide is then sent to an underground geological formation (such as a saline aquifer), a depleted oil and gas field, or an unminable coal seam. The remaining combustion gases are emitted to the air or filtered further. Geological formations worldwide may theoretically store up to 2 trillion metric tonnes of carbon dioxide, which compares with a fossil-fuel carbon dioxide emission rate in 2019 of about 36.4 billion metric tonnes.

Another proposed CCS method is to inject the carbon dioxide into the deep ocean. Because carbon dioxide is an acid, this method acidifies the ocean. Some carbon dioxide injected into the deep ocean also eventually works its way back to the surface, where it is released back to the air.

A third type of sequestration method is to mix captured carbon dioxide with concrete material, trapping the gas inside the concrete.

Carbon capture and use (CCU) is the same as CCS, except that the carbon dioxide isolated during carbon capture is sold to industry to pay back the cost of the carbon capture equipment. To date, the major application of CCU has been **enhanced oil recovery**. With this process, captured carbon dioxide is piped to a nearby oil field where it is pumped underground. There, it binds with the oil, reducing oil's density and allowing the oil to rise to the surface faster. Once the oil rises up, the carbon dioxide is separated from it and sent back into the reservoir. However, during the transport and separation process, up to 40 percent of all carbon dioxide originally captured is released back to the air.[150,151] Enhanced oil recovery also results in about two additional barrels of oil for every tonne of carbon dioxide injected into the ground. When this is burned, it produces even more air pollution and carbon dioxide emissions.

Another proposed use of carbon dioxide has been to create carbon-based fuels to replace gasoline and diesel. The first problem with this proposal is that it allows combustion to continue in vehicles.

Combustion creates air pollution, only some of which can be stopped by emission control technologies. In addition, producing the fuels requires machines, materials, and energy, all of which require money and additional energy and result in emissions. The money is more effectively spent on electric vehicles, which eliminate all tailpipe emissions of air pollutants and carbon dioxide.

8.4.1   Air Pollution and Climate Impacts of Natural Gas or Coal with Carbon Capture

Proponents of carbon capture claim that the equipment captures 90 percent or more of carbon dioxide from a fossil-fuel exhaust stream. However, several factors cause the yearly average carbon dioxide capture efficiency to be only 20 to 80 percent.

The capture efficiency accounts only for carbon dioxide capture from the smoke stack. But running carbon capture equipment takes energy, and that energy usually comes from fossil fuels as well, and these fossil fuels emit carbon dioxide and methane during their mining, transport, and use. In addition, mining for fossil fuels used in the power plant itself takes energy, and that energy use results in emissions. None of these additional emissions are captured. When such emissions are accounted for, the overall capture efficiency declines to between zero and 25 percent.

In addition, the capture equipment captures only carbon dioxide, not air pollutants. Also, no air pollutants are captured during the mining of fossil fuels, including of the additional fossil fuels needed to run the capture equipment. As such, carbon capture increases air pollution relative to no capture. Similarly, carbon capture increases fuel mining relative to no capture.

To summarize,

(1) Carbon capture equipment captures 75 to 90 percent of the carbon dioxide in a concentrated exhaust stream when operating at its fullest and best.[152,153]

(2) However, the carbon capture equipment may be shut down temporarily, either owing to a lack of demand for the captured carbon dioxide or owing to scheduled or unplanned maintenance of the capture equipment. These factors result in the yearly averaged capture rate of the carbon dioxide exhaust stream declining to 20 to 80 percent.[152,153,154,155,156]

(3) Carbon capture equipment does not capture the upstream carbon dioxide or other greenhouse gas emissions resulting from mining, transporting, or processing the fossil fuel used in the plant emitting carbon dioxide. Accounting for these emissions reduces the overall carbon dioxide capture rate more.

(4) Carbon capture equipment does not capture any of the health-affecting air pollutants from the fossil-fuel exhaust or upstream mining, transporting, or processing of the fossil fuel. Such pollutants include carbon monoxide, nitrogen oxides, sulfur dioxide, organic gases, mercury, toxins, black carbon, brown carbon, and fly ash, all of which affect health.

(5) A plant with carbon capture needs 25 to 50 percent more energy to run the carbon capture equipment than does a plant without the equipment.[137,154,157] If that energy comes from a fossil fuel, the plant then emits even more uncaptured carbon dioxide. If that energy comes from a renewable source, the use of the renewable energy for carbon capture prevents the energy from displacing fossil-fuel energy. In both cases, fossil-fuel mining and infrastructure and air pollution increase 25 to 50 percent relative to no capture.

(6) If the captured carbon dioxide is used for enhanced oil recovery or to produce synthetic fuels, up to 40 percent of the captured carbon dioxide is released back to the air.

(7) If the captured carbon dioxide is sequestered underground, some of it leaks back to the air over time.

Overall, coal-CCS and coal-CCU are estimated to result in 33 to 211 times the CO_2-equivalent emissions of onshore wind per unit electricity generated. Natural gas-CCS and natural gas-CCU are estimated to result in 27 to 100 times the emissions of onshore wind (Figure 8.1).

In addition, air pollution emissions from coal and natural gas without carbon capture are 100 to 400 times those of onshore wind per unit energy. Adding carbon capture to a coal or gas plant increases non-CO_2 air pollution emissions another 25 to 50 percent.

The high air pollution and climate-relevant emission rates of coal and natural gas with carbon capture suggest that spending money on them represents an opportunity cost relative to spending money on lower-emitting technologies.

8.4.2 Carbon Capture Projects

To date, carbon dioxide has been captured and separated primarily from natural gas processing facilities, coal-fired power plants, a plant that gasifies coal to produce natural gas, ethanol refineries, and a facility producing hydrogen. In most cases, the carbon dioxide has been used to enhance oil recovery. In rare cases, the carbon dioxide was sequestered underground.

As of 2021, only two fossil-fuel electric power-generating plants have operated with carbon capture equipment. In both cases, the separated carbon dioxide was used for enhanced oil recovery. These projects, plus a third, the largest carbon capture project in the world, in which carbon dioxide from a liquefied natural gas processing facility is being captured and stored, are discussed. All three captured little carbon and increased air pollution and fossil-fuel mining.

8.4.2.1 Boundary Dam Project

The first electric power plant with CCU equipment was the Boundary Dam power station in Estevan, Saskatchewan, Canada. This plant has been operating with CCU equipment on one coal boiler connected to a steam turbine since October 2014. The cost of the retrofit project was 1.47 billion Canadian dollars, which included a C$240 million subsidy from the Canadian government. The remaining C$1.23 billion was paid for by coal plant electricity customers as an additional charge added to their normal bills. From 2015 through the first quarter of 2021, the average carbon dioxide capture efficiency from the exhaust stream of this plant was 55 percent.[155] The captured carbon dioxide was sold for enhanced oil recovery. Of the captured carbon dioxide, up to 40 percent may have been lost during its transport and processing.[150,151] That means that only 33 percent of coal plant emissions were stored underground. However, even if true, most of that was offset by the fact that the carbon capture equipment increased the electricity required by the coal plant by at least 25 percent, which resulted in 25 percent more coal mining, combustion, and emissions. In addition, none of the carbon dioxide or air pollution from the mining of the coal has been captured.

8.4.2.2 Petra Nova Project

The second coal plant with carbon capture equipment was the W.A. Parish coal power plant near Thompsons, Texas. The plant was retrofitted with carbon capture equipment as part of the Petra Nova project and began using the equipment during January 2017. The carbon capture equipment (240 megawatts) received 36.7 percent of the emissions from a 654-megawatt boiler at the coal plant. The equipment required almost 0.5 kilowatt-hours of electricity to run per kilowatt-hour of electricity produced by the coal plant. An entire natural gas turbine with a heat recovery boiler was built to provide this electricity. A cooling tower and water treatment facility were also added. The retrofit cost $1 billion ($4,200 per kilowatt) beyond the coal plant cost.[158] Owing to poor economics, the carbon capture plant ceased operating in December 2019.

During operation, captured carbon dioxide was compressed and piped to an oil field, where it was used to enhance oil recovery. Carbon dioxide from the natural gas turbine was not captured. The mining and transport of natural gas also emitted uncaptured carbon dioxide, methane, and air pollutants. Upstream carbon dioxide, methane, and air pollution emissions from the coal plant itself were also uncaptured. Finally, 40 percent of the captured carbon dioxide was released back to the air during enhanced oil recovery.

Even after ignoring upstream emissions and carbon dioxide emissions during enhanced oil recovery, the capture equipment captured only 55.4 percent of carbon dioxide from the coal combustion exhaust stream and 33.9 percent of the coal plus natural gas combustion carbon dioxide during 2017. When upstream emissions from the mining and processing of the coal and natural gas were accounted for, the capture equipment reduced coal and gas combustion plus upstream CO_2-equivalent emissions a net of only 10.8 percent over a 20-year time frame and 20 percent over a 100-year time frame.[154]

Hypothetically, using wind, instead of natural gas, to power the carbon capture equipment would reduce CO_2-equivalent emissions more, by a total of 37.4 percent over 20 years and 44.2 percent over 100 years versus no carbon capture. The decrease is greater because wind does not result in any natural gas mining or combustion emissions.

However, hypothetically using the wind electricity that powered the carbon capture equipment instead to replace coal electricity directly would reduce CO_2-equivalent emissions over both 20 and 100 years by even more, 49.7 percent, than in any of the other cases.

The climate benefit of using WWS electricity to replace fossil-fuel electricity instead of to power carbon capture equipment is only part of the story. Carbon capture does not reduce any air pollution from the coal plant or coal mine. Instead, it increases air pollution when a fossil fuel is used to provide power for the capture equipment. Even when wind powers the capture equipment, air pollution continues from the coal plant and coal mine. Only when wind replaces the coal plant itself does air pollution from both the coal plant and coal mine decrease.

In sum, the total social cost (equipment plus health plus climate cost) of coal-CCU powered by natural gas is over twice that of wind replacing coal directly and even higher than that of doing nothing. The social cost of using wind to power CCU equipment is also higher than using the same wind to replace a coal plant. In other words, the best strategy is to use WWS to replace fossil fuels.[154] This conclusion is independent of the fate of the carbon dioxide after it leaves the carbon capture equipment, and it applies to CCS or CCU with bioenergy or cement manufacturing as well.

When the fate of captured carbon dioxide is considered, the problem often deepens. If carbon dioxide is sealed underground, little additional emission may occur. If the captured carbon dioxide is used to enhance oil recovery, its current major application, up to 40 percent of the captured carbon dioxide is released back to the air[150,151] and more oil is extracted and burned. If the captured carbon dioxide is used to a create carbon-based fuel to replace gasoline and diesel, energy is still required to produce the fuel, the fuel is still burned in vehicles (creating pollution), and little carbon dioxide is captured to produce the fuel with. A fourth proposal is to use the carbon dioxide to produce carbonated drinks. This is already done. However, most carbon dioxide in carbonated drinks is released to the air during consumption.

8.4.2.3 Gorgon Project

A third project is the world's largest CCS plant, which is attached to the Gorgon liquefied natural gas (LNG) facility on Barrow Island, which is 130 kilometers off the northwest coast of Australia. On March 21, 2016, the facility started converting a portion of the raw natural gas extracted to LNG for export and using the rest for local consumption. The raw natural gas stream from the Gorgon field near Barrow Island contains about 15 percent carbon dioxide. The government of Australia permitted the LNG facility on the condition that 80 percent of the carbon dioxide from the natural gas be captured and injected 2 kilometers below Barrow Island, starting on the day that LNG exports commenced. The CCS equipment cost $3 billion.[156]

Delays prevented capture and injection until August 2019. Inefficiencies and problems limited capture rates thereafter. As a result, over the 5 years between 2016 and 2021, only 20 percent of the carbon dioxide emitted from the LNG facility exhaust stream was captured. The rest was released to the air. Also, a good portion of natural gas for domestic consumption was used to provide electricity to run the CCS equipment. The burning of this gas resulted in enough carbon dioxide emissions to cause the net carbon dioxide emitted by the CCS facility to be positive. Thus, instead of reducing carbon dioxide release to the air, the Gorgon CCS plant likely increased carbon dioxide emissions and at a cost of $3 billion.

In sum, carbon capture is not close to a zero-carbon technology. For the same energy cost, wind turbines and solar panels replacing fossil fuels reduce much more carbon dioxide and also eliminate fossil-fuel air pollution and mining, which carbon capture increases. Using renewables to replace fossils also reduces pipelines, refineries, gas stations, tanker trucks, oil tankers, and coal trains, oil spills, oil fires, gas leaks, gas explosions, and international conflicts over energy. Carbon capture increases these by increasing energy use.

8.5 Why Not Nuclear Power

In evaluating solutions to global warming, air pollution, and energy security, two important questions that arise are (1) should new nuclear electricity-producing plants be built to help solve these problems, and (2) should existing, aged nuclear plants be kept open as long as possible to help solve the problems? This section discusses these issues after nuclear power is explained.

All nuclear electricity today is generated by nuclear fission. **Nuclear fission** is the process by which tiny neutrons bombard and split certain fissile heavy elements, such as uranium-235 or plutonium-239 in a **nuclear reactor**. The 235 and 239 refer to the isotope – that is, the number of protons plus neutrons in the nucleus of a uranium or plutonium atom, respectively. A **fissile** element is one that can be split during fission upon neutron bombardment and whose neutrons released during splitting can split other fissile atoms in a chain reaction. Fissile elements do not spontaneously release neutrons, creating a chain reaction. Instead, they require outside neutrons to bombard them, thereby initiating the chain reaction. Uranium-235 is the only fissile element found in nature. Plutonium-239 is also a fissile element, but it is produced artificially in a nuclear reactor. It is the product of uranium-238 and a free neutron. This produces uranium-239, which decays to plutonium-239.

In a nuclear reactor, a moving neutron may either pass through or be absorbed by uranium-235. Slow-moving neutrons have a higher probability than fast-moving neutrons of being absorbed. If a neutron is absorbed, the resulting uranium atom's total energy is spread among its 236 protons and neutrons now in its nucleus. The nucleus is now unstable, and some uranium atoms fragment into two smaller elements, whereas the remaining atoms form uranium-236. A variety of element pairs arise from fragmentation. Two of the most common are krypton-92 and barium-141. The fragmentation, with this product pair, also produces gamma rays and three free neutrons. The new neutrons may then collide with other uranium-235 atoms or with plutonium-239 atoms, splitting them in a chain reaction. When the fragments and the gamma rays collide with water, the collision converts kinetic energy and electromagnetic energy, respectively, to massive amounts of heat.

In a **boiling water reactor nuclear power plant**, the heat boils water directly. The high-pressure steam turns a turbine connected to a generator to produces electricity. The steam is then recondensed to liquid water in a condenser, and the liquid water is returned back to the reactor core. In the condenser, heat from the steam is transferred to a separate (in an enclosed pipe) stream of cooling water that originates from a lake, a river, or the coastal ocean. The heated water is then returned to where it originated from, warming the outdoor water body, creating thermal pollution. Other thermal power plants, such as those running on coal, oil, or gas, similarly heat water bodies.

In a **pressurized water reactor** plant, the air pressure in the reactor is increased substantially, up to 155 bar. For comparison, air pressure at the Earth's surface is 1 bar. Because the boiling point of water increases with increasing pressure, water in the reactor does not boil, even when the reactor temperature reaches 282 degrees Celsius (at sea level, water usually boils at 100 degrees Celsius). The hot water in the reactor, which is radioactive, passes through a pipe and exchanges its heat with a different batch of water maintained at normal air pressure, causing the latter water to boil. The boiling water creates steam that is pushed through a steam turbine to generate electricity. The water batches are kept separate to ensure radioactive material in the high-pressure reactor does not pass through to the water vapor running through the steam turbine. Boiling water reactors and pressurized water reactors are both called **light water reactors**, which are reactors that use normal water.

Uranium in a nuclear power plant is originally stored in small ceramic pellets within a metal fuel rod, often 3.7 meters long. A conventional light water reactor will go through one rod during about 6 years, and the rod and remaining material in it become radioactive waste. Reactors that use rods once are referred to as **once-through** reactors. The radioactive waste in the fuel rod must be stored for hundreds of thousands of years.

A fuel rod that has gone through a fission reactor once still has 99 percent of its uranium left over, including more uranium-235 than natural uranium. This remaining uranium and its fission product, plutonium, can be extracted and reprocessed for use in a **breeder reactor**, extending the life of a given mass of uranium and reducing waste significantly. However, the reprocessing increases the cost of energy. It also increases the production of plutonium-239 by the collision of uranium-238 with fast-moving neutrons. Breeder reactors can thus be optimized to produce plutonium-239 for use in nuclear weapons.[159] As such, they are a concern with respect to nuclear weapons proliferation.

Nuclear fission became a source of electricity starting in the 1950s. The first nuclear power plant to produce electricity was an experimental reactor in Arco, Idaho. On December 20, 1951, it powered four light bulbs. On June 26, 1954, a 5-megawatt nuclear reactor was connected to the electric power grid for industrial use in Obninsk, Russia. Subsequently, on August 27, 1956, a 50-megawatt reactor was connected to the grid for commercial use in Windscale, England.

As of 2021, over 400 active nuclear reactors still provide electricity in 30 countries. Only two of these reactors are breeder reactors. For all active reactors, uranium mines produce about 54,200 tonnes of uranium per year. Uranium reserves (aside from hard-to-extract uranium in seawater) as of 2019 were about 8.1 million tonnes.[160] This suggests that about 147 years of uranium are available for the number of once-through fuel cycle reactors open in 2021. As such, even if the issues discussed below were not issues, uranium is a limited resource, and growing nuclear power will deplete uranium faster.

An alternative fuel to uranium in nuclear reactors is thorium. **Thorium**, like uranium, can be used to produce nuclear fuel in a breeder reactor. The advantage of thorium is that it produces less radioactive waste with a long half-life than does uranium. Its products are also more difficult to convert into nuclear weapons material. However, thorium still produces ^{232}U, which was used in one nuclear bomb core produced during the **Operation Teapot** bomb tests in 1955. Thus, thorium is not free of nuclear weapons proliferation risk. In addition, thorium reactors require the same long time lag between planning and operation as do uranium reactors and likely longer because hardly any contractors or scientists have experience building or running thorium reactors. Thus, thorium reactors will produce greater emissions from the background electric grid than WWS technologies, which have a shorter time lag. Finally, lifecycle emissions of carbon from a thorium reactor are similar to those from a uranium reactor.

A proposed alternative to the large once-through reactor and the breeder reactor is the **small modular reactor**. Small modular reactors are nuclear fission reactors that are on the order of one-third the size of traditional reactor. They have some parts that could be prefabricated in a factory, which could help to reduce construction time, costs, and mistakes during construction.

Many types of small modular reactors have been proposed, including miniature versions of current reactors. One type of new design is a **fast reactor**, in which the fuel is reformulated to allow fast-moving neutrons, rather than slow-moving neutrons, to split an atom. One way to do this is to increase the quantity of plutonium-239, which absorbs more fast-moving neutrons than does uranium-235. This is done by surrounding the core with uranium-238, which absorbs a fast-moving neutron to become uranium-239, which then decays to plutonium-239. By this mechanism, though, fast reactors become breeder reactors, increasing weapons proliferation risk.

In sum, whereas slow reactors produce significant radioactive waste, fast reactors would produce less waste but would increase nuclear weapons proliferation risk by producing more plutonium-239. Because all small modular reactors are small and many are proposed to be transportable, numerous countries that do not currently have nuclear energy facilities will have an incentive to purchase them, increasing weapons proliferation risk. Most small modular reactors also have meltdown risk. In addition, they require uranium, which must be mined. Slow reactors have the same uranium mining resource limitation, underground mining lung cancer risk, and land despoilment risk as do large reactors.

No small modular reactor being designed in 2022 is expected to be commercial until about the year 2030. However, the world needs to eliminate 80 percent of climate-relevant emissions by 2030 and the rest soon after to avoid the harshest consequences of global warming. In addition, the world needs to eliminate 100 percent of its air pollution emissions today to avoid the 7 million air pollution deaths that occur yearly. As such, small modular reactors will not be able to help address global warning or air pollution in a rapid or meaningful way. Instead, money spent on them will prevent faster and less expensive solutions from being implemented, exacerbating the climate and air pollution problems the world faces.

Historically, small nuclear reactors were planned and developed before large ones were conceived. However, nuclear plant developers abandoned small reactors in favor of large reactors because building one larger reactor was much less expensive than building three small ones. Even today, the cost of new small modular reactors is expected to be higher per unit energy than the cost of large reactors. Further, the cost per unit energy of a large reactor, in 2022, is 5 times that of new onshore wind or utility PV.[161] As such, it is expected that small modular reactors will be much more expensive than new WWS devices.

The sun and all stars in the universe are powered by a different type of nuclear power, nuclear fusion. **Nuclear fusion** involves the fusing together of light atomic nuclei (protium, deuterium, or tritium) into heavier elements. In theory, nuclear fusion could supply all power on Earth indefinitely. It could also do so without long-lived radioactive waste because its products are isotopes of helium, which are not harmful. Nuclear fusion has been explored for decades. However, its commercialization has always been about 30 to 100 years away. Recently, technical advances have been made in the development of nuclear fusion. But even the strongest proponents of fusion admit that a commercial reactor could not be available until at least the mid-2030s, and even that date is highly uncertain. Given that we need to eliminate 80 percent of world emissions by 2030, that proposed date is too far away for fusion to be useful. As such, nuclear fusion is unlikely to address global warning in a meaningful way.

8.5.1 Risks Affecting Nuclear's Ability to Address Global Warming and Air Pollution

The risks associated with nuclear power can be broken down into two categories: (1) risks affecting nuclear's ability to reduce global warming and air pollution, and (2) risks affecting its ability to provide environmental security.

Risks under Category 1 include delays between planning and operation of a nuclear power plant, emissions contributing to global warming and air pollution, and cost. Risks under Category 2 include weapons proliferation risk, reactor meltdown risk, radioactive waste risk, underground mining lung cancer risk, and land despoilment risk. These risks are discussed herein.

8.5.1.1 Delays Between Planning and Operation and Due to Refurbishing Reactors

The longer the time lag between the planning and operation of an energy facility, the more the emissions of air pollutants and climate-damaging chemicals from the background electric grid. Similarly, the longer the time required to refurbish a nuclear plant for continued use at the end of its life, the greater the emissions from the background grid while the nuclear plant is down.

The time lag between planning and operation of a nuclear plant includes the times to secure a construction site, a construction permit, an operating permit, financing, and insurance; the time between construction permit approval and issue; and the time to build the plant, which includes the time between the end of construction and grid connection.

In March 2007, the United States Nuclear Regulatory Commission approved the first request for a site permit in 30 years. This process took 3.5 years. The time to review and approve a construction permit is another 2 years and the time between the construction permit approval and issue is about half a year. Thus, the minimum time for preconstruction approvals (and financing) in the United States is 6 years. An estimated maximum time is 10 years. The time to construct a nuclear reactor depends significantly on regulatory requirements and costs. Although nuclear reactor construction times worldwide are often shorter than the 9-year median construction times in the United States since 1970,[162] they averaged 7.4 years worldwide in 2015.[163] As such, a reasonable estimated range for construction time is 4 to 9 years, bringing the overall time between planning and operation of a nuclear power plant worldwide to 10 to 19 years.

An examination of some recent nuclear plant developments confirms that this range is not only reasonable, but an underestimate in at least one case. The **Olkiluoto 3** reactor in Finland was proposed to the Finnish cabinet in December 2000 to be added to an existing nuclear power plant. Its construction started in 2005, and it first generated electricity in 2022, giving it a construction time of 17 years and a **planning-to-operation** time of 22 years. The final cost was at least 3 times the original estimate.

The **Hinkley Point C** nuclear power station was planned starting in 2008. Construction began only on December 11, 2018. It has an estimated completion year of 2026 to 2027, giving it an estimate construction time of 8 to 9 years and a planning-to-operation time of 18 to 19 years. The projected cost in 2022 was twice the original projected cost.

The **Vogtle 3 and 4** reactors in the U.S. state of Georgia were first proposed in August 2006 to be added to an existing nuclear power station site. Construction started for both in 2013. The anticipated completion dates are 2023 and 2024, respectively, giving them estimated construction times of 10 and 11 years, respectively, and planning-to-operation times of 17 and 18 years, respectively. The estimated cost in 2022 was about twice the original estimated cost.

The **Flamanville**, France, Unit 3 reactor was planned on an existing nuclear power station site starting in 2004. A contract was awarded in 2005. Construction started in 2007. The reactor is not expected to be completed until 2023, for an estimated construction time of 16 years and planning-to-operation time of 19 years. The project is over budget by a factor of 5.

The **Haiyang 1 and 2** reactors in China were planned in 2005. Construction started in 2009 and 2010, respectively. Haiyang 1 began commercial operation on October 22, 2018. Haiyang 2 began operation on January 9, 2019, giving them construction times of 9 years and planning-to-operation times of 13 and 14 years, respectively.

The **Taishan 1 and 2** reactors in China were planned starting in 2006. Construction began in 2008. Taishan 1 began commercial operation on December 13, 2018. Taishan 2 began operation on September 9, 2019, giving them construction times of 10 and 11 years and planning-to-operation times of 12 and 13 years, respectively.

The **Shidao Bay** nuclear power plant in China includes the development of a 200-megawatt high-temperature gas-cooled (as opposed to water-cooled) reactor. Planning started in 2005. Construction began in December 2012. Grid connection occurred in early 2022. Thus, the construction time was 10 years and the planning-to-operation time was 17 years.

Planning and procurement for four reactors in **Ringhals**, Sweden, started in 1965. One took 10 years, the second took 11 years, the third took 16 years, and the fourth took 18 years to complete.

Some contend that France's 1974 Messmer Plan resulted in the building of its 58 reactors in 15 years. The **Messmer Plan** was a proposal, enacted without public or parliamentary debate, by the Prime Minister of France, Pierre Messmer, to build 80 nuclear reactors by 1985 and 170 by 2000. In fact, the plan had been in the works for years prior and was only proposed publicly following the international oil crisis of 1973.[164] For example, the Fessenheim nuclear reactor obtained its construction permit in 1967 and was planned before that. In addition, 10 of the reactors were completed only between 1991 and 2000. As such, the whole planning-to-operation time for the 58 reactors was at least 33 years, not 15. That of any individual reactor was 10 to 19 years.

In sum, planning-to-operation times for both recent and historic nuclear power plants have been in the range of 10 to 22 years, but with most between 10 and 19 years.

Planning-to-operation delays are also caused by downtime during reactor refurbishment. Nuclear reactors have an expected lifetime on the order of 40 years. To run longer, they need to be refurbished. An estimate of the time to refurbish a nuclear reactor is 3 to 4 years. Refurbishment of the Darlington 2, Ontario nuclear reactor, for example, began in October 2016 and was completed in June 2020, taking just less than 4 years.

The background grid, which consists primarily of fossil fuels in most places worldwide, emits pollution when a nuclear plant is being constructed or down for refurbishing. The total opportunity-cost emissions of nuclear due to both factors average out to be about 64 to 102 grams of CO_2-equivalent per kilowatt-hour of electricity generated (Figure 8.1). These emissions are higher than the lifecycle emissions of nuclear.

Transition highlight

China's investment in nuclear plants, which have long planning-to-operation times, instead of in wind or solar power, may have resulted in China's carbon dioxide emissions rising 1.3 percent from 2016 to 2017 rather than declining by an estimated 3 percent during that period.

The health impacts of such delays in China are substantial. In 2016, 1.9 million people died from air pollution particles and gases in China. Assuming that air pollution emissions are proportional to carbon dioxide emissions, 82,000 more people may have died in 2016 alone due to China's investment in nuclear power instead of wind or solar. Thus, opportunity-cost emissions affect both climate and health.

8.5.1.2 Air Pollution and Global Warming Relevant Emissions from Nuclear

Nuclear power contributes to global warming and air pollution in several ways: through (1) emissions of air pollutants and climate-damaging pollutants from the background grid owing to nuclear's long planning-to-operation and refurbishment times; (2) emissions of health- and climate-damaging air pollutants during construction, operation, and decommissioning of a nuclear plant; (3) heat and water vapor emissions during the operation of a nuclear plant; (4) carbon dioxide emissions due to the covering of soil or clearing of vegetation during the construction of a nuclear plant, uranium mine, and nuclear waste site; and (5) the risk of emissions arising from nuclear weapons proliferation.

Each of these categories represents an actual emission or emission risk, yet all of these emissions, except for lifecycle emissions, are incorrectly ignored in virtually all studies of nuclear power impacts on climate. Studies that ignore these real emissions distort the impacts of nuclear power on climate and air pollution health.

The estimated range of lifecycle emissions of nuclear (9 to 70 grams of CO_2-equivalent per kilowatt-hour of electricity) in Figure 8.1 is well within the range (4 to 110) from studies examined by the Intergovernmental Panel on Climate Change.[165] On top of those emissions are opportunity cost emissions (64 to 102); emissions due to heat and water vapor fluxes (4.4); emissions due to covering and clearing soil (0.17 to 0.28); and emissions due to the risk of nuclear weapons use arising from the spread of nuclear energy (0 to 1.4). The total is 78 to 178 grams of CO_2-equivalent per kilowatt-hour of electricity. These emissions are 9 to 37 times the emissions from onshore wind (Figure 8.1).

Although the emissions from nuclear are lower than those from coal or natural gas with carbon capture, nuclear power's high CO_2-equivalent emissions coupled with its long planning-to-operation time render it an opportunity cost relative to the faster-to-operate and lower-emitting WWS technologies.

8.5.1.3 Nuclear Costs

The third risk of nuclear power that affects its ability to reduce global warming and air pollution is its high cost. In addition, the cost of running existing nuclear reactors has increased so much that many existing reactors are shutting down early. Owners of others have requested large subsidies to stay open. This section discusses nuclear costs.

The 2021 mean levelized cost of energy for a new nuclear plant in the United States is 16.8 cents per kilowatt-hour.[161] This compares with 3.8 and 3.5 cents per kilowatt-hour for onshore wind and utility-scale solar PV, respectively. Thus, new nuclear electricity is almost 5 times the cost per unit electricity of new wind and solar. A good portion of the high cost of nuclear is related to its long planning-to-operation time, which in turn is partly due to construction delays.

This levelized cost of nuclear does not account for the future cost of storing radioactive waste. For example, in the United States alone, about $500 million is spent yearly to safeguard nuclear waste from about 100 civilian nuclear energy plants.[166] Such waste must be stored for hundreds of thousands of years.

The cost also does not account for the damage due to nuclear reactor meltdowns. For example, the estimated cost to clean the damage from three Fukushima Dai-ichi nuclear reactor core meltdowns in 2011 was $460 to $640 billion.[168] This is equivalent to 10 to 18.5 percent of the capital cost of every nuclear reactor that exists worldwide.

The spiraling cost of new nuclear plants in recent years has resulted in the canceling of several nuclear reactors under construction. For example, two reactors in South Carolina were canceled during July 2017. The high cost of nuclear has also resulted in threats by nuclear reactor owners that they would shut plants unless they received a subsidy. The risk of shutting a functioning nuclear plant is that its energy will be replaced by energy from high-emitting fossil fuels. However, the risk of subsidizing the plant is that the subsidies would otherwise be used to replace the nuclear plant output with lower-cost and lower-emitting wind and solar electricity. Because the nuclear plant usually needs to be replaced within a decade of a subsidy request in any case, simply incurring the cost of new renewables immediately will almost always be less expensive than both spending the same money on renewables in 10 years and paying nuclear a subsidy each year for 10 years.

Transition highlight

In 2016, three upstate New York nuclear plants requested and received subsidies to stay open until 2028, using the argument that the plants were needed to keep emissions low. However, subsidizing such plants may have increased carbon dioxide emissions and costs relative to replacing the nuclear plants quickly with wind or solar.[169]

High costs have reduced the number of new nuclear reactors under construction to a crawl in liberalized markets. However, in some countries, such as China, nuclear reactor growth continues because of large government subsidies.

In sum, before accounting for waste storage or meltdown damage costs, a new nuclear power plant costs about 5 times as much as a new onshore wind farm, takes 5 to 17 years longer between planning and operation than a wind farm, and produces 9 to 37 times the emissions per unit electricity generated as a wind farm. Thus, a fixed amount of money spent on a new nuclear plant means much less power generation, a much longer wait for power, and a much greater emission rate than the same money spent on WWS.

The Intergovernmental Panel on Climate Change similarly concludes that the economic, social, and technical feasibility of nuclear power have not improved over time:[170]

The political, economic, social and technical feasibility of solar energy, wind energy and electricity storage technologies has improved dramatically over the past few years, while that of nuclear energy and Carbon Dioxide Capture and Storage (CCS) in the electricity sector has not shown similar improvements.

8.5.2   Risks Affecting Nuclear's Ability to Address Environmental Security

The second category of risk related to nuclear power is the risk of a nuclear plant not being able to provide environmental security. One reason is the risk of weapons proliferation. Others are the risks of meltdown, radioactive waste leakage, and cancer and land degradation due to uranium mining. WWS technologies do not create such risks.

8.5.2.1 Weapons Proliferation Risk

The first risk of nuclear power related to environmental security is weapons proliferation risk. The growth of nuclear energy has historically increased the ability of nations to obtain or harvest plutonium or enrich uranium to manufacture nuclear weapons:[171]

> *Peaceful nuclear cooperation and nuclear weapons are related in two key respects. First, all technology and materials related to a nuclear weapons program have legitimate civilian applications. For example, uranium enrichment and plutonium reprocessing facilities are dual-use in nature because they can be used to produce fuel for power reactors or fissile material for nuclear weapons. Second, civilian nuclear cooperation builds-up a knowledge-base in nuclear matters.*

The Intergovernmental Panel on Climate Change recognizes this fact. They conclude, with *"robust evidence and high agreement"*, that nuclear weapons proliferation concern is a barrier and risk to the increasing development of nuclear energy:[165]

> *Barriers to and risks associated with an increasing use of nuclear energy include operational risks and the associated safety concerns, uranium mining risks, financial and regulatory risks, unresolved waste management issues, nuclear weapons proliferation concerns, and adverse public opinion.*

The building of a nuclear reactor for energy in a country that does not currently have a reactor increases the risk of nuclear weapons being developed in that country. It allows the country to import uranium for use in the nuclear energy facility. If the country so chooses, it can secretly enrich the uranium to create weapons-grade uranium and harvest plutonium for nuclear weapons from uranium fuel rods used in a nuclear reactor. This does not mean any or every country will do this, but historically some have.

Nuclear weapons can be produced from nuclear energy infrastructure as follows. Uranium ore is mined in an open pit or underground and contains 0.1 to 1 percent uranium by mass. The ore is milled to concentrate the uranium in the form of a yellow power called **yellowcake**, which contains about 80 percent uranium oxide. Uranium is then processed further into uranium dioxide or uranium hexafluoride for use in nuclear reactors. However, before the uranium can be used in a reactor, it must be enriched.

Of all uranium on Earth, 99.2745 percent is uranium-238, 0.72 percent is uranium-235, and 0.0055 percent is uranium-234. Uranium-238 has a half-life of 4.5 billion years. Most commercial light water nuclear reactors use uranium consisting of 3 to 5 percent uranium-235. As such, the concentration of uranium-235 in a fuel rod must be increased to 4 to 7 times its ore concentration. This is done by enrichment. **Uranium enrichment** is the process of separating the isotopes of uranium to increase the percent of uranium-235 in a batch. Enriched uranium is useful for both nuclear energy and nuclear weapons.

Enrichment is done by gas diffusion, centrifugal diffusion, or mass separation by magnetic field. Only gas diffusion and centrifugal diffusion are commercial processes, and most enrichment today is by **centrifugal diffusion** because it consumes only 2 to 2.5 percent as much energy as gas diffusion. Nevertheless, centrifugal diffusion still requires many centrifuges running for long periods, thus lots of energy. Centrifugal diffusion works by spinning a cylinder containing uranium. The heavier uranium-238 atoms collect toward the outside edge of the cylinder and the lighter uranium-235 atoms collect toward the inside.

Uranium with less than 20 percent uranium-235 is called **low enriched uranium. Highly enriched uranium** contains 20 to 90 percent uranium-235. A nuclear weapon can be made with highly enriched uranium. However, nuclear weapons increase their destructiveness with even more enrichment. Thus, 90 percent or more uranium-235 is considered **weapons-grade uranium** and is generally used together with enriched plutonium in a nuclear bomb. An estimated 9,000 centrifuges can produce enough weapons-grade uranium-235 for one nuclear weapon from natural uranium in about 7 months. With 5,000 centrifuges, the process takes about one year.[172] Because uranium in a fuel rod used for nuclear energy has only 3 to 5 percent uranium-235 and even less once it goes through a nuclear reactor, spent fuel rods are not considered a useful source of weapons-grade uranium.

Plutonium is also used in nuclear weapons. Ten kilograms of plutonium-239 were used in the bomb dropped on Nagasaki. Plutonium can be obtained from a once-through nuclear reactor running on a uranium fuel rod. When uranium-235 decays and releases neutrons in a nuclear reactor, one of the neutrons can bind with a uranium-238 atom to produce uranium-239, which decays to plutonium-239. Plutonium that contains 93 percent or more plutonium-239 is considered weapons-grade plutonium. Plutonium with less than 80 percent plutonium-239 is reactor grade. Because plutonium can be used to make a bomb and is easier to obtain than is enriching uranium (since plutonium can be harvested from a fuel rod running once through a nuclear reactor), plutonium is considered the element of even greater concern than uranium with respect to nuclear weapons proliferation.

A large-scale worldwide increase in nuclear energy facilities would exacerbate the risk of nuclear weapons proliferation. In fact, producing material for a weapon requires merely operating a civilian nuclear power plant together with a sophisticated plutonium separation facility. The historic link between nuclear energy facilities and nuclear weapons is evidenced by the development or attempted development of weapons capabilities secretly under the guise of peaceful civilian nuclear energy or nuclear energy research programs in Pakistan, India, Iraq (prior to 1981), Syria (prior to 2007), Iran, and North Korea, among other countries.

If the world's all-purpose energy were converted to electricity and electrolytic hydrogen by 2050, the 9 trillion watts (TW) in resulting annual average end-use electric power demand would require about 12,500 850-megawatt nuclear reactors (31 times the number of active reactors today), or one installed every day for 34 years. Not only is this construction timeline impossible given the long planning-to-operation times of nuclear, but it would also result in all known reserves of uranium worldwide for once-through reactors running out in about 3 years. As such, there is no possibility the world will run solely on once-through nuclear energy by 2050.

Even if only 6.4 percent of the world's energy were supplied with nuclear, the number of active nuclear reactors worldwide would nearly double to around 800. Many more countries would possess nuclear reactors, increasing the risk that some of these countries would use the facilities to mask the development of nuclear weapons, as has occurred historically.

If a country were to develop a weapon as a result of its acquisition of one or more nuclear energy facilities, the risk that it would use the weapons is not zero. Here, the emissions associated with a limited nuclear exchange are estimated.

The explosion of one hundred 15-kilotonne nuclear bombs (a total of 1.5 megatonnes, or 0.1 percent of the yield of a full-scale nuclear war) during a limited nuclear exchange in a megacity could kill 2.6 to 16.7 million people from the explosion and burn 63 to 313 million tonnes of city infrastructure, adding 1 to 5 million tonnes of warming and cooling aerosol particles to the atmosphere, including much of it to the stratosphere.[133] The particle emissions would cause significant short- and medium-term regional temperature changes. The carbon dioxide emissions, estimated at 92 to 690 million tonnes, would enhance long-term global warming. The warming impact of one such nuclear exchange over 100 years is equivalent to 1.4 grams of CO_2-equivalent emissions per kilowatt-hour of electricity produced by nuclear during that period. It arises from nuclear weapons development facilitated by the spread of nuclear energy. That is the high-end risk of carbon dioxide emissions from nuclear weapons proliferation due to nuclear energy in Figure 8.1. The low-end estimate is zero emissions because no exchange occurs.

8.5.2.2 Meltdown Risk

The second risk of nuclear power related to environmental security is reactor core meltdown risk. The Intergovernmental Panel on Climate Change points to operational risks (meltdown) as a barrier and risk associated with nuclear power.

About 1.5 percent of all nuclear reactors operating in history have had a partial or significant core meltdown. Meltdowns have been either catastrophic (Chernobyl, Russia, in 1986; three reactors at Fukushima Dai-ichi, Japan, in 2011) or damaging (Three-Mile Island, Pennsylvania, in 1979; Saint-Laurent, France, in 1980). The nuclear industry has suggested that new reactor designs are safer. However, these designs are generally untested, and there is no guarantee that the reactors will be designed, built and operated correctly or that a natural disaster or act of terrorism, such as an airplane flown into a reactor, will not cause the reactor to fail, resulting in a major disaster.

On March 11, 2011, an earthquake measuring 9.0 on the Richter scale, and a subsequent tsunami knocked out backup power to a cooling system, causing six nuclear reactors at the **Fukushima 1 Dai-ichi plant** in northeastern Japan to shut down. Three reactors experienced a significant meltdown of nuclear fuel rods and multiple explosions of hydrogen gas that formed during efforts to cool the rods with seawater. Uranium fuel rods in a fourth reactor also lost their cooling. As a result, cesium-137, iodine-131, and other radioactive particles and gases were released into the air. Locally, tens of thousands of people were exposed to the radiation, and 170,000 to 200,000 people were evacuated from their homes. 1,600 to 3,700 people perished during the evacuation alone.[168,173] At least one nuclear plant worker died from lung cancer from direct radiation exposure.[174]

The radiation release created a dead zone around the reactors that may not be safe to inhabit for decades to centuries. The radiation also poisoned the water and food supplies in and around Tokyo. The radiation plume from the plant spread worldwide within a week. Although radioactivity levels in Japan within 100 kilometers of the plant were extremely high, those in the rest of Japan and eastern China were lower, and those in North America and Europe were even lower. It is estimated that 130 (15 to 1,100) radiation-related deaths and 180 (24 to 1,800) radiation-related illnesses will occur worldwide, primarily in eastern Asia, during the decades after the meltdown.[167] The cost of the cleanup of the Fukushima reactors and the surrounding area is estimated at $460 to $640 billion.[168]

The 1.5 percent risk of a nuclear reactor meltdown is a high risk. Catastrophic risks with all WWS technologies aside from the risk of a large hydropower dam collapsing are zero. WWS roadmaps do not call for an increase in the number of large hydropower dams worldwide, only the more effective use of existing ones.

8.5.2.3 Radioactive Waste Risks

Another risk associated with nuclear power is the risk of human and animal exposure to radioactivity from fuel rods consumed by once-through reactors. Used fuel rods are considered **radioactive waste**. Currently, most used fuel rods are stored at the reactor site. This has given rise to hundreds of radioactive waste sites in many countries that must be maintained for hundreds of thousands of years, far beyond the lifetime of any nuclear power plant. The United States houses about one-quarter of all nuclear reactors worldwide. Plans to store the waste of all U.S. reactors inside of Yucca Mountain, Nevada, never passed into law, so waste will continue to accumulate at reactor sites. The more that waste accumulates, the greater the risk that a radioactive leak will damage water supply, crops, animals, and humans.

8.5.2.4 Uranium Mining Health Risks and Land Degradation

Nuclear power increases the risk that underground miners contract lung cancer and that open-pit uranium mines degrade land. Such risks continue so long as nuclear power plants operate because the plants need uranium to produce electricity. WWS technologies, on the other hand, do not require the continuous mining of any material, only one-time mining to produce the WWS equipment.

In 2019, 13 countries mined uranium. Of these, Kazakhstan, Canada, Australia, Namibia, and Uzbekistan, Niger, and Russia produced the most uranium. Mines can be open-pit or underground. Open-pit mines cause the most land degradation. Underground mines cause the greatest lung cancer risk.

Underground uranium mining causes lung cancer in large numbers because uranium mines contain natural radon gas, some of whose decay products are carcinogenic. Several studies have found a link between high radon levels and cancer.[175,176] A study of 4,000 underground uranium miners between 1950 and 2000[177] found that about 10 percent died of lung cancer, a rate 6 times that expected based on smoking rates alone. Another 1.5 percent died of mining-related lung diseases, supporting the hypothesis that uranium mining is unhealthy. In fact, the combination of radon and cigarette smoking increases lung cancer risks above the normal risk associated with smoking.[178]

Clean, renewable energy does not have this risk because (1) it does not require the continuous mining of any material, only one-time mining to produce energy generators and storage; and (2) the mining for materials related to WWS does not carry the same lung cancer risk as does uranium mining.

8.6   Why Not Biomass for Electricity or Heat?

Biofuels are solid, liquid, or gaseous fuels derived from organic matter. Most biofuels are derived from dead plants or from animal excrement. **Solid biofuels**, such as wood, grass, agricultural waste, and dung, are burned directly for home heating and cooking in developing countries and for electric power generation in most all countries. **Liquid biofuels** are generally used for transportation as a substitute for gasoline, diesel, jet fuel, or bunker fuel. **Gaseous biofuels**, such as methane, are used for electricity, heat, and transportation.

Here, **biomass** (a subset of **bioenergy**) is defined to be a gaseous or solid biofuel that is used for electricity or heat generation.

Biomass combustion for electricity or heat is not recommended in a 100 percent WWS world for several reasons, discussed herein. Similarly, **biomass with carbon capture and storage (BECCS)** also represents an opportunity cost in comparison with WWS options, so is not recommended. Biomass combustion without and with carbon capture is discussed here.

8.6.1   Biomass Combustion without Carbon Capture

The main sources of solid biomass combusted for energy are as follows:[179]

agriculture residues, which include dry crop residue (such as straw and sugar beet leaves), and livestock waste (such as solid or liquid manure);

forestry residues, which include bark, wood blocks, wood chips from treetops and branches, and logs from forest thinning;

energy crops, which include dry wood crops (such as willow, poplar, eucalyptus, and short-rotation coppice), dry herbaceous crops (such as miscanthus, switchgrass, reed, canary grass, cynara, and Indian shrub), oil energy crops (such as sugar beet, cane beet, sweet sorghum, Jerusalem artichoke, sugar millet), starch energy crops (such as wheat, potato, maize, barley, triticale, corn, and amaranth), and other energy crops (such as flax, hemp, tobacco stems, aquatic plants, cotton stalks, and kenaf);

industry residues, which include wood industry residues (such as bark, sawdust, wood chips, and cutoffs from saw mills), food industry residues (such as beet root tails, used cooking oils, tallow, yellow grease, and slaughterhouse waste), and industrial products (such as pellets from sawdust and wood shavings, bio-oil, ethanol, and bio-diesel);

park and garden wastes, which include grass and pruned wood; and

contaminated wastes, which include demolition wood, municipal waste, sewage sludge, sewage gas, and landfill gas.

The primary reason biomass combustion is not proposed for use in a WWS world is that biomass combustion, like coal and natural gas combustion, produces air pollution. A 100 percent WWS energy infrastructure is designed to eliminate air pollution. The air pollution from biomass combustion is similar to that from liquid biofuel combustion. The problem is even worse with the burning of municipal waste, which contains toxic chemicals. In sum, whereas biomass is partly renewable, it is not clean. A 100 percent WWS world requires both clean and renewable energy rather than just renewable energy.

The second reason for not including biomass is that it causes more global warming per unit electricity produced than does WWS. Biomass grows by converting carbon dioxide and water vapor from the air into organic material and oxygen (photosynthesis). Although growing biomass takes carbon dioxide out of the air, almost all of that carbon dioxide is returned to the air when the biomass is burned. Biomass causes additional carbon dioxide to go into the air due to the fertilizing, watering, growing, collecting, transporting, separating, and/or incinerating the biomass. These processes all require fossil-fuel energy and emissions.

The overall emissions from biomass used for electricity production ranges from 86 to 1,788 grams of CO_2-equivalent per kilowatt-hour of electricity, or 10 to 373 times the emissions per unit electricity of onshore wind (Figure 8.1). These emissions are due mostly to lifecycle emissions (43 to 1,730 grams of CO_2-equivalent per kilowatt-hour of electricity). A review[179] suggests that the combustion of forestry and industry residues may result in the lowest emissions (43 to 46 grams) among biomass fuels. Combustion of agricultural residues and energy crops may cause higher emissions (200 to 300 grams). Combustion of municipal solid waste may result in the highest emissions (mean at 1,730 grams). The low emissions from forestry and industry residues are due to the fact that the feedstock in those cases does not need to be produced actively as it does with agricultural residues or energy crops. The high emissions from burning municipal solid waste are due to emissions from the energy required to collect, segregate, sort, transport, and incinerate the waste.

Biomass energy facilities have opportunity cost emissions of 36 to 51 grams of CO_2-equivalent per kilowatt-hour of electricity generated because they take 4 to 9 years between planning and operation versus 2 to 5 years for onshore wind or utility PV. During the additional time, the background grid is emitting.

Other sources of climate-affecting emissions due to using biomass for electricity are heat and water vapor emitted during biomass combustion. Because biomass combustion is less efficient than is even coal combustion, biomass combustion releases more heat per unit electricity than does coal combustion.

A third problem with some types of biomass, particularly energy crops, is the much greater land requirement for them than for WWS. Given that photosynthesis is only 1 percent efficient at converting sunlight to biomass energy, whereas solar PV panels are now 20 to 47 percent efficient at converting sunlight to electricity, a solar panel needs less than 1/20 the land to produce the same energy as a biomass crop.

An alternative to burning biomass for electricity or heat is to extract landfill gas, which contains mostly methane, and use the methane to produce hydrogen by steam reforming. The use of methane captured from landfills and from methane digesters to produce hydrogen is one method of consuming methane that would otherwise leak to the air. The steam reformer allows carbon dioxide, a much less potent greenhouse gas per molecule than methane, to leak to the air but also produces hydrogen for use in a fuel-cell vehicle, displacing the need for gasoline. Some might suggest the carbon dioxide from this process should be captured, but the energy and additional fuel required to capture that carbon dioxide result in little reduction in CO_2-equivalent emissions compared with no capture.[117]

In sum, burning forest and industry residue and other forms of biomass to provide electricity and heat results in higher CO_2-equivalent emissions and more air pollution than does using WWS. Some forms of biomass also require much more land than does WWS. As such, using biomass for energy represents an opportunity cost. The exception is to use landfill and digester methane to produce hydrogen by steam reforming, where the hydrogen is subsequently used in a fuel cell.

8.6.2   Biomass Combustion with Carbon Capture

A proposed method of reducing biomass CO_2-equivalent emissions, and even creating negative carbon emissions, is to combine biomass combustion with carbon capture and storage to give **biomass with carbon capture and storage or use (BECCS/U)**. Negative carbon emissions arise if a process removes more carbon from the air than it adds to the air.

BECCS would theoretically result in negative carbon emissions if, for example, forest wood residue (containing carbon dioxide from the air) were collected, little energy were used to collect, transport, and incinerate the wood, and the carbon dioxide were captured from the exhaust stream of the biomass electricity-generating facility and pumped underground. If successful, this method would be a one-way conduit for carbon dioxide to go from the air to underground, thereby resulting in negative carbon emissions.

The problems, however, are several-fold. As with natural gas and coal with carbon capture, the carbon capture system with BECCS/U requires 25 to 55 percent more energy than without it. If that energy comes from natural gas, coal, or biomass, 25 to 55 percent more air pollution occurs with BECCS/U than with biomass combustion without capture. Biomass combustion without carbon capture already produces substantial air pollution, whereas WWS produces none.

Similarly, as with carbon capture for coal and natural gas, the carbon dioxide reductions with BECCS/U are much lower than anticipated owing to the high energy requirements of carbon capture equipment. Leakage of carbon dioxide from underground storage or from industrial use is also an issue.

Second, like with gas and coal, few reliable underground storage facilities exist for BECCS. Because of the high cost of carbon capture, biomass with carbon capture facilities are likely to be coupled with for-profit uses of the carbon dioxide, such as enhanced oil recovery or synthetic fuel production. Thus, BECCU will be favored over BECCS. This will encourage more combustion of and emissions from oil products.

Third, the efficiency of biomass combustion for electricity (electricity output per unit energy in the fuel) is low (20 to 27 percent), even compared with the efficiency of coal combustion (33 to 40 percent). Thus, a large mass of biomass is needed to produce a small amount of electricity. As such, if BECCS/U were to provide negative emissions on a large scale, substantial land areas dedicated to biomass crops would be needed to maintain a continuous energy supply. Consequently, a share of agricultural land would be used for fuel instead of food, increasing the price of food. Higher food prices trigger deforestation by incentivizing people to turn high-carbon-storage forestland into low-carbon-storage agricultural land.

Fourth, removing agricultural residues usually means crops need to be fertilized more since the residues contain nutrients that are no longer available once they are removed. Fertilizers contain nitrous oxide, a greenhouse gas, and ammonia, a major air pollutant. Both nitrous oxide and ammonia are emitted to the air.

Finally, the cost of BECCS/U is high, even compared with CCS for fossil fuels. As of 2020, only six BECCS/U facilities have survived, and each has been at a high cost. One captures carbon dioxide at an ethanol refinery. The others have captured carbon dioxide at municipal solid waste plants.

In sum, paying for BECCS/U instead of WWS means less energy production, a longer time lag between planning and operation, more air pollution, greater land use (for some crops), and less carbon removal. As such, BECCS/U is not recommended.

8.7   Why Not Liquid Biofuels for Transportation?

Liquid biofuels are generally used for transportation as a substitute for gasoline or diesel. The most common transportation biofuels are ethanol, used in passenger cars and other light-duty vehicles, and biodiesel, used in many heavy-duty vehicles. Liquid biofuels should not be used as part of a 100 percent WWS infrastructure since they result in substantial air pollution death and illness, climate damage, land consumption, and water use. This section discusses these issues.

Ethanol is produced in a factory, generally from corn, sugarcane, wheat, sugar beet, or molasses. The most common among these sources are corn and sugarcane, resulting in the production of **corn ethanol** and **sugarcane ethanol**, respectively. Microorganisms and enzyme ferment sugars or starches in these crops to produce ethanol.

Fermentation of cellulose in switchgrass, wood waste, wheat, stalks, corn stalks, or *Miscanthus* also produces ethanol. However, the process is more energy intensive than fermentation of sugar and starches because the breakdown of cellulose by natural enzymes (as it occurs in the digestive tracts of cattle) is slow. Faster breakdown of cellulose requires genetic engineering of enzymes. The ethanol resulting from these sources is referred to as **cellulosic ethanol**.

Ethanol may be used on its own, as it is frequently in Brazil, or blended with gasoline. A blend of 6 percent ethanol and 94 percent gasoline is referred to as **E6**. Other typical blends are **E10**, **E15**, **E30**, **E60**, **E70**, **E85**, and **E100**. In many countries, including the United States, **E100** is required to contain 5 percent gasoline as a **denaturant**, which is a poisonous or foul-tasting chemical added to a fuel to prevent people from drinking it. As such, E85, for example, contains about 81 percent ethanol and 19 percent gasoline.

A proposed alternative to ethanol for transportation fuel is **butanol**. It can be produced by fermenting the same crops used to produce ethanol but with a different bacterium, *Clostridium acetobutylicum*. Butanol contains more energy per unit volume of fuel than does ethanol. However, unburned butanol also reacts more quickly in the atmosphere than does unburned ethanol, speeding up ozone production relative to ethanol. Ozone is harmful to those who breathe it. On average, ethanol used in internal combustion engine vehicles produces more ozone than does gasoline used in vehicles in most regions of the United States.[180,181,182]

Biodiesel is a liquid diesel-like fuel derived from vegetable oil or animal fat. Major edible vegetable oil sources of biodiesel include soybean, rapeseed, mustard, false flax, sunflower, palm, peanut, coconut, castor, corn, cottonseed, and hemp oils. Inedible vegetable oil sources include jatropha, algae, and jojoba oils. Animal fat sources include lard, tallow, yellow grease, fish oil, and chicken fat. Soybean oil accounts for about 90 percent of biodiesel production in the United States. Biodiesel derived from soybean oil is referred to as **soy biodiesel**.

Biodiesel is produced by the chemical reaction of a vegetable oil or animal fat (both lipids) with an alcohol. It is a standardized fuel designed to replace diesel in standard diesel engines. It can be used as pure biodiesel or blended with regular diesel. Blends range from 2 percent biodiesel and 98 percent diesel (B2) to 100 percent biodiesel (B100). Generally, only blends B20 and lower can be used in a diesel engine without engine modification. The use of vegetable oil or animal fat directly (without conversion to biodiesel) in diesel engines is also possible; however, such use results in more incomplete combustion, and thus more air pollution byproducts, as well as a greater build-up of carbon residue in, and damage to, the engine, than biodiesel.

Significant efforts have been made to produce **algae biodiesel**, which is biodiesel from algae grown from waste material, such as sewage. However, these efforts have been hampered by the fact that algae can grow quickly only when exposed to the sun. As such, algae cannot grow quickly when piled with one layer on top of the other. Instead, algae must be spread in a single layer over a large, unshaded area. Each volume of oil produced from algae also requires about 100 times that volume of water. Both factors have limited the growth of the algae biodiesel industry.

Liquid biofuels (corn ethanol, cellulosic ethanol, butanol, and biodiesel) are not recommended as part of a 100 percent WWS energy infrastructure. The main reasons are that (1) nearly all biofuels are burned to generate energy for vehicle motion, resulting in air pollution similar to that from fossil fuels; (2) liquid biofuels do not reduce CO_2-equivalent emissions nearly to the extent that WWS-powered battery-electric or hydrogen fuel-cell vehicles do; (3) some liquid biofuels increase CO_2-equivalent emissions relative to fossil fuels; (4) many biofuels require rapacious amounts of land; (5) many biofuels require excessive quantities of water; and (6) many biofuels are derived from food sources, increasing food shortages, food prices, and starvation.[133,183,184] Because liquid biofuels cause greater climate, pollution, land, water, and food problems than do WWS technologies, biofuels represent opportunity costs.

The main issues with liquid biofuels can be illustrated by comparing the impacts of using ethanol to power internal combustion engine vehicles with the impacts of using wind or solar to power battery-electric or hydrogen fuel-cell vehicles.

In the United States, about 30.2 percent of all CO_2-equivalent emissions in 2017 were from vehicle exhaust and 10.1 percent were from the upstream production of vehicle fuel. As such, eliminating vehicle-related emissions would reduce U.S. CO_2-equivalent emissions by a maximum of 40.3 percent.

Replacing gasoline vehicles with battery-electric vehicles powered by wind reduces CO_2-equivalent emissions by 99.3 to 99.8 percent of the maximum possible (40.3 percent) reduction. The remaining emissions are due to the fossil energy required to build and decommission the wind turbines.

Similarly, replacing gasoline cars with hydrogen fuel-cell cars, where the hydrogen is produced from wind electricity, reduces CO_2-equivalent emissions by 98 to 99 percent of the maximum possible reduction. This reduction is less than if wind-powered battery-electric vehicles are used because producing, compressing, and storing hydrogen then using the hydrogen in a fuel cell requires about 3 times as much wind electricity as simply putting wind electricity in a battery then withdrawing the electricity. Because more wind turbines are needed, more carbon dioxide is emitted when building wind turbines with hydrogen fuel vehicles than with battery-electric vehicles.

Nevertheless, wind powering battery-electric and hydrogen fuel-cell vehicles reduces carbon emissions and energy requirements substantially versus gasoline vehicles. On the other hand, using corn E85 or cellulosic E85 vehicles does not change carbon emissions significantly relative to using gasoline vehicles. The reason is that ethanol production results in a lot of fossil-fuel energy used, and thus emissions, to grow, fertilize, water, and cultivate the crop; transport the crop; refine the crop into fuel; and transport the fuel to market. It also results in non-energy emissions, such as of nitrous oxide from fertilizer. Ethanol is too corrosive to be used in pipes, so it must also be transported by train, truck, or barge, all of which emit diesel exhaust.

The air pollution mortality associated with either corn or cellulosic ethanol vehicles also significantly exceeds that associated with WWS-powered battery-electric and hydrogen fuel-cell vehicles. The reason is that battery-electric vehicles have zero tailpipe emissions, and hydrogen fuel-cell vehicles emit only water vapor. Thus, the only pollutant emissions from either vehicles are from the upstream production of wind turbines or solar panels and emissions from tire wear and some brake pad wear. Because electric vehicles use regenerative breaking, their brake pads are hardly engaged and brake pad emissions are rare. Production of the vehicles themselves also results in emissions, but such emissions are similar for all vehicles in this analysis.

On the other hand, E85 vehicles have high tailpipe emissions, high air pollution emissions from ethanol production and transport, tire particle emissions, and brake pad particle emissions. E85's tailpipe emissions cause air pollution health impacts that are often greater than those of gasoline vehicles[180,182], especially at low temperatures.[181]

Another problem with using ethanol as a fuel is water consumption. For example, irrigating only 13.2 percent of U.S. corn (the U.S. average irrigation rate for corn) that is needed to power a U.S. on-road vehicle fleet would require about 10 percent of the U.S. water supply.[133]

Finally, because of the substantial land required for corn or cellulosic ethanol, neither can provide enough energy for more than a few percent of the U.S. vehicle fleet.

In sum, liquid biofuels are not recommended as part of a 100 percent WWS energy infrastructure because of the climate, health, water supply, and land opportunity costs they incur.

8.8   Why Not Gray, Blue, or Brown Hydrogen?

In a WWS world, hydrogen will be used primarily for long-distance, heavy transport; steel production; remote microgrids for electricity and heat; and to produce ammonia. Such hydrogen should be produced only by electrolysis or photoelectrochemical water splitting, where WWS produces the electricity in both cases. This section discusses why hydrogen should not be so-called gray, blue, or brown hydrogen.

Two methods of producing hydrogen from natural gas are steam reforming of methane and autothermal reforming. When not coupled with carbon capture, the hydrogen resulting from these methods is referred to as **gray hydrogen**. When the methods are coupled with carbon capture, the hydrogen is referred to as **blue hydrogen**. When hydrogen is produced from coal gasification, the hydrogen is referred to as **brown hydrogen**. Each of these techniques is discussed, in turn.

Steam reforming of methane is a method of producing hydrogen by mixing methane with steam at a high temperature (700 to 1,000 degrees Celsius) and pressure (3 to 25 times atmospheric pressure). Natural gas is needed for two purposes during steam reforming. The first is to provide the energy needed for high temperatures and pressures. The second is to provide the methane used in the chemical reactions that produce hydrogen. During the steam reforming process, methane in natural gas reacts with two water vapor (H_2O) molecules to form four hydrogen molecules (H_2) and a molecule of carbon dioxide. About 96 percent of hydrogen production worldwide in 2019 was from steam reforming. This method of hydrogen production consumes about 6 percent of all natural gas globally.[117]

During the mining, transporting, and processing of the natural gas that arrives at a steam reforming plant, air pollutants, carbon dioxide, and methane are released to the air. Air pollutants and carbon dioxide are emitted because mining requires construction equipment, which runs on diesel fuel. Mining also requires concrete production for wells, which results in pollution. Trees are also cleared and land is trenched for pipes. Those processes require diesel and gasoline machinery and vehicles, resulting in even more pollution. Tree-clearing releases stored carbon to the air as carbon dioxide.

Methane leaks from both active and abandoned natural gas wells. The United States has 1.3 million active wells and 3.2 million abandoned wells. Worldwide, 29 million oil and gas wells have been abandoned, and two-thirds are estimated to leak.[148] Many of these leaks occur because 5 percent of steel casings or cement in natural gas wells leak immediately and 50 percent or more leak during their lifetimes. These underground leaks are not sealed even when the well is capped at its top.

Methane leak rates from two of the world's largest natural gas fields, in Turkmenistan, are estimated by satellite to be 4.1 percent.[142] Leak rates in the United States are generally between 1.2 and 9.4 percent. Methane also leaks from pipes running between gas wells and the natural gas processing center (where impurities are removed from natural gas). More leaks occur at the processing center, from pipes running between the processing center and the hydrogen plant, and at the hydrogen plant itself. In addition, electricity from natural gas is used to compress and transport natural gas through pipes and to process the natural gas at the processing station. Generating that electricity results in more emissions. An overall estimate of the worldwide average methane emission rate from natural gas mining and transport is 3.5 (1.5 to 4.3) percent.[117]

In sum, the production of hydrogen from steam reforming results in air pollution, carbon dioxide, and methane emissions from mining, transporting, and processing natural gas and from using the mined natural gas for energy and as a feedstock.[117]

Another method of producing hydrogen from natural gas is **autothermal reforming**. With this method, natural gas is brought to an autothermal reforming facility. However, instead of methane reacting with water vapor, it reacts with pure oxygen to produce a carbon dioxide molecule and two hydrogen molecules. This reaction releases heat, which reduces the need for additional fossil-fuel energy to produce heat for this process

One shortcoming of autothermal reforming is that it produces only two molecules of hydrogen per molecule of methane, whereas steam reforming produces four molecules of hydrogen per molecule of methane. As such, more methane is needed with autothermal reforming than with steam reforming to produce the same quantity of hydrogen. Autothermal reforming also produces carbon monoxide (CO), so additional equipment is needed to react carbon monoxide with steam (H_2O) at high temperature to produce carbon dioxide and one more hydrogen molecule. Even when this is done, though, a maximum of 2.89 molecules of hydrogen can be obtained per molecule of methane.[147] Thus, autothermal reforming always requires at least 38 percent more methane (and associated mining and leaks) than steam reforming to produce the same quantity of hydrogen. A third disadvantage of autothermal reforming is that it requires pure oxygen, which must be separated from air. Separation requires equipment and energy. For these three reasons, steam reforming is generally preferred by industry over autothermal reforming.[186]

The inefficiency and emissions from autothermal reforming can be reduced slightly by using some of the hydrogen produced by the autothermal reforming process in a fuel cell to generate the electricity for oxygen separation from air. The fuel cell also releases heat that can be used for the autothermal reforming process. However, this means a fuel cell must be purchased and more methane is needed to produce the additional hydrogen needed for the fuel cell. More methane means more natural gas mining, transporting, and processing, thus more upstream air pollution, carbon dioxide, and methane emissions. More methane consumption also means more carbon dioxide emitted during the autothermal reforming process than during the steam reforming process.

Some have proposed to capture the carbon dioxide from steam reforming and autothermal reforming, resulting in the hydrogen produced becoming "blue" hydrogen. During steam reforming, carbon dioxide emissions occur during combustion to produce electricity and heat and during chemical reactions to produce hydrogen. Additional emissions arise from powering carbon capture equipment.

During autothermal reforming coupled with a hydrogen fuel cell, carbon dioxide emissions arise from the chemical reactions that produce hydrogen but not from combustion to produce energy. That is because heat is provided by both the chemical reaction producing hydrogen and by the fuel cell. So only carbon dioxide from chemical reaction needs to be captured. However, far more carbon dioxide is produced and needs capturing per molecule of hydrogen with autothermal reforming than with steam reforming. Also, much more equipment is needed with autothermal reforming than with steam reforming.

Whereas carbon dioxide capture rates from chemical reaction during steam reforming or autothermal reforming can be 90 percent or more when the capture equipment is fully operational, real annual average capture rates of pure carbon dioxide streams from steam reforming equipment have been reported as only an average of 78.8 percent.[152] The reasons are that capture equipment is often down for scheduled or unscheduled maintenance; the demand for carbon dioxide is temporarily low; or the capture equipment is less efficient than expected. Capture rates of carbon dioxide from energy production (relevant only to the steam reforming process) are even lower, ranging from 20 to 70 percent.

The carbon dioxide that is captured with both steam reforming and autothermal reforming must be piped to either an underground storage facility or an industrial facility, where it is used. Piping requires additional energy and results in additional emissions. Of the carbon dioxide that is captured and piped for enhanced oil recovery, for example, 40 percent leaks to the air.[150,151] In addition, neither steam reforming nor autothermal reforming reduces upstream emissions or leaks of air pollutants, carbon dioxide, or methane.

Because of the need for additional equipment and energy required for the carbon capture equipment, steam reforming with carbon capture (blue hydrogen) may reduce the warming impact of steam reforming with no carbon capture (gray hydrogen) by only 9 to 12 percent over a 20-year time frame.[117] Blue hydrogen, of course, always costs more than gray hydrogen because blue hydrogen requires carbon capture equipment, pipes, and additional energy not required by gray hydrogen.

Hydrogen can also be produced by **coal gasification**. With this process, coal reacts with oxygen gas and steam under high temperature and pressure to form a mixture of hydrogen gas, carbon monoxide, carbon dioxide, and other chemicals. After impurities are removed, the carbon monoxide reacts with steam to form more hydrogen. Coal gasification results in more carbon dioxide and other pollutant emissions than does steam reforming of methane. Continuous coal mining and transport result in additional fossil-fuel combustion emissions, leaked methane, and land degradation. Additional energy is needed to create the high temperatures and pressures for the gasification process. Hydrogen from coal gasification is referred to as **brown hydrogen.**

In sum, because the steam reforming, autothermal reforming, and coal gasification processes emit carbon dioxide, methane, and other pollutants and degrade land, they are not suitable candidates for producing hydrogen in a 100 percent WWS world. In a WWS world, all carbon and pollution emissions and continuous fuel mining are eliminated. The simplest and cleanest way to produce hydrogen in a WWS world is with green hydrogen.

8.9   Why Not Synthetic Direct Air Carbon Capture?

Synthetic direct air carbon capture and storage (SDACCS) is the direct removal of carbon dioxide from the air by its chemical reaction with other chemicals followed by sequestration of the carbon dioxide, either underground or in a material. **Synthetic direct air carbon capture and use (SDACCU)** is the same as SDACCS, except that in this case, the captured carbon dioxide is sold for use in industry.

SDACCS/U should not be confused with **natural direct air carbon capture and storage (NDACCS)**, which is the natural removal of carbon dioxide from the air by either planting trees or reducing permanent deforestation (by reducing open biomass burning). Growing a tree removes carbon dioxide naturally by photosynthesis and sequesters the carbon within organic material in the tree for decades to centuries. Reducing open biomass burning similarly sequesters carbon in trees and eliminates emissions of health-affecting air pollutants and climate-affecting global warming agents aside from carbon dioxide at the same time. Trees also absorb air pollutants, helping to filter them from the air.

Whereas NDACCS is recommended in a 100 percent WWS world, SDACCS/U is not. SDACCS/U is basically a cost, or tax, added to the cost of fossil-fuel generation. As such, it raises the cost of using fossil fuels while increasing air pollution. It also enables the fossil-fuel industry to expand its damage to the environment and human health by allowing mining and air pollution to continue at an even higher cost to consumers than with no carbon capture.

Based on data from an existing SDACCU facility powered by natural gas, 90 percent (averaged over 20 years) or 69 percent (averaged over 100 years) of the carbon dioxide captured is returned to the air owing to the production of electricity needed to run the equipment, before even considering the fate of the carbon dioxide after capture. Even if SDACCU were powered by renewable electricity, it would still reduce CO_2-equivalent emissions less than does using the same renewable electricity to replace a coal or natural gas power plant.

Because air capture allows air pollution to continue, reduces little carbon, and incurs an equipment cost, spending on it rather than on WWS replacing fossil fuels or bioenergy always increases total social cost (equipment plus health plus climate cost). No improvement in air capture equipment can change this conclusion, since air capture always incurs an equipment cost never incurred by WWS. Also, air capture never reduces, but instead mostly increases, air pollution and mining.

8.9.1   How Does Air Capture Remove Carbon Dioxide from the Air?

In 1754, Joseph Black isolated carbon dioxide, which he named fixed air. He found that heating the odorless white powder magnesium carbonate ($MgCO_3$) released a gas (carbon dioxide, CO_2) that could not sustain life or fire. The remaining solid, magnesium oxide (MgO), weighed less than the original magnesium carbonate. He similarly found that adding potassium carbonate (K_2CO_3) to magnesium oxide (MgO) in solution resulted in the production of solid magnesium carbonate ($MgCO_3$). The mass of $MgCO_3$ exceeded that of MgO by the same amount that was lost when $MgCO_3$ was heated to form MgO. The difference in mass in both cases was the mass of carbon dioxide. As such, Black quantified the mass of carbon dioxide for the first time.

Today, an important air capture technique is to react carbon dioxide from the air with alkali metal oxides (Na_2O or K_2O, among others), alkali metal hydroxides (NaOH or KOH), alkaline Earth metal oxides (MgO or CaO), or alkaline Earth metal hydroxides [$Mg(OH)_2$ or $Ca(OH)_2$].[187]

For example, a classic method of removing carbon dioxide from the air while recycling the material removing it is to expose the carbon dioxide to a large pool of calcium hydroxide [$Ca(OH)_2$,], also called slaked lime. The products of the reaction are calcium carbonate ($CaCO_3$) and water vapor. Heating the calcium carbonate to 700 K releases a concentrated stream of carbon dioxide that can be captured and used. The leftover calcium oxide (CaO) is then reacted with water to reform calcium hydroxide. The problem with this process is that it needs a continuous electricity input.

8.9.2   Opportunity Cost of Direct Air Capture

By removing carbon dioxide from the air, air capture does exactly what WWS generators, such as wind turbines and solar panels, do. This is because WWS generators replace fossil generators, preventing carbon dioxide from getting into the air in the first place. The impact on climate of removing one molecule of carbon dioxide from the air is the same as the impact of preventing one molecule from getting into the air in the first place.

However, WWS generators also (1) eliminate air pollutants aside from carbon dioxide from fossil-fuel combustion; (2) eliminate the upstream mining, transport, and refining of fossil fuels and the corresponding emissions; (3) reduce the pipeline, refinery, gas station, tanker truck, oil tanker, and coal train infrastructure of fossil fuels; (4) reduce oil spills, oil fires, gas leaks, and gas explosions; (5) reduce international conflicts over energy; and (6) reduce the large-scale blackout risk associated with centralized power plants by decentralizing/distributing power.

Air capture does none of that. Its sole job is to remove carbon dioxide from the air, but at a higher cost than using renewable energy to do the same thing. In fact, air capture is just a cost added onto the cost of using fossil fuels.

Based on data from a real plant, carbon dioxide removal from the air by an air capture facility is an inefficient process.[154] In this plant, electricity for the air capture equipment is provided by a natural gas combined cycle turbine. Mining, transporting, processing, and burning the natural gas that is used to power the air capture equipment produces air pollutants, methane, and carbon dioxide. The methane and carbon dioxide emissions from the natural gas use offset much of the benefit of capturing carbon dioxide from the air. In fact, averaged over 20 and 100 years, 89.5 percent and 69 percent, respectively, of all carbon dioxide captured by the air capture equipment is estimated to be returned to the air.

The social cost of air capture includes the carbon capture equipment cost, the natural gas electricity cost, and the social cost of air pollution associated with capture, less the cost benefit of removing carbon dioxide from the air. The social cost of air pollution is the health-related cost of air pollution. For example, air pollution increases death and illness, both of which increase hospitalization costs, emergency room visit costs, lost workdays, lost school days, insurance rates, taxes, workers' compensation payments, and loss of companionship, among other costs. The social cost of running the air capture plant in this case is about 8 times that of using the same amount of electricity, but powered by wind, to replace part of a coal plant. The reason is that wind replacing coal reduces substantial air pollution and more carbon dioxide than does the direct air capture plant, and wind replacing coal does not incur a carbon capture equipment cost.

Even when wind, instead of natural gas, powers the air capture equipment, the social cost of using air capture is still 5 times that of using wind to replace part of a coal plant.

In fact, there is no case where wind powering an air capture plant has a social cost below that of replacing any fossil-fuel or biomass power plant with wind electricity.[154] The reasons are that wind-powering air capture always incurs an equipment cost that wind replacing a fossil plant never incurs. Air capture also allows air pollution and mining to continue, whereas wind electricity always eliminates air pollution and mining from a fossil plant.

An argument for using air capture is that it will be needed to remove carbon dioxide from the air once all fossil fuels are replaced with 100 percent WWS. If all energy is provided by renewables at that point, air capture should reduce carbon dioxide without increasing air pollution. However, the question at that point is whether growing more trees, reducing open biomass burning; reducing agriculture and waste burning; or reducing halogen, nitrous oxide, or non-energy methane emissions are more cost-effective methods of limiting global warming. Until the time that such an evaluation can be made, air capture will always be an opportunity cost.

In sum, like with carbon capture, direct air capture is not close to a zero-carbon technology. For the same energy cost, wind turbines and solar panels replacing fossil or biomass power plants reduce much more carbon dioxide while also eliminating fossil air pollution, mining, and infrastructure, which air capture increases.

8.10 Why Not Geoengineering?

Geoengineering is the large-scale alteration of the natural properties of the Earth or the atmosphere in an attempt to reduce global near-surface temperatures. The two primary categories of geoengineering that have been proposed are techniques to remove carbon from the air (**carbon capture techniques**) and techniques to increase the reflectivity of the Earth or its atmosphere in order to decrease sunlight hitting the Earth's surface (**solar radiation management** techniques).

Carbon capture techniques have already been discussed. These are geoengineering techniques because they are intended to reduce the amount of carbon dioxide in the air to modulate the Earth's average temperature. Of the carbon capture techniques, only natural direct air carbon capture is recommended in a 100 percent WWS world.

The main solar radiation management techniques that have been proposed include (1) injecting reflective aerosol particles into the stratosphere to reflect sunlight directly, (2) injecting fine sea spray particles into the air just above the ocean surface to increase the number and decrease the average size of cloud drops, thereby increasing the overall cross-sectional area of cloud drops to increase their reflectivity, and (3) installing white roofs or roads.

The first problem with all these techniques is that none reduces fossil-fuel or bioenergy emissions of gases or particles that cause global warming and 7 million deaths per year. To the contrary, with geoengineering, the public and policymakers become complacent, no longer feeling the urgency to reduce global temperatures or fossil-fuel emissions. As such, pollutant gases and particles continue to cause damage and, in fact, increase.

Second, geoengineering may temporarily mask some warming regionally. However, because long-lived greenhouse gases continue to accumulate with geoengineering, even more investment in geoengineering is needed to keep up with the increase in emissions. Any interruption or stoppage of the geoengineering results in an immediate worsening of the climate problem because of the accumulation of even more greenhouse gases during the period of geoengineering.

Third, since geoengineering does nothing to stop air pollution, death and illness continue to occur without abatement compared with no geoengineering. Such health impacts worsen since complacency allows more fossil fuel and bioenergy to be burned. Health problems may also worsen because of the particles injected into the atmosphere to assist with the geoengineering. Such particles, when breathed, are harmful for health.

Fourth, since geoengineering does not reduce fossil-fuel or nuclear use, it does nothing to help reduce energy insecurity associated with those energy sources.

Fifth, if the money spent on geoengineering were spent instead on WWS, not only would WWS eliminate greenhouse gas emissions (thus reducing temperatures, as geoengineering does), but WWS would also eliminate air pollution emissions, death and illness resulting from the air pollution emissions, mining of fossil fuels and uranium, and energy insecurity. As such, geoengineering is an opportunity cost compared with WWS.

A sixth problem with geoengineering is its unintended consequences. For example, reducing sunlight reaching the ground reduces crop yields, which can cause starvation in some parts of the world. Injecting aerosol particles into the stratosphere catalyzes stratospheric ozone loss in the presence of halogens currently in the stratosphere. Injecting particles into the stratosphere or into the air above the ocean results in changes in weather patterns. Injecting particles into marine air increases the concentration of particles entering populated coastal cities, increasing death and illnesses from air pollution. Particles injected into the stratosphere ultimately deposit to lower levels and the ground, increasing air pollution health and acid rain problems as well.

An example of the unintended consequences of a geoengineering proposal is the potential impact of **white roofs** and roads on global climate. Although white roofs and roads reflect radiation, cooling buildings and the ground in cities locally, they may cause large-scale global warming.[188]

The first reason is that, because white roofs cool the ground relative to the air locally, they reduce the ability of air to rise, and thus of clouds to form. Since clouds are reflective, reducing cloudiness increases sunlight to the surface, offsetting some of the reduced sunlight to the surface caused by the white roofs. However, since clouds travel and spread beyond a city, their reduction increases sunlight reaching the ground and temperatures outside of the city.

Second, black and brown carbon pollution particles in the air absorb sunlight, then convert that sunlight directly to heat, which is released to the air. In the presence of white roofs or roads, black and brown carbon absorb not only downward sunlight but also sunlight reflected upward by the white surfaces, warming the air further.

Finally, whereas white roofs cool buildings and thus reduce air conditioning energy requirements at low latitudes and during summers, their reflectivity increases heating requirements during winters. In many places worldwide, heating requirements exceed cooling requirements, so adding a white roof to a building simply increases the fossil-fuel use required to heat the building.

A better solution than using a white roof is to install solar PV panels on a rooftop. The primary purpose of installing a PV panel is to generate electricity; however, panels also have several side benefits. Not only do rooftop PV panels remove 20 percent or more of incoming sunlight, converting it to electricity and cooling the underlying building, but the electricity they produce also displaces fossil-fuel use and its emissions. In addition, because solar panels do not reflect sunlight upward as white roofs do, solar panels do not allow absorption of upward-reflected sunlight by black and brown carbon pollution particles. Similarly, because PV panels are warmer than a white roof, PV panels do not increase air stability and thus do not reduce cloudiness like white roofs do.

In sum, with the exception of natural direct air capture by trees and reducing deforestation, geoengineering is not recommended in a WWS world.

9 ELECTRICITY GRIDS

A 100 percent wind–water–solar (WWS) energy infrastructure involves electrifying or providing direct heat for all energy sectors and then providing the electricity or heat with WWS. The solution also requires interconnecting geographically dispersed WWS generators on the grid. Because electricity and the grid are such a large part of the solution, understanding how both work is important. This chapter discusses these issues along with the battle between George Westinghouse/Nikola Tesla and Thomas Edison to determine whether alternating current (AC) or direct current (DC) would predominate worldwide. Finally, the chapter discusses how transformers, motors, and generators work.

9.1   Types of Electricity

Electricity generally means the free flowing movement of charged particles, usually either negatively charged electrons, negatively charged ions, or positively charged ions. The flow of electric charge constitutes an **electric current**. Charged particles moving in an electric current are called **charge carriers**. Thus, charge carriers can be electrons, negative ions, or positive ions.

Electricity can travel in a wire or another medium, including the air. This chapter focuses on electricity through wires with application to the electric grid. Two other types of electricity are static electricity and lightning.

Static electricity arises on rubbing two materials, such as silk and glass, together so that one material strips electrons off of the other. The result is that one material becomes negatively charged and the other, positively charged. If the charge difference is small, and if neither material is a strong conductor of electricity, such as with glass and silk, the two materials may merely stick to each other because of the attraction of the excess electrons in one material to the excess positive ions in the other. On the other hand, if one material is a strong conductor of electricity, a spark may occur. For example, a person walking on a wool rug scrapes electrons off the rug, giving his or her body a net negative charge. If the person's hand approaches a metal doorknob, which is a good conductor of electricity, the excess electrons in the person's hand induce positive charges in the metal doorknob to rise to the knob's surface. Just before contact, millions of electrons that are attracted to the positive charges on the doorknob fly from the person's hand, through the air, smashing into the doorknob and creating a spark.

Lightning occurs when a large charge difference occurs between a cloud and the ground, between two parts of the same cloud, or between two clouds. If the charge difference builds up beyond a threshold, lightning strikes. Before cloud-to-ground lightning occurs, the bottom of a cloud builds up a net negative charge and the top, a net positive charge. This happens because small ice crystals bounce off of heavy hail particles near the middle of the cloud, stripping the surface of hail particles of positive charges. The positively charged small crystals are then blown by updrafts to the top of the cloud. The negatively charged hail falls to the base of the cloud.

The net negative charge at the base of the cloud induces positive ions at high points along the ground to migrate to the ground surface. Once the charge difference between the base of the cloud and the ground exceeds a threshold, a group of electrons streams from the base of the cloud toward the ground. As the electrons bash into air molecules, the electrons split the air molecules into positive ions and additional electrons. The additional electrons join the downward parade of electrons, but the positively charged ions slow down the parade by attracting some of the electrons back upward. This temporary upward motion followed by the continued downward motion of electrons contributes to lightning's jagged shape.

As the electrons approach the ground, their attraction is so strong that a **streamer** of positive ions flows up from high points along the ground into the air to meet them. The connection creates a channel in which electrons in the air near the ground first rush to the ground. These electrons bash into air molecules, forming a hot (30,000 to 60,000 degrees Celsius) **plasma** of ions and electrons. **Lightning** occurs because cooler molecules (3,700 to 7,300 degrees Celsius) on the outer edge of the flash emit photons of visible light. The thermal expansion of the air creates a sonic boom, or **thunder**.

Electrons backed up behind the first batch then rush to the ground, just as a line of cars released from a stoplight start moving from the front to the rear of the line. In this manner, lightning starts near the ground and propagates up to the cloud as each successive batch of electrons behind the previous one accelerates toward the ground. This first major flash is the most luminous. Once the channel is open, additional electrons from the cloud come through it in a downward stroke of lighting.

Wired electricity is electricity that travels through a metal wire. In a metal wire, the metal's atomic nuclei are fixed in position and electrons carry charge freely through the wire to produce a current. The flowing electrons are sufficiently far from the nucleus of any specific atom in the wire that the attraction of the electrons to a nucleus is easily overcome. These conduction electrons wander from atom to atom, creating a current.

Negatively charged electrons balance positive charges in a wire. So, when an electron moves, it leaves behind a positive charge. Thus, the positive charges appear to move in the opposite direction from the negative charges. The direction of a current is defined to be the direction of positive flow (or opposite the direction of negative flow). So, if electrons are moving to the right, the current (positive flow) is moving to the left. In a circuit, a flow of positive charge in one direction has the same impact as a flow of negative charge in the opposite direction.

An **electric current** is the flow rate of electrons or other charged particles through a wire and is the same at any point along a closed circuit. Electric current is quantified as the number of electric charges transported per second.

Direct current is current flowing at a constant rate over time in one direction. An example of a DC electricity source is a battery. In a wire conducting DC electricity, electrons as a whole move relatively slowly. Electrons move in random directions, but in the presence of an **electric field**, which is an electric force per unit charge, more will move in one direction than another, resulting in a net flow of electrons in one direction. Although electrons as a whole move slowly in a wire, each electron carries a lot of energy. As electrons collide with each other, they transfer energy to each other, sending a stream of electricity down a wire at nearly the speed of light. As such a wire can carry a lot of current.

Alternating current is current that flows back and forth, in a sinusoidal manner, over time. In one sinusoidal cycle, electrons first accelerate their flow to the right, reach a peak speed, decelerate to zero speed, accelerate to the left, reach a peak speed, then decelerate to zero speed. The number of complete cycles in a second is called the **AC frequency**. One cycle per second is defined as 1 **hertz**. Three hertz means that three complete cycles of flow and counter flow occur in 1 second. In the United States and some other countries, the AC frequency for the electric power grid is set at 60 cycles per second, or 60 hertz. **Nikola Tesla** (1856 to 1943) calculated that 60 hertz was optimal for the new U.S. electric grid. In Europe and in most other countries, the AC frequency is set at 50 hertz. In a wire conducting alternating current at 60 hertz, electrons reverse themselves 60 times a second, so hardly move at all. Yet, they transfer energy to each other, sending a wave of electricity down a wire at nearly the speed of light.

9.2   Voltage

Voltage is the energy per unit charge needed to move a single positive charge between two points in an electric field. Voltage can be understood using an analogy with potential energy, which equals mass multiplied by gravity and height. As a mass is lifted against gravity, it gains potential energy. Analogously, as the voltage of a charge is raised, the charge also gains potential energy, referred to as electrical energy.

A way to think of the difference between voltage and current is by using an analogy of water in a hose used to wash a sidewalk. The amount of water passing through the hose is analogous to current. The thrust of the water coming out of the hose is analogous to voltage.

A simple electrical circuit can also be used to illustrate voltage and current. An **electrical circuit** is a closed-loop electrical network with a return path through which a current runs. A simple circuit may consist of a battery, a light bulb, a wire from one end of the battery going to the light bulb, a wire going from the other side of the light bulb to the other end of the battery, and a switch that connects and disconnects the circuit.

According to **Kirchhoff's voltage law**, *"the sum of the voltages around any loop of a circuit at any instant is zero."* Kirchhoff's voltage law is analogous to the changes in elevation along a circular path. A hiker walking up in elevation on a trail eventually descends back to his or her original position. As such, the sum of elevation changes along the path always equal zero. Similarly, the sum of voltage changes around a circuit always equals zero.

Suppose a battery in a circuit is a 12-volt battery. Voltage is measured across components. This means that, if the voltage at one end of the battery is zero volts, the voltage at the other is 12 volts, so the overall voltage across the battery is 12 volts. If the battery is connected to a light bulb in the circuit, the voltage drops 12 volts when measured from the wire going into the light bulb to the wire leaving the light bulb. Thus, the light bulb reduces the voltage by 12 volts, converting electrical energy into light energy through a resistance element. Because the voltage increases by 12 volts from one end of the battery to the other then decreases by 12 volts from the wire going into the light bulb to the wire going out of the light bulb and back to the battery, the net change in voltage through the circuit is zero volts, satisfying Kirchhoff's voltage law. Meanwhile, the current through the circuit is constant everywhere.

9.3   Power

Power is the change in energy per unit time. With respect to electricity, since energy per unit time equals energy per unit charge (voltage) multiplied by charge per unit time (current), then power equals voltage multiplied by current. Power has units of watts, where 1 watt equals 1 joule of energy per second.

Electric current flows through wires in order to power an electric load. An **electric load** (or electric demand) is a component or part of an electrical circuit that consumes electrical energy, thus reduces voltage. A light bulb, for example, is a load. Loads can also be heaters, cookers, dishwashers, car chargers, arc furnaces, etc. When loads consume energy, they reduce the voltage across the load.

Voltage is also reduced along a circuit if the current passes through an electrical insulator. An **electrical insulator** is a material that impedes the flow of a current by preventing electric charges from flowing freely through the material. Good insulators are rubber, glass, air, plastic, paper, wood, wax, and wool. The opposite of an electrical insulator is an **electrical conductor**, through which charges flow freely. Good conductors are copper, aluminum, gold, and silver.

When a current passes through an insulator, electric energy carried by the current turns into heat, and voltage decreases, just as it does when current passes through an electrical load.

Electrical resistance arises when an electric load or insulator impedes the flow of an electrical current. Both loads and electrical insulators are forms of electrical resistance.

The electrical resistance of an object is quantified as the voltage across the object divided by the current through the object. Alternatively, the voltage across an object equals the current through it multiplied by the resistance. In other words, resistors drop the voltage across an object proportionally to the current through the object. If no voltage change occurs across a load, the load's resistance is zero.

9.4   Electromagnetism and AC Electricity

Magnets, in addition to batteries, can produce an electric current through a wire. For example, a magnet moving toward or away from a coil of wire that is part of a circuit induces an electric current in the wire. Similarly, an electric current flowing along a straight wire creates a circular magnetic field around the wire. The relationship between electricity and magnetism is referred to as **electromagnetism**. Electromagnetism is the key physical process occurring in inductors, transformers, motors, and generators.

On April 21, 1820, the Danish physicist and chemist, **Hans Christian Orsted** (1777 to 1851) discovered electromagnetism when he noticed that the needle of a compass was deflected in the presence of an electric current originating from a battery that he switched on and off.

He investigated and published this phenomenon the same year, concluding that an electric current flowing through a wire produces a circular magnetic field around the wire.

In 1825, **André-Marie Ampère** (1775 to 1836) found that a wire carrying a current exerted an attracting force on a second wire carrying a current in the same direction and a repelling force on a second wire carrying a current in the opposite direction.

Michael Faraday (1791 to 1867) made a third important advancement in electromagnetism. On August 29, 1831, he coiled two unconnected wires around opposite ends of a circular iron ring. He then connected one of the wires to a **galvanometer**, which measures current. Upon connecting the second wire to a battery, he noticed that the galvanometer briefly measured an electric current through the first coiled wire. Disconnecting the battery from the second wire also induced a brief current in the first wire.

What happened was that connecting the battery to the second coiled wire created a current through that wire. The sudden increase in current created a pulse magnetic field around the second coil. The pulse increase in magnetic field induced a current through and a voltage across the first coiled wire. Disconnecting the battery similarly decreased the current through the second wire, and thus the magnetic field around it. The sudden drop in magnetic field induced a current through and a voltage across the first wire.

Faraday similarly showed that moving a magnet toward or away from a coiled wire on a circuit induces a current through the circuit and a voltage across the coil.

In a third experiment, he showed that moving a coiled wire connected to a battery in and out of the center of a larger, second coiled wire induces a current in the larger coil. The current through the first coil creates a magnetic field, and the movement of that magnetic field induces a brief current through the larger coil. When the two coils are held still, no current flows through the second coil. A current can be induced only when a magnetic field fluctuates with time, not when it stays constant with time.

The principle of electromagnetism was essential for the development of AC electricity. An AC current arises when current flows in one direction for a short time, changes direction for a short time, then changes direction back again in a repeating cycle. With DC electricity, on the other hand, current flows constantly in one direction.

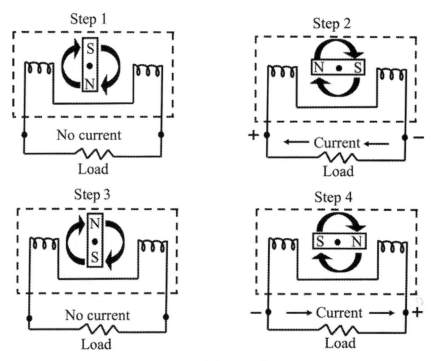

Figure 9.1 Production of single-phase AC current by rotating a magnet in the presence of stationary wire coils. In the diagram, S identifies the magnet's south pole and N identifies the magnet's north pole. The text describes the process.

AC electricity can be produced with a magnet that spins clockwise inside of two stationary coiled wires placed on opposite sides of the magnet (Figure 9.1). The two stationary coiled wires are part of a closed circuit that has a load connected to it. When the magnet is at rest and its south and north ends are away from the coils, no current flows (Step 1 in Figure 9.1). As the magnet rotates clockwise, its south pole approaches the coiled wire on its right side (Step 2). The increase in magnetic flux experienced by the coiled wire induces a clockwise current in the circuit. As the south pole continues rotating clockwise away from the coiled wire on the right side, the current in the circuit decreases until it reaches zero (Step 3). As the magnet continues rotating clockwise, its south pole approaches the coiled wire on the left side, again inducing a current, but in the counterclockwise direction (Step 4). As the magnet continues rotating clockwise, the current decreases again, ultimately to zero again (Step 1). Thus, one clockwise rotation by the magnet creates an alternating-in-direction current through the circuit. The faster the magnet turns, the greater the frequency that the current alternates direction in the wire.

The current generated with one pair of coiled wires is called **single-phase AC electricity**. Such electricity is produced in an alternator, which is a type of generator. **Generators** use magnets and coiled wires to convert the rotational energy in a spinning shaft to electricity. Almost all rotational energy for generators worldwide is obtained from a fluid (steam, air, combustion gases, or liquid water) passing through turbine blades connected to a rotating shaft. **Motors** are generators run in reverse. They convert electricity into rotational energy.

The problem with single-phase AC electricity on a circuit is that the oscillating AC electricity current causes a continuous alternating increase and decrease in current. Thus, appliances, such as light bulbs, continuously increase and decrease in brightness. This flickering is ameliorated by the fact that the frequency of the increases and decreases is so rapid, 50 or 60 cycles per second, that most people cannot detect the flickering.

Three-phase AC electricity increases the apparent constancy of the current. It can be accomplished in a generator simply by rotating a rectangular magnet inside of six (three pairs of) instead of two (one pair of) equally spaced coiled wires. Another way to generate three-phase AC electricity is to rotate three equally spaced coiled wires inside of a circular magnet that has a north end and a south end. The advantage of three-phase AC electricity is that it results in the maximum positive current occurring three times per rotation rather than one time per rotation, decreasing flickering. Having more than three phases increases equipment cost with little additional benefit.

Three-phase generators smooth current compared with single-phase generators and produce virtually all commercial AC electricity worldwide. Most three-phase AC electricity is sent along three-phase transmission lines. These have three wires to transmit each phase independently to the end load.

Since each phase in a three-phase AC system is transmitted along its own wire, different phases can be delivered to a household for different purposes. Because of the high frequency (60 or 50 hertz) of the waves, flickering is not noticeable by most people using even a single phase for home appliances.

9.5   Capacitors and Inductors

Capacitors store electric charge in an electric circuit. They are made of two flat conducting metal plates in parallel, separated by a thin nonconducting insulator in between. The simplest capacitor consists of two sheets of aluminum foil separated by air or paper. If a battery is connected to a circuit with nothing else connected to it but a capacitor, then one metal plate becomes positively charged and the other, negatively charged. The difference in charge across the plates creates an electric field in which electrostatic energy is stored. Once fully charged, the capacitor holds the same voltage as the battery.

If a circuit contains a battery, a light bulb, and a large capacitor, the light bulb will first light up but become dimmer as the current flows to the capacitor, charging the capacitor. When the capacitor is fully charged, the light bulb will darken completely since the voltage drop across the capacitor is now the same as the voltage increase across the battery, so no more voltage is available for the light bulb. If the battery is then disconnected from the circuit, the capacitor will act like a battery and instantly provide current to the light bulb, lighting it, until the stored energy in the capacitor is depleted.

In sum, capacitors are like batteries in that they discharge stored electricity quickly. However, capacitors charge much faster than do batteries. They also weigh less, require fewer materials, cost less, and last many more cycles than do batteries. The disadvantage of a capacitor is that it stores much less total energy than does a battery. Thus, capacitors are used only when a small amount of energy is needed quickly, such as to maintain power in electronic devices when batteries are being charged or when voltage drops.

Capacitors are also used to smooth voltage in DC power lines because they resist rapid changes in voltage. In other words, voltage cannot change instantaneously in a capacitor so capacitors smooth out sudden variations in voltage on a power line.

Because capacitors can be charged quickly and last long but store little energy, efforts were made starting in the late 1950s to improve them so that they could hold more energy. These efforts resulted in the first **supercapacitors**, which are capacitors with much larger plates, much smaller distance between plates, and made of different materials than capacitors. Even the best supercapacitors today store only 10 to 20 percent of the energy that a battery stores. However, supercapacitors do have important applications. For example, a cordless drill powered by a bank of supercapacitors allows astronauts in space to charge the drill quickly. Banks of supercapacitors are used in electrical and electronic equipment to smooth out power supplies. They are also used to store temporarily the electricity from regenerative braking in vehicles. The electricity is used immediately when the vehicle starts moving again.

Whereas capacitors store energy in an electric field, **inductors** store energy in a magnetic field. An inductor consists of an insulated wire coiled around an iron core. When a current passes through the coil of an inductor, the inductor creates a magnetic field in which voltage is stored.

Just as capacitors smooth sudden changes in voltage across them, inductors smooth sudden changes in current through them. When current that flows through an inductor suddenly increases, for example, the magnetic field becomes stronger. As the magnetic field strength increases, electrical energy is drawn from the current and converted to magnetic potential energy, increasing the voltage. Because electrical energy is taken from the current, the current decreases. In sum, inductors smooth sudden changes in current. As such, inductors are useful in the electric power grid for limiting the damage due to short circuits or other abnormal currents. Inductors are also integral components of transformers.

9.6   The Battle of DC versus AC Electricity

When electricity travels down a transmission line, it loses power proportional to the square of the current through the line. Thus, increasing current through a transmission line by a factor of 10 increases power loss by a factor of 100. As such, current through a transmission line should be kept low to minimize power loss. Since power equals voltage multiplied by current, decreasing current means increasing voltage to maintain the same overall power. Thus, transmitting electricity long distance can be accomplished without significant transmission line loss only if the current through the line is low and the voltage transmitted is high. This principle is the main reason the world's electricity grids today are AC grids, as described next.

9.6.1 Thomas Edison and the DC Grid

On September 4, 1882, **Thomas Edison** (1847 to 1931) and the Edison Illuminating Company began operating the first investor-owned electric utility in the world, the Pearl Street Power Station, in New York City. The electricity was produced by burning coal to boil water to produce steam for use in a steam engine. In the steam chamber of a steam engine, steam pushes a piston up. Liquid water is then squirted into the steam chamber to recondense the vapor, creating a vacuum that pulls the piston down. The up and down motion of the piston allows a lever attached to the piston to create a rotating motion. The rotating lever is connected to a **dynamo**, which is a generator that produces DC electricity when a coiled wire rotates continuously inside of a magnetic field.

Edison had invented the incandescent light bulb in 1879. His 1882 electric utility initially provided power for 400 light bulbs owned by 82 customers. By 1884, the number of light bulbs had increased to 10,164, spread among 508 customers.

The DC power provided to customers was at 110 volts. If the voltage delivered to customers were substantially higher, it would be dangerous. Because the voltage through the transmission lines was relatively low, the current needed to be high in order for the total power (energy per unit time) delivered to the homes to be sufficient. The high current required thick copper wires, resulting in high transmission line losses and costs. The line losses were so high that the distance between the power generator and customers was limited to 2 kilometers. As such, the first coal-fired electricity-generating station in the world emitted pollution in the middle of one of the most densely populated cities in the world. However, if the coal plant were sited outside of the city, transmission line losses would have been so high that customers would have received virtually no power.

The Pearl Street Station power plant burned down in 1890. In the 1880s, however, Edison opened up several other power stations and local grids based on DC electricity. These include utilities in Sunbury, Pennsylvania (1883); Shamokin, Pennsylvania (1883); Brockton, Massachusetts (1883); Mount Carmel, Pennsylvania (1883); Fall River, Massachusetts (1883); Cumberland, Maryland (1884); Tamaqua, Pennsylvania (1885); and Boston, Massachusetts (1886). By the end of the 1880s, in fact, Edison had built or licensed over 1,500 electricity-generating stations worldwide, many of which provided dedicated power to factories. The rest provided grid electricity for homes and commercial buildings.

9.6.2   Nikola Tesla

Several of Edison's utilities were in Europe. In 1882, the French branch of Edison Electric Light Co., Continental Edison Company, which was building power stations and grids in Paris and other European cities, hired **Nikola Tesla** (1856 to 1943) (Figure 9.2). Tesla, born in Croatia, previously worked in Budapest, Hungary, as an electrical engineer. Tesla's new job was to install incandescent lights in a Paris electric grid and to troubleshoot problems in grids being built by Edison in France and Germany.

Tesla's skill as a technician and troubleshooter became apparent quickly, so much so that Tesla was called on to investigate and repair what could have been the greatest disaster to date resulting from Edison's DC power grid. In early 1883, during the grand opening of a DC power station at the Strasbourg (in Germany at the time) rail station, the ceremonial switching on of power caused an explosion that nearly killed Emperor William I of Germany. Tesla was asked to find out what happened and to fix the problem. He determined that the explosion was due to a major short circuit resulting from poor dynamos and faulty wiring. He spent a year redesigning the system, particularly the dynamos. The new power system worked well. Tesla's talent as an electrical engineer was recognized. However, Tesla's boss and Edison's confidant in Paris, Charles Batchelor, who had orally promised Tesla a $25,000 bonus for fixing the problems in Strasbourg, backed out from paying him. Tesla was not happy but continued with his job.

While in Paris, Tesla recognized the problem with Edison's DC electricity grids – namely that they were limited to low voltages and short distances and required thick copper-wire transmission lines to carry the high current. This meant that a power plant was needed every 2-kilometer radius to provide electricity for a whole city. In addition, because thick copper wires were needed to carry the high current, even over short distances, copper requirements for transmission lines would be enormous for powering a large population. These issues created a practical problem that limited the spread of DC electricity. In 1882, just prior to moving to Paris, Tesla designed a prototype AC motor.

Figure 9.2 Nikola Tesla in 1890 at age 34. Photo by Napoleon Sarony (1821 to 1896), public domain, via Wikimedia Commons.

Tesla felt that a solution to the DC grid problem might be AC electricity. However, his superiors at Edison's company in Paris did not want to hear about AC electricity, as they were heavily invested in DC.

Batchelor, though, recognized Tesla's talent. He also knew that Tesla should speak directly to Edison about Tesla's AC grid idea so that the two could work together on the idea rather than Tesla going off on his own. Batchelor thus convinced Tesla to move to the United States, where Tesla would have a better opportunity to interact directly with Edison and design and improve dynamos and motors. Tesla agreed to go.

Having been robbed of money and luggage at the shipyard in Calais, France, Tesla travelled by ship to New York City with no money or clothes beyond what he was wearing. He did have a letter of introduction to Edison from Batchelor. Soon after arriving, on June 8, 1884, Tesla began working at the Edison Machine Works. His first job was to repair broken dynamos powering lights on the ship S.S. *Oregon*, which was docked in New York City. Edison was impressed with Tesla's successful repairs, which took less than a day. Thereafter, Edison had even more confidence in Tesla and decided to give him a more difficult task.

Edison asked Tesla to improve the design of the dynamo so that multiple dynamos could be linked together to provide electricity for large buildings without causing light flickering or short circuits. Edison verbally offered Tesla a $50,000 bonus for this job. Not learning a lesson from his previous mistake, Tesla agreed and, over the next year, developed 24 different types of improved dynamos. Edison applied for patents on the dynamos, and his company reaped substantial financial rewards from them. However, upon completion of Tesla's work in 1885, Edison refused to pay the bonus, telling Tesla, "*You don't understand our American humor.*" This was the last straw. Tesla immediately resigned.

9.6.3   Transformers

Meanwhile, in 1882, **Sebastian Ziani de Ferranti**, working at the Siemens brothers' firm in London with Lord Kelvin (William Thompson), designed a prototype AC power system. In the process, he developed an early transformer. Working off the design of Ferranti, **Lucien Gaulard** of France, with the financial backing of John Gibbs, invented a step-down transformer in 1883. Gaulard and Gibbs demonstrated the transformer, which they called a *secondary generator*, that year in London's Royal Aquarium. In 1884, Gaulard used his transformer in a demonstration 40-kilometer AC transmission system between Lanzo and Turin, Italy. The transformer reduced high-voltage AC current to low-voltage AC current to provide power for incandescent lights and arc lamps along the path. He also showcased the transformer at the Turin Italian National Exhibition that year, where **Otto Blathy**, a Hungarian engineer, saw it. In 1885, Blathy and two other Hungarian engineers, Miksa Deri and Karoly Zipernowsky, improved and patented a similar transformer, which Blathy named the *ZBD transformer*, where ZBD are the initials of the three inventors.

A **transformer** consists of two or more inductors in proximity to each other. A transformer increases or decreases voltage between a powered inductor coil and an unpowered inductor coil. It works when an AC current winding through one of the inductor coils creates a fluctuating magnetic field around the second inductor coil not connected to the first. The fluctuating magnetic field induces a current through and voltage across the second coil. The AC voltage induced across the unpowered coil equals that across the powered coil multiplied by the ratio of the number of coil turns in the unpowered coil to those in the powered coil. So, if the unpowered coil has twice as many coils as the powered coil, the voltage in the unpowered coil will be twice that in the powered coil. The current in the unpowered coil is the opposite. So, in this example, the current through the unpowered coil will be half that in the powered coil. In other words, if the number of coils in the unpowered coil is higher than in the powered coil, a transformer will increase voltage and decrease current (step-up transformer). If the number is lower, the transformer will decrease voltage and increase current (step-down transformer). The power (voltage multiplied by current) stays the same between the two coils.

A transformer does not work with DC current. Although DC electricity passing through one coil creates a magnetic field around the second coil, the magnetic field stays constant over time because DC current in the first coil stays constant over time. A current occurs in the second coil only if the magnetic field fluctuates with time, and this requires the current in the first coil to fluctuate with time. Because AC electricity through the first coil continuously alternates direction, the magnetic field produced by it continuously fluctuates as well, inducing a current in the second coil.

Transformers are used not only to increase or decrease voltage along a transmission line but also, because they are made of inductors, to limit abnormal electrical currents.

While still working with Edison, Tesla had offered to redesign Edison's grid from DC to AC. Tesla recognized that a transformer would allow AC electricity voltage to be increased (stepped up) for long-distance transmission, and a second transformer would allow the voltage to be decreased (stepped down) again for end use. Increasing the voltage of AC electricity to produce high-voltage AC (HVAC) electricity would allow current to be reduced proportionately, decreasing line loss tremendously over a long distance, thereby increasing power output at the end of the line relative to with no transformers.

The main implication of an AC grid was that AC electricity could be produced far away from where it was used. This would allow a single electricity generating station to serve far more customers than Edison's generator stations, which had to be near customers. A single, large generator station could operate at much lower cost than many small stations. In addition, although this was not a consideration at the time, a single large coal-fired power plant outside of a city would cause much less air pollution exposure to a city's population than would many smaller plants within the city

Edison called Tesla's ideas *"splendid"* but *"utterly impractical"*.[189] After Tesla resigned from Edison's company, Tesla set out to start his own AC utility, Tesla Light & Manufacturing.

9.6.4 George Westinghouse and the AC Electricity Grid

Coincidentally, in 1885, **George Westinghouse** (1846 to 1914), who started a DC lighting business in 1884, became interested in AC electricity. He was motivated and had the financial resources to invest in an AC grid. When he read an 1885 account in a newspaper about Gaulard's transformer, which had been used in the 1884 Turin Exhibition, Westinghouse quickly purchased the American rights for Gaulard's design and ordered several of the transformers and a Siemens alternator (single-phase AC generator) to be shipped to Westinghouse's factory in Pittsburgh, Pennsylvania.

Westinghouse then asked an engineer working for him, **William Stanley Jr.**, to develop an electric AC lighting system with the alternator and transformers. In 1885, Stanley began experimenting in Pittsburg. He also improved the transformer enough so that it became the world's first practical transformer.

In 1886, Stanley used his new expertise to build, with Westinghouse's funding, the world's first AC power system using both a step-up and a step-down transformer, in Great Barrington, Massachusetts. In 1886, Stanley, Westinghouse, and Oliver B. Shallenberger then built the first commercial AC power system, in Buffalo, New York. Westinghouse installed several other AC grid systems that year around the United States, particularly in remote locations that Edison could not reach economically with his DC grid. By 1887, after only a year, Westinghouse had half as many generators as Edison. The same year, C. S. Bradley invented the first three-phase AC generator. Only single-phase generators were available until then. A problem at the time, though, was that an efficient AC motor had not yet been invented, so AC electricity could not be used to power any equipment that needed a motor.

In 1886, Tesla attempted to sell his AC power system idea to investors in New York City but found little interest, since New York City already had some of Edison's DC power. Nevertheless, Tesla kept working to develop AC technology. In an 1888 breakthrough, Tesla invented a three-phase AC induction motor, which was critical for powering equipment on an AC grid. After Westinghouse heard about Tesla's invention, Westinghouse visited Tesla's laboratory. The two began a collaboration to design a better AC grid system with three-phase generators and motors. Westinghouse paid Tesla well.

Transition highlight

George Westinghouse poured financial resources into developing an AC power system with step-up and step-down transformers, AC generators, and AC motors. In 1891, he built the first power plant (a hydroelectric plant) to supply AC power over a long distance (5.6 kilometers) for a gold mine near Ophir, Colorado.

9.6.5   The Fight Begins

Edison considered going into the AC transmission business. In fact, he purchased options on Blathy's transformer. However, he ultimately decided against going forward with AC.

Instead, the storm of development by Westinghouse caused Edison anxiety, which he soon reacted to. In November 1887, a dentist in Buffalo, New York, Alfred P. Southwick, wrote to Edison asking Edison to support the use of electricity as a humane way of executing criminals. Since beoming aware of the accidental August 1881 electrocution of a drunken dockworker by a dynamo, Southwick began experimenting with electrocuting dogs to euthanize them. He published work on his experiments in scientific journals and wanted to develop a way to adopt electrocution for executing criminals. In an early design, Southwick used his dental chair as a way to restrain a person during an execution. This gave rise to the name *electric chair*.

Edison did not believe in capital punishment but did believe that his competition should be punished. When pressed by Southwick about what kind of equipment should be used for executions, Edison eventually replied, "*The most effective of these are known as alternating machines manufactured principally in this country by Mr. Geo. Westinghouse, Pittsburgh*".[189] Edison subsequently engaged Harold Brown, a senior engineer working for him, to publicly stoke fears about AC electricity.

Brown, who pretended he was not paid by Edison, stated correctly in front of audiences, for example, that the AC voltage required to electrocute a dog was much less than the DC voltage. However, to incite fear, he then went on to demonstrate the electrocution of dogs with AC versus DC electricity. In one stunt, he connected a sheet of tin to an AC dynamo and led a dog onto the tin to drink from a dish. The dog was electrocuted and died. This and other demonstrations often resulted in the smell of burned flesh and hair permeating a room. Despite protests, Brown ramped up electrocutions for show, but now of a horse and calf in addition to dogs. Edison and his helpers also leaked photographs of other animals they electrocuted to the press. They even attempted, unsuccessfully, to get a law banning AC electricity introduced.

When Westinghouse challenged Brown and Edison's advertising campaign, Brown came back proposing that Westinghouse *"take through his body the alternating current while I take through mine a continuous current."* Westinghouse refused. In 1889, Brown, funded by Edison, published a pamphlet stoking fears about the alleged dangers of AC electricity, calling it *"the executioner's current."* In fact, through Brown, Edison lobbied hard for the first electric chair in the United States to run on AC electricity. The lobbying was successful. Brown and Edison convinced the New York superintendent of prisons to adopt AC electricity for this purpose. The first use of the electric chair for capital punishment occurred on August 6, 1890, at Auburn Correctional Facility, New York. The criminal, William Kremler, was executed with AC electricity. The execution went awry, taking two and a half minutes and resulting in a charred and smoking body.

9.6.6 AC Rises Victoriously

Nevertheless, within a year of the execution, AC electricity had all but taken over. DC electricity could operate only a few appliances, whereas AC could operate more. AC was also much less expensive and could run on larger, more distant power supplies. With the adoption of AC electricity at the Chicago World's Fair in 1893 and at New York's Niagara Falls power station in 1895, AC electricity crushed any hope of a large-scale DC grid. AC was ultimately adopted worldwide.

The major reason for AC's victory was that AC power could be transmitted through long-distance HVAC lines without suffering the same line losses as DC power. AC line losses were minimized by using a transformer to step up the voltage (reducing the current) at the start of transmission and another transformer to step down the voltage (increasing the current) at the end of transmission. Transformers did not work with DC current. Thus, transformers settled the battle of DC versus AC electricity.

Since then, a method has been developed to transmit DC current long distances. However, even this method requires an AC power grid. The method is to boost AC voltage with a step-up transformer, then convert the HVAC current to high-voltage DC (HVDC) current. The HVDC current is then sent through a long-distance HVDC transmission line. At the end of the line, the HVDC current is converted back to an HVAC current. The AC voltage is then reduced with a step-down transformer. The world's first HVDC transmission line was installed in 1954 between mainland Sweden and the island of Gotland.

An advantage of using HVDC instead of HVAC transmission over long distances (greater than 600 kilometers) is that, although current is low and line losses are minimized in both cases, line losses are even less for HVDC than for HVAC over long distances. In addition, over long distances, HVDC is less expensive than is HVAC. For distances less than 600 kilometers, HVDC transmission lines are more costly than are HVAC lines because expensive AC-to-DC conversion equipment comprises a larger share of the overall HVDC cost than it does for long-distance transmission. The greater line losses of HVAC are also less of an issue over short distances.

Today, generators produce electricity with voltages of between 12,000 and 25,000 volts. Step-up transformers boost the voltage up to between 100,000 volts and 1 million volts for HVAC transmission. If long-distance HVDC is used, voltage is stepped up further to between 100,000 volts and 1.5 million volts. Step-down transformers decrease the voltage down to between 4,160 and 34,500 volts for distribution to the local utility. Local utilities then distribute AC power to neighborhood power poles. Transformers on power poles step voltage down further to 120 or 240 volts.

In the United States and many countries, home wall receptors receive 60 hertz AC power at 120 (110 to 125) volts. Dryers, electric car chargers, and electric water heaters, and some other appliances require 240 volts. In Europe and most countries, the power to homes is 50 hertz AC power at 240 volts.

9.7 Transmission and Distribution Losses

In an AC transmission and distribution system, AC electricity flows along a transmission line between an electric power generating facility and a step-up transformer station. The station boosts the voltage to produce HVAC electricity in order to reduce long-distance AC transmission losses. Along the HVAC line, AC electricity may or may not be converted to DC electricity for extra-long-distance HVDC transmission. At the end of an HVDC line, the DC electricity is converted back to HVAC electricity. The HVAC electricity is then transmitted to a step-down transformer station in a neighborhood, where the voltage is decreased, and the electricity is sent to local distribution lines. Electricity then goes to a transformer near buildings, where the AC voltage is dropped further for use in the buildings.

Losses along transmission and distribution lines arise through five factors. First, resistance along the lines converts some electricity to heat. Second, losses arise from step-up and step-down transformers that convert low-voltage to high-voltage AC electricity and back to low voltage again. Losses similarly arise in local transformers, which reduce voltage further at the end of distribution lines for electricity use in buildings. Third, losses occur in equipment converting HVAC electricity to HVDC electricity and back again. Fourth, losses arise from downed power lines. Fifth, in countries with transmission and distribution loss rates above 15 percent, electricity theft from power lines is a major source of loss.[190]

Of all transmission and distribution losses in the current energy system in countries without power theft, about 16 to 33 percent are short- and long-distance transmission losses from the electricity generator station to the step-down transformer substation, 32 to 40 percent are distribution losses between the step-down substation and the end user, and 27 to 52 percent are transformer losses.[191]

When HVAC electricity is converted to HVDC electricity for extra-long-distance transmission, transmission losses are reduced compared with HVAC transmission. For example, the overall loss of electricity along an 800,000-volt HVDC line ranges from 2.5 to 4 percent per 1,000 kilometers compared with twice that for an HVAC line.[192,193] However, a portion of the benefit of long-distance HVDC transmission is offset by losses arising from the HVAC-to-HVDC-to-HVAC conversion process. Such converter station losses are about 0.6 percent of energy transmitted.[192]

Overall transmission and distribution losses worldwide in 2014 ranged from lows of 2 percent in Singapore, 2.3 percent in Trinidad/Tobago, 2.5 percent in the Slovak Republic, and 2.7 percent in Iceland to highs of 60.1 percent in Haiti, 69.7 percent in Libya, and 72.5 percent in Togo.[194] Fifty-four percent of countries had transmission and distribution losses that were 10 percent or higher. Losses in some large countries and regions were as follows: China (4 percent), the United States (5.9), the European Union (6.4), the Russian Federation (10), Brazil (15.8), and India (19.4).

A lot of room exists for reducing transmission and distribution losses in many countries, particularly countries, such as Brazil and India, with losses exceeding 10 percent. Reducing losses decreases energy needs substantially. For example, if a country's current transmission and distribution loss rate is 10 percent, reducing the loss rate by one percentage point down to 9 percent reduces electricity generation requirements by 1.1 percent. If the current loss rate is 20 percent, a one-percentage-point reduction reduces generation requirements by 1.23 percent. If the loss rate is 72.5 percent, as in Togo, simply reducing the loss rate to 71.5 percent reduces generation requirement by 3.5 percent.

Finally, in a 100 percent WWS world, a portion of new electricity generation, including from offshore wind turbines, tidal turbines, ocean current devices, wave devices, and floating solar panels, will be offshore. Offshore renewables often require short transmission distances to load centers because most people in the world live along the coasts, and most offshore resources will be sited less than 200 kilometers offshore. As such, the growth of offshore renewables will increase the efficiency of the transmission and distribution system.

10 PHOTOVOLTAICS AND SOLAR RADIATION

Solar and wind will make up the bulk of a 100 percent wind–water–solar energy generation infrastructure worldwide. The main types of solar generation are from photovoltaics on rooftops and in utility-scale power plants, concentrated solar power plants, and solar thermal collectors for water heating. The sun produces enough energy worldwide to power the world with PV for all purposes in 2050 about 2,200 times over if all energy were electrified. Over land, PV can power all energy about 640 times over. Needless to say, the world needs only a small portion of this. For example, if half of the world's all-purpose power were from solar PV, only about 0.08 percent of the world's solar resource over land would be needed. Given the large potential of solar PV in particular for powering all energy needs, it is useful to understand PV panels and solar resources better. This chapter discusses both. The chapter starts with a detailed description of solar photovoltaic cells, panels, and arrays, and their efficiencies. It also discusses solar resource availability and optimal tilt angles for solar panels worldwide.

10.1   Solar Photovoltaics

A solar PV is a material or device that converts photons of sunlight to DC electricity. Each photon of sunlight breaks an electron in a PV material free of the atom that holds it. If an electric field is present, it sweeps the free electron toward a metallic contact, where the electron becomes part of an electric current in a circuit.

PV cells are made of semiconductor materials that convert sunlight to electricity. A **semiconductor** material has an electrical conductivity between that of a metal and a nonmetal. A **metal** is material in which electrons can readily break free from atoms, and thus conduct electricity, at any temperature but less so with increasing temperature. A **nonmetal** has very low electrical conductivity at all temperatures and thus is an insulator. A semiconductor has very low conductivity at low temperature but much higher conductivity at moderate and high temperatures. At near zero degrees Kelvin (−273.15 degrees Celsius), for example, silicon is a perfect insulator (it has no free electrons). As the temperature rises, some electrons can break free of the nucleus and flow in an electric current. The primary semiconductor materials used today are silicon (Si), germanium (Ge), gallium (Ga), and arsenic (As).

10.1.1 Brief History of the Solar PV Cell

Today's PV cells evolved from early discoveries of the properties of light interacting with materials and the invention of early PV cells that used the semiconductor materials selenium (Se) and cuprous oxide (Cu_2O).

In 1839, the 19-year-old French physicist **Edmond Becquerel** (1820 to 1891), working in his father's laboratory, discovered that shining sunlight onto one of two parallel plates of platinum, gold, or silver placed in an electricity-conducting solution produced an electric current and voltage.[195] Visible or ultraviolet light hitting electrons on one of the metal plates caused the electrons to absorb the light, energizing them. The added energy allowed them to escape their attraction to the atomic nuclei they were bound to. The free electrons flowed into the conducting solution, where they joined other electrons moving to the second plate, creating electricity in a circuit attached to the second plate. The generation of electricity by exposing one material to visible or ultraviolet sunlight, thereby causing the release of electrons into a different material, is called the **photovoltaic effect**. The term photovoltaic was named after the Italian physicist and chemist **Alessandro Volta** (1745 to 1827), who made substantial contributions to understanding electricity.

Capitalizing on Becquerel's discovery, **Augustin Mouchot** (1825 to 1912) of France registered patents for solar-powered engines in the 1860s. He also unveiled a solar-powered printing press at the 1878 University Exhibition in Paris.

In 1873, the English engineer **Willoughby Smith** found that selenium conducted electricity produced by sunlight. Shortly after, in 1876, the English engineers **William Adams** and **Richard Day** discovered that exposing a cylinder of solid selenium with platinum wires on each end to sunlight generated electricity through the photovoltaic effect.[196] This was the first time a material was able to convert sunlight to electricity without heat or moving parts.

In 1883, the American **Charles Fritts** (1850 to 1903) developed the first solar cell to generate electricity.[197] To do this, he pressed a thin film of selenium against a brass metal plate, then laid an even thinner layer of gold on top. The gold layer was so thin that sunlight could penetrate through it. Exposing the gold leaf to sunlight resulted in electricity generation. The efficiency of the cell, however, was only 1 to 2 percent. Werner Siemens, who subsequently confirmed Fritts' experiment, commented,[198]

> *In conclusion, I would say that however great the scientific importance of this discovery may be, its practical value will be no less obvious when we reflect that the supply of solar energy is both without limit and without cost, and that it will continue to pour down upon us for countless ages after all the coal deposits of the earth have been exhausted and forgotten.*

Fritts' invention spurred others to develop solar photovoltaic cells and panels as well as ways to improve exposure of panels to sunlight. In 1894, for example, the American inventor **Melvin Severy** (1863 to 1951) received separate patents for a solar photovoltaic panel that generates electricity and for a two-axis tracker for the panel.

In 1904, the German physicist **Wilhelm Hallwachs** (1859 to 1922) found that combining a plate of pure copper with a thin coating of cuprous oxide (which can be obtained by heating the copper plate in the presence of oxygen), and exposing the cuprous oxide/copper cell to sunlight, produced a current. Visible and ultraviolet light passing through the cuprous oxide (a semiconductor material) excited and released electrons that were then conducted away to a load by a wire. Thus, Hallwachs developed a semiconductor junction cell.

Dozens of photovoltaic cells containing copper and cuprous oxide were experimented on through 1932 by multiple groups.[199] However, because of the low efficiency of copper/cuprous oxide and the degradation of cuprous oxide upon its long-term exposure to sunlight, selenium became the material of choice in commercial cells during the 1930s. In 1932, Audobert and Stora discovered that cadmium sulfide, another semiconductor material, could be used to generate electricity through the photovoltaic effect.

The first silicon-based photovoltaic cell was patented in 1941 by the American engineer **Russell Ohl** (1898 to 1987). In 1939, Ohl discovered the p–n junction and how it worked by adding different impurities to 99.85-percent-pure monocrystalline silicon on each side of the junction. Ohl's PV cell was only 1 to 2 percent efficient.

A method of growing monocrystalline (single-crystal) silicon was first discovered in 1916 by **Jan Czochralski** (1885 to 1953), a Polish chemist. Czochralski made his discovery after accidentally dipping his pen in molten tin. This resulted in solid tin hanging from the nib of his pen. He verified that the crystal was **monocrystalline**, which is a solid material in which the crystal lattice is continuous and unbroken to the edges.

It was not until 1954 that **Daryl Chapin, Calvin Fuller,** and **Gerald Pearson**, working for Bell Labs, developed a more efficient silicon PV cell. This cell was 4 percent efficient and was the first that could provide sufficient electricity to run basic electrical equipment. Hoffman Electronics improved the efficiency of silicon cells to 8 percent in 1957 and 14 percent in 1960. In 1985, the University of New South Wales, Australia, increased silicon PV cell efficiencies to 20 percent. In 2021, the record efficiency for a monocrystalline silicon solar cell was 26.7 percent. That of a multicrystalline silicon cell was 24.4 percent.[200]

In 1955, Western Electric in the United States began selling licenses for products that ran on silicon PV cells. Such products included a dollar bill changer and a device that decoded computer punch cards and tape. In 1958, several space satellites, including the Vanguard I and II, Explorer III, and Sputnik-3, used solar PV cells to power some equipment. Satellites after that had larger arrays. In 1980 ARCO Solar produced more than 1 megawatt nameplate capacity of solar PV modules, the first time a company had done that.

In 1981 the *Solar Challenger*, an aircraft with over 16,000 solar cells, was built by Paul MacCready. It was the first solar aircraft and flew from France to England. In 1982, Hans Tholstrup drove the first solar car 4,500 kilometers from Sydney to Perth, Australia, in 20 days. In 1982, the first megawatt-scale PV power station went online in Hisperia, California. It had 108 dual-axis trackers.

10.1.2 How Silicon PV Cells Work

Pure silicon crystals consist of silicon (Si) atoms bonded together. Silicon has 14 total electrons, including 4 in its outer shell, called the **valence shell**. **Valence electrons** are the electrons that occupy the outermost shell of an atom and determine the charge of the nucleus. With four electrons (−4 charge) in its valence shell, silicon's nucleus has a +4 charge, so the overall atom is neutrally charged (zero net charge).

Each electron in a valence shell can bond with another electron in the valence shell of another atom. Thus, one silicon atom can bond with four other adjacent silicon atoms.

In 1888, Wilhelm Hallwachs, who later invented the copper/cuprous oxide solar cell, found that exposing a conducting metal to ultraviolet radiation caused the metal to obtain a net positive charge by gouging out an electron. This concept was the basis for what is called the **photoelectric effect**, which arises when photons of visible or ultraviolet radiation of sufficiently short wavelength hit electrons in metals, nonmetallic solids, liquids, or gases and release the electrons to the air or a vacuum. The electrons emitted in this way are called **photoelectrons.**

The photoelectric effect is closely related to but differs from the photovoltaic effect in that, with the photoelectric effect, electrons are ejected out of the material into the air or a vacuum; with the photovoltaic effect, electrons are ejected into a new material to produce a current of electricity. Modern photovoltaics are based on the photovoltaic effect rather than the photoelectric effect, although the main concept of electron release is the same in both cases.

In 1905, **Albert Einstein** (1879 to 1955) published a theoretical paper explaining the photoelectric effect. After **Robert Millikan** (1868 to 1953) proved the effect experimentally in 1916, Einstein won the Nobel Prize for his theory in 1921.

The photoelectric effect is explained with respect to semiconductors as follows. Semiconductor atoms have three main energy bands: a valence band, a forbidden band, and a conduction band. The **valence band** of an atom is the energy band occupied by the valence electrons. If excited with enough energy from sunlight, an electron in the valence band can jump into the conduction band, which is a band of energy that is otherwise vacant. Thus, the valence and conduction bands hold electrons of different energy levels, with the valence band holding electrons with less energy and the **conduction band** holding electrons with more energy.

The **forbidden band** is an energy gap between the conduction band and the valence band. When an electron obtains sufficient energy, it jumps from the valence band into the conduction band, bypassing the forbidden band. As such, no electrons reside in the forbidden band. In order to jump to the conduction band, an electron in the valence band must obtain enough energy, called **band-gap energy**. It is the energy required for a valence electron to free itself from the electrostatic force holding it to its own nucleus to jump into the conduction band. The band-gap energy is the energy needed for the photoelectric effect to occur.

Electrons in the conduction band contribute to current flow. For metals, the conduction band is partly filled naturally, even at low temperature, owing to the fact that some electrons in metals are thermally excited enough to break free of their nucleus at low temperature. For semiconductors, the conduction band is empty at zero degrees Kelvin (−273.15 degrees Celsius), the coldest temperature possible in the universe. At room temperature, only one out of 10 billion electrons in the valence band of a semiconductor is thermally excited enough to jump into the conduction band.

When an electron jumps out of the valence band of a silicon atom, it leaves a +4 nucleus with only three electrons, so the silicon atom has a net +1 positive charge, or a **hole**. Soon, another electron will recombine with the positively charged silicon atom to fill the hole. Since that electron comes from another silicon atom in the lattice of silicon atoms, the hole (location of positive charge) shifts to the other atom. As such, holes in the lattice appear to move around.

The **band-gap energy**, as noted above, is the energy needed for a semiconductor's electron to jump from its valence band to its conduction band The unit of band-gap energy is the **electron volt**, which is the energy an electron acquires when voltage increases by 1 volt. Silicon has a band-gap energy of 1.12 electron volts. Thus, a wavelength of sunlight must supply at least 1.12 electron volts to propel an electron from silicon's valence band into its conduction band.

The shorter a wavelength of sunlight, the more energy the wavelength contains. Thus, at some long wavelength, the energy contained in the wavelength is less than the band-gap energy. The solar wavelength above which the energy contained in the wavelength is less than the band-gap energy is called the **band-gap wavelength**. For silicon, the band-gap wavelength is 1.11 micrometers (millionths of a meter). Thus, every wavelength of sunlight below 1.11 micrometers can promote an electron from the valence band to the conduction band. Every wavelength above this size contains too little energy to promote an electron.

At each wavelength below the band-gap wavelength, only 1.12 electron volts is used for promotion. Since wavelengths below the band-gap wavelength contain more energy than the band-gap energy, the extra energy beyond the band-gap energy is wasted. This is one reason why solar panels can never be 100 percent efficient at converting sunlight to electricity: only a fraction of the energy at each wavelength below the band-gap wavelength is used to promote an electron into the conduction band. In addition, none of the wavelengths above the band-gap wavelength have enough energy to promote an electron.

The band-gap wavelength of 1.11 micrometers is a solar infrared wavelength. The solar spectrum includes ultraviolet wavelengths (below 0.38 micrometers), visible wavelengths (0.38 to 0.75 micrometers), and solar infrared wavelengths (0.75 to 10 micrometers). Thus, silicon PV cells generate electricity when they are exposed to ultraviolet, visible, and some solar infrared wavelengths. Ultraviolet wavelengths comprise only about 5 percent of total solar energy.

10.1.3   Maximum Possible PV Cell Efficiency

Solar cells use the band-gap energy in solar wavelengths below the band-gap wavelength to propel valence band electrons into the conduction band. Only wavelengths below the band-gap wavelength are used, and at each wavelength, only the band-gap energy is used. Thus, the amount of solar energy that is useful for generating electricity in a PV cell is limited.

For silicon, a maximum of only 49.6 percent of total sunlight can be converted to useful band-gap energy. This is because 20.2 percent of radiation is above the band-gap wavelength and thus unavailable, and 30.2 percent of radiation that is below the band-gap wavelength is above the band-gap energy. Of the remaining 49.6 percent, seven percentage points are lost because of high temperatures, which result in heat release, and another nine percentage points are lost because some of the electrons recombine with positively charged ions.

The resulting maximum possible efficiency of a single junction PV cell is, therefore, about 33.7 percent, which is called the **Shockley–Queisser limit.**

The solar cell efficiency is the actual power output of a solar cell obtained under standard test conditions divided by the maximum solar power available under those test conditions. **Standard test conditions** are conditions under which solar PV panels industry-wide are tested and compared with each other to evaluate their efficiency. As of 2021, the record efficiency for a single-junction solar cell, which is made of monocrystalline silicon, was 26.7 percent,[200] thus about 79.2 percent of the maximum possible efficiency.

10.1.4   Creating Electricity in a PV Cell

Electric fields are built into PV cells to carry away electrons boosted from the valence band to the conduction band, preventing them from recombining with silicon atom holes. Electric fields push electrons in one direction and holes in the other. To create an electric field in today's silicon-based PV cells, one side of the cell is contaminated with one atom of phosphorus (P) per thousand atoms of silicon and the other side is contaminated with one atom of boron (B) per thousand atoms of silicon. Phosphorus has five electrons in its valence shell. Boron has three electrons in its valence shell.

Each phosphorus and boron atom substitutes for one atom of silicon in the lattice of the PV cell (Figure 10.1). Since phosphorus has five electrons in its valence band, it can form five bonds. Silicon can form only four bonds. As such, when phosphorus replaces a silicon atom in a silicon lattice, phosphorus has a free fifth electron in its valence band. At room temperature, this free electron escapes the atom and roams, giving the immobile phosphorus atom a positive charge. Phosphorus, however, is called an **n-type** (negative-type) material since it donates a negatively charged roaming electron.

Since boron has three electrons in its valence shell, it can form only three bonds with silicon. When boron is substituted into the lattice of silicon, boron forms bonds with three out of its four silicon neighbors. The boron tries to bond to the fourth neighbor, first by moving its own three valence band electrons around, then by borrowing an electron from a nearby silicon atom. This results in the immobile boron becoming net negatively charged and a positive hole forming in

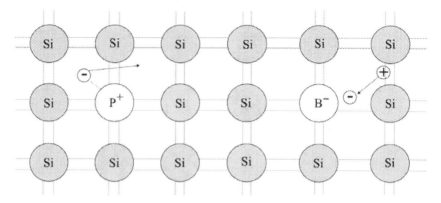

Figure 10.1 Diagram showing the substitution of a phosphorus (P) atom (left) and boron (B) atom (right) for a silicon (Si) atom in the lattice of a PV cell. The P atom, which attains an immobile positive charge, is an n-type material since it donates a negatively charged roaming electron. The B atom, which attains an immobile negative charge, is a p-type material since it results in a positively charged roaming hole. Adapted from Karolkalna at the English Wikipedia.

a nearby silicon atom where the electron came from. Since a roaming electron from another silicon atom soon fills the hole, the hole roams among silicon atoms. Because the boron creates a roaming hole (positive charge), boron is called a **p-type** material.

Mobile electrons from the n-type (phosphorus) side of the PV cell drift by diffusion toward the p-type side. Similarly, mobile holes from the p-type side roam toward the n-type side. In the middle, mobile electrons fill mobile holes, creating uncharged, electrically neutral atoms. This no-charge **depletion region** is about 1 micrometer wide and is called a **p–n junction**.

In the depletion region, the immobile positively charged phosphorus atoms on the n-type side of the p–n junction are no longer balanced by roaming negative charges. As such, that region takes on a net positive charge. Similarly, the immobile negatively charged boron atoms on the p-type side of the p–n junction are no longer balanced by roaming holes, so that side takes on a net negative charge. As such, the n-type side of the depletion region now has immobile positive charges and the p-type side now has immobile negative charges, giving rise to a charge gradient. When an external wire is connected from a cap at the end of the p-type region to a cap at the end of the n-type region, the charge gradient gives rise to an electric field that can sweep away electrons in the conduction band.

In a PV cell, the n-type layer can be on top (closest to the sun) or on bottom. The most efficient solar cells generally have a p-type layer on top and an n-type layer on bottom, with a p–n junction in between. In either case, photons of sunlight energize electrons in the valence band of silicon atoms in the layer on top, moving the electrons into the conduction band. If these drift into the depletion region, the electric field sends the electrons to the n-type layer. If the wire attached to the PV cell is connected to a load, electrons flow out of the n-type side into the wire and through the load and back to the p-type side. Electrons reaching the p-type side recombine with holes there, completing the circuit.

10.1.5   Types of PV Cells and Their Materials

Silicon is the second most abundant element on Earth, comprising 20 percent of Earth's crust. The source of silicon for PV cells is high-quality silicon dioxide (SiO_2), also called silica and quartz, that comes from sand or a mine. Silicon processing begins with a high-temperature electric arc furnace that currently uses carbon to reduce silicon dioxide to metallic grade (99 percent pure) silicon. Silicon is purified further until it is more than 99.9999 percent pure for use in solar PV cells.

First-generation PV cells, which are still widely used, are 160 to 240 micrometers thick. They are made of either **single-crystal silicon** (also called monocrystalline silicon) or **polycrystalline silicon** (also called multicrystalline silicon).

Single-crystal silicon cells are uniform in structure and appear square but with clipped corners, so are really octagonal. They have a distinct pattern of small white diamonds. Polycrystalline silicon cells resemble rock-like chunks of a multifaceted metal. Because single-crystal silicon cells are more uniform in structure, they are more efficient but also more expensive to manufacture than are polycrystalline cells.

Second-generation PV cells, called **thin-film PV** cells, are 1 to 10 micrometers thick. Thin-film cells are much thinner than are first-generation cells, reducing the amount of expensive material needed. Thin-film PV cells usually consist of either **amorphous silicon** or a material with two to four elements, such as gallium arsenide, cadmium telluride, copper sulfide, cadmium sulfide, or copper–indium–gallium–selenium. Thin-film materials are wedged between two panes of glass. Since glass is heavy and first-generation cells use only one pane of glass, second-generation cells are heavier than are first-generation cells. However, the energy required to produce first-generation cells is greater.

Copper–indium–gallium–selenium and gallium arsenide cells are both efficient. However, they are also expensive. Thus, despite their higher efficiency, they result in higher overall energy cost than do first-generation silicon-based cells. Cadmium telluride cells, on the other hand, are less efficient than are first-generation cells but cost less per cell and have an overall energy cost comparable with that of first-generation cells.

Thin-film solar cells were invented to power the first solar calculators in the late 1970s. They contained a small strip of amorphous silicon. In 1980, the University of Delaware developed the first thin-film solar cell for electricity generation. The efficiency exceeded 10 percent with a cell made of copper sulfide and cadmium sulfide. In 1992, the University of South Florida increased the efficiency of thin film to 15.9 percent with a cell made of cadmium telluride. As of 2021, the record efficiency of a thin-film cadmium telluride cell was 21 percent. The record for thin-film copper–indium–gallium–selenium cells was 23.34 percent. That for thin-film gallium arsenide cells was 29.1 percent.[200]

Amorphous silicon is made by chemical vapor deposition of silane gas and hydrogen gas. As such, it is noncrystalline or has tiny crystals about a micrometer in size (thus called microcrystalline). Amorphous silicon has a much greater band-gap energy than does monocrystalline silicon. Thus, amorphous silicon absorbs across a much smaller portion of the solar spectrum (has a lower band-gap wavelength) than does monocrystalline silicon. As a result, amorphous silicon also has a much lower efficiency (10.2 percent) than does monocrystalline silicon (26.7 percent).[200]

Third-generation cells are thin-film cells that are either emerging or still too expensive for widespread commercial use. Three of these are as follows:

Organic cells (cells that contain organic material). These cells tend to be less expensive per cell than silicon-based cells but also have lower efficiency. As of 2021, the record efficiency of a thin-film organic cell was 18.2 percent.[200] In addition, sunlight photochemically degrades organic-based cells over time.

Multijunction (tandem) cells are either individual thin-film PV cells containing multiple materials or a stack of thin-film PV cells, each made of a different material, connected in series. In both cases, each material absorbs a different portion of the solar spectrum (thus has a different band-gap energy and band-gap wavelength). For example, with a two-band-gap stack of cells, the top cell may have a high band-gap energy, and thus absorb more energy per photon, but only up to a limited wavelength. The bottom material may have a low band-gap energy, and thus absorb less energy per photon, but over more wavelengths in the solar spectrum.

Whereas the theoretical peak efficiency of a single-junction PV cell is about 33.7 percent, that of a stack of two cells with different band gaps is around 47 percent. The theoretical peak efficiency of a stack of three cells is around 53 percent; and that of a stack of eight cells is about 62 percent.[201] To date, tandem cells have been used primarily in satellite PV arrays and in technologies that use lenses and mirrors to focus light onto a cell to increase the intensity of sunlight hitting the cell.

In 1994, the U.S. National Renewable Energy Laboratory developed a gallium–indium phosphide and gallium arsenide double-junction cell that was the first cell with an efficiency exceeding 30 percent. As of 2021, the record for this double-junction cell was 32.8 percent and that for a double-junction perovskite–silicon cell was 29.5 percent.[200] A **perovskite** is a material with a crystal structure similar to the crystal perovskite, which contains calcium titanium oxide. Common perovskite materials used in solar cells are hybrid organic–inorganic lead or tin halide-based materials. The record for a triple-junction cell was 37.9 percent and that for a five-junction cell was 38.8 percent.[200]

Concentrator cells are solar PV cells in which lenses or curved mirrors are used to focus light onto multijunction cells. The efficiency of a concentrator cell is larger than that of any other cell type. The record efficiency for any cell measured in 2021, for example, was 47.1 percent, which was for a four-junction concentrator cell.[200] Concentrator cells, however, require direct sunlight, just like concentrated solar power does, so do not work well under cloudy or hazy conditions where direct light is scattered to become diffuse light. Thus, concentrator cells are limited to desert-like areas with high direct sunlight. Also, the concentrated light increases cell temperature more than in conventional cells, so a heat sink is needed. Concentrator cells may also need more frequent cleaning than conventional cells.

Because so many alternative PV cell types are available, it is unlikely that materials will constrain the large-scale growth of PV. For example, in multijunction cells, the limiting material is germanium; however, substituting gallium, which is more abundant, would allow terawatt expansion. In addition, the production of silicon-based PV cells is limited not by crystalline silicon (because silicon is widely abundant) but by silver, which is used as an electrode. Reducing the use of silver as an electrode would allow the virtually unlimited production of silicon-based solar cells. In sum, the development of a large global PV system will not be limited by the scarcity or cost of raw materials.

10.1.6   PV Panels and Arrays

A **PV panel** consists of 32, 36, 48, 60, 72, 96, or 128 pre-wired PV cells in series, fitted into a rectangular package. A **PV array** is a group of PV panels wired in series to increase the voltage, or in parallel to increase the current. For panels wired in series, the total voltage is the sum of the voltages across all individual panels, and the total current is the current running through any one panel. For panels wired in parallel, the total current is the sum of currents through each panel, and the total voltage is the voltage across one panel.

Alternatively, an array can be wired partly in series and partly in parallel to optimize power (voltage multiplied by current). In this case, panels are first wired in series to increase the voltage by as much as is safe, then each series string is wired in parallel to maximize power. Maximizing voltage also minimizes current, minimizing power losses along wires. When arrays are set up in this manner, the total voltage is the sum of voltages across one string of arrays. The current through the string is the current through one panel in the string. The total current through the array is the sum of currents through each parallel string.

10.2   Solar Resources and How to Use Them Efficiently

10.2.1   Solar Resources

In the annual average, incident solar radiation reaching the Earth's surface peaks near the equator and generally decreases between the equator and the north and south poles. However, a slight increase in surface solar radiation occurs at the poles relative to at 60 degrees north and 60 degrees south of the equator, owing to lower cloud cover in polar regions than at those latitudes. Similarly, in the United States, more sunlight reaches the ground in the southwest than in the southeast owing to the greater cloud cover in the southeast. The Sahara desert also receives a substantial amount of sunlight at the ground owing to low cloud cover there.

Worldwide, about 97,000 terawatts of solar power hit the Earth's surface in the annual average, after accounting for clouds, gases, and particles in the air, but before accounting for any reflection by the ground. Of this downward sunlight, about 28.7 percent hits land.

If the Earth's land and ocean were covered completely with horizontal solar PV panels with an efficiency of 20 percent (with no spacing between panels and ignoring any feedbacks of the panels to climate), the global electricity available from solar PV worldwide would then be about 19,400 terawatts. That over land would be about 5,600 terawatts.

If the world's all-purpose energy demand were electrified, the annual average end-use power demand in 2050 would be about 9 terawatts. As such, enough solar theoretically exists over land alone to provide the world's 2050 end-use power 620 times over. However, many locations over land, such as Antarctica and the Himalayan Mountains, are not easily accessible to transmission. In addition, panels require walking space between rows and space for inverters, which convert PV panel DC electricity to AC electricity that is sent to the grid. In addition, space is needed for agriculture and forests. These factors reduce the practical solar installation potential. The world's likely developable solar PV resource over land is thus about 1,300 terawatts. As such, about 144 times the world's all-purpose end-use demand is available from accessible solar PV over land.

Fortunately, the small amount of sunlight incident on flat (horizontal) panels in high northern and southern latitudes does not limit the installation of solar PV there. The reason is that incident sunlight on a panel is magnified by up to a factor of 2.6 if the panel tracks the sun or is tilted. As such, solar PV can be used to generate substantial electric power almost anywhere on Earth.

Floating solar PV can also be used effectively over inland waters, which are relatively calm, to generate electricity. Ocean surfaces are more volatile because of high waves, and corrode panels more because of salt. Nonetheless, breakwaters and jetties can be constructed to create areas of calm water, just like in a harbor. Floating solar PV can be installed over these calm waters. Since most people live near the coast, offshore floating solar arrays are expected to grow significantly. This is particularly the case in countries such as Gibraltar, Malta, and Singapore, for example, that have high populations and are land constrained.

10.2.2   Sunlight Reaching Solar Panels

PV panel output depends a lot on the sunlight reaching the panel. The radiation reaching a panel at a given time of day and year depends on the amount of sunlight available on that day; the quantity of gases, aerosol particles, and clouds in the atmosphere; shading by hills, mountains, trees, and buildings; and the orientation of the panel relative to the direct solar beam. In this section, these factors are discussed, aside from panel orientation, which is discussed in the next section.

The amount of sunlight reaching the outside of the Earth's atmosphere varies with day of the year because the Earth's distance from the sun varies each day. The average Earth–sun distance is about 150 million kilometers. But that distance is 147.1 million kilometers on December 22 (Northern Hemisphere winter solstice) and 152.1 million kilometers on June 22 (Northern Hemisphere summer solstice) because the Earth rotates around the sun in an elliptical orbit with the sun at one focus. This 3.4 percent difference in Earth–sun distance between December and June corresponds to a 6.9 percent difference in sunlight reaching the outside of Earth's atmosphere between these months. In other words, 6.9 percent more radiation falls on the Earth in December than in June. Although more sunlight falls on the Earth during the Northern Hemisphere winter than summer, the north pole is tilted away from the sun during December, so most of that sunlight falls on the Southern Hemisphere, keeping the Northern Hemisphere cold during December.

Sunlight is relatively unimpeded as it travels from the sun to the outside of the Earth's atmosphere. However, when it passes through our atmosphere, it interacts with gases, aerosol particles, and clouds. These obstacles may either scatter or absorb the light or allow the light to transmit through. **Scattering** is the process by which gases, aerosol particles, or cloud particles deflect radiation in a random direction, just like a tree deflects a rock thrown at it. All gases, aerosol particles, and cloud particles in the atmosphere scatter radiation to some degree. The amount of scattering depends on several factors and varies with wavelength of light. **Absorption** is the process by which sunlight enters a gas or particle in the air, and the light is converted to invisible heat, which is then re-radiated to the surrounding air. Only a few gases, such as nitrogen dioxide and ozone, absorb sunlight. Very few particle types, such as black carbon, brown carbon, and iron-containing particles, absorb sunlight.

Sunlight that is unimpeded by scattering and absorption, so makes its way directly to a solar panel, is called **direct sunlight**. Sunlight that is scattered out of the direct solar beam by gases and particles, but then encounters additional gases and particles that scatter some of the light to the solar panel, is called **diffuse sunlight**. For example, if the sun is close to the horizon, the direct solar beam may intersect the underside of a cloud, which scatters the light. Some of that scattered diffuse sunlight may then hit a solar panel.

Cloud particles, aerosol particles, and gases all decrease direct sunlight by scattering light out of the direct beam or absorbing light along the beam, but they all increase diffuse radiation as well. Hills, mountains, trees, and buildings also decrease direct sunlight. So long as the surfaces are not black, they also reflect light, increasing diffuse radiation. On average, the quantity of diffuse radiation to the ground is about 10 percent of the amount of direct sunlight reaching the ground. However when the sun is blocked by heavy clouds, pollution, or shading by hills, mountains, trees, or buildings, nearly 100 percent of sunlight reaching a panel is diffusion radiation.

10.2.3   Solar Panel Tilting and Tracking

If a solar PV panel sits flat (lies horizontally) on the ground, then it receives the maximum possible radiation from the sun only if the sun is directly overhead. The sun can shine directly over a horizontal panel only if the panel lies between 23.5 degrees north of the equator (Tropic of Cancer) and 23.5 degrees south of the equator (Tropic of Capricorn). The reason is simply that the Earth's axis of rotation is tilted at 23.5 degrees relative to a line perpendicular to the sun, so the sun is directly overhead the Tropic of Cancer on June 22, the Tropic of Capricorn on December 22, and all other latitudes in-between on every other day. If the Earth had no tilt, the sun's light would be directly over the equator every day.

However, if the face of a solar panel follows the sun by swiveling east–west and tilting up and down, the panel will receive the maximum possible direct beam sunlight, in addition to receiving diffuse sunlight, every minute of every day (but not at night).

During sunset or sunrise, a panel sitting flat on the ground receives no direct sunlight since direct radiation flies right over the panel, parallel to it. Instead, the panel receives only diffuse radiation. If the panel instead tracks the sun, the sunlight, even at sunset, will hit the panel straight on. As such, tracking the sun with a panel maximizes solar PV panel output.

Even tilting panels at an optimal tilt angle, without swiveling the panels east–west, improves solar output relative to flat panels. An **optimal tilt angle** of a solar panel is the estimated fixed tilt angle of the panel relative to the horizon that gives the greatest annual average incident sunlight on a panel compared with any other fixed tilt angle. The optimal tilt angle depends not only on latitude, but also on cloud cover, elevation, and air pollution levels.

Rooftop panels are mostly placed at a fixed tilt angle relative to the sun although some lie flat. Utility-scale solar panels are almost always at a fixed tilt angle or move to track the sun. Tilted panels generally, but not always, face south, southwest, or southeast in the Northern Hemisphere or north, northwest, or northeast in the Southern Hemisphere.

Panels that track the sun can track either on one axis vertically, one axis horizontally, or two axes. One-axis vertical-tracked PV panels are panels that face south or north and swivel vertically (up and down) around a horizontal axis. Thus, they follow the sun as it goes from the horizon to its highest point in the sky, but not from east to west. One-axis horizontal tracked panels are tilted at an optimal tilt angle and swivel horizontally between east and west around a vertical axis. Thus, they follow the sun from east to west but not from the horizon to the highest point in the sky. Two-axis tracked PV panels combine one-axis vertical and one-axis horizontal tracking capabilities to follow the sun perfectly during the day.

Whether panel tracking or tilting is used depends not only on how much sunlight the panel receives but also on the cost of tracking equipment and of the land or roof area needed to avoid shading.[202] For example, two-axis tracking of solar panels in a utility PV farm requires more land area to avoid shading panels behind the front row of panels than does one-axis tracking or optimal tilting. Two-axis tracking equipment is also more expensive than is one-axis tracking equipment or optimal tilting equipment

Shading depends not only on panel tilting, but also on the height that panels are placed relative to each other. For example, panels that track the sun placed on a south-facing hillside will likely see less shading than will panels on uniformly elevated ground. Shading further depends on the number of panels placed on each single platform that tracks the sun.

10.2.3.1 Optimal Tilt Angles

Optimal tilt angles increase with increasing latitude from equator to pole in both hemispheres. The higher the latitude, the less that sunlight is perpendicular to the surface of the Earth. Thus, panels need to be tilted more and more toward the low-lying sun with increasing latitude. The simplest estimate of the optimal tilt angle of a PV panel at a given latitude is the latitude itself. Thus, at 45 degrees north latitude, a PV panel should face south with a tilt angle of 45 degrees relative to the horizon. However, at high latitudes, where clouds are prevalent, optimal tilt angles are not determined so simply.

Transition highlight
At the same high latitude but at different longitudes, optimal tilt angles can differ significantly from each other. For example, Calgary (51.12 degrees north latitude) has a higher optimal tilt angle (45 versus 34 degrees) than does Beek, the Netherlands (50.92 degrees north latitude), which is at a similar latitude. The reason is that Calgary is exposed to less cloud cover than is Beek so panels can take advantage of the overhead sun more efficiently in Calgary. In Beek, panels are exposed to less direct sunlight owing to heavier cloud cover so must take advantage of diffuse light scattered by clouds above them. The lower the optimal tilt angle, the more such panels can receive diffuse light under clouds.

10.2.3.2 Impacts of Tilting and Tracking versus Horizontal Panels on Solar Output

In the global and annual average, two-axis tracked panels receive about 39 percent more sunlight than do horizontal panels. Similarly, one-axis horizontal-tracked panels and one-axis vertical-tracked panels receive about 35 percent and 22 percent, respectively, more sunlight than do horizontal panels. Optimally tilted panels receive about 19 percent more sunlight than do horizontal panels.[203] These percentages vary with latitude.

At virtually all latitudes, one-axis horizontal-tracking panels receive almost the same (within 1 to 3 percent) amount of sunlight as two-axis tracked panels. Because one-axis tracked panels require less land and cost less for nearly the same output as two-axis panels, one-axis horizontal-tracking panels are preferred over two-axis tracking panels. Moreover, one-axis horizontal-tracking panels are more efficient than, and thus preferred over, one-axis vertical-tracking panels. Optimally tilted panels receive less sunlight than do one-axis horizontal-tracking panels, but optimal tilting is still preferred on rooftops and in many solar arrays to minimize space needs and to eliminate the cost of tracking equipment.

In sum, ignoring land needs or cost of tracking equipment, the default recommendation for utility-scale PV is for one-axis horizontal tracking for all except the highest latitudes, where optimal tilting appears sufficient.[203] For rooftop PV, optimal tilting is recommended.

Transition highlight

Where is the sunniest place on Earth from a solar panel's perspective? Over the South Pole, horizontal panels receive relatively little radiation. However, optimally tilted and tracked panels there receive more sunlight, averaged annually, than anywhere else on Earth. This is primarily due to the fact that the South Pole receives sunlight 24 hours per day during the Southern Hemisphere summer. Also, much of the Antarctic is at a high altitude, thus above more air and clouds than over the Arctic or other latitudes. Tilted and tracked panels over the Arctic receive more radiation than between 40 to 80 degrees north latitude, but less than over the Antarctic, owing to the lower altitude and greater cloudiness above the surface of the Arctic.

11 ONSHORE AND OFFSHORE WIND ENERGY

After solar energy, onshore and offshore wind energy have potential to supply the greatest portion of the world's all-purpose energy demand. Not only are wind resources abundant in almost every country of the world, but the cost of onshore wind energy has also declined so much that wind in 2022 is the least expensive form of new electricity in many countries. Its low cost has resulted in massive wind installations to replace fossil-fuel power plants and to supply new energy demand. This chapter discusses the history of windmills and wind turbines followed by a discussion of the different types and components of wind turbines. The chapter also discusses how wind turbines work and how generators convert the energy from a rotating blade into electricity. Finally, the chapter discusses wind farm footprint and spacing areas needed to power the world, wind resources available worldwide, and the impacts of wind turbines on global wind speeds and temperatures as well as on hurricanes, birds, and bats.

11.1   Brief History of Windmills and Wind Turbines

People often confuse windmills with wind turbines, but they are different. A **windmill** converts kinetic energy from the wind into mechanical energy. A **wind turbine** converts the wind's kinetic energy into electricity.

The first known use of a circular wheel driven by the wind to provide mechanical energy was in the first century A.D. The engineer,

Heron of Alexandria (circa 10 A.D. to circa 70 A.D.) in Roman Egypt, invented such a windmill to provide mechanical power for a musical instrument, an organ. Windmills were also developed in the fourth century A.D. in India, Tibet, and China to rotate a prayer wheel. Between the seventh and ninth centuries A.D., the Persians, near the border of present-day Iran and Afghanistan, developed a windmill with 6 to 12 rectangular sails rotating horizontally around a **vertical axis** (an axis that is perpendicular to the ground) to pump water and to grind grain. The use of windmills for these purposes spread across the Middle East, Central Asia, India, and China. By 1000 A.D., the Chinese and Sicilians used windmills to pump seawater to make salt. Windmills rotating around a **horizontal axis** (parallel to the ground) were developed to grind flour in northwestern Europe beginning around 1180 A.D.

Windmills were popularized by Miguel de Cervantes in his 1604 novel, *Don Quixote*. In it, a man believes he is a knight named Don Quixote. He travels around Spain by horse with his squire, Sancho Panza. In one adventure, Don Quixote encounters 30 to 40 windmills on the plain in front of him but imagines they are giants. He tells Sancho he wants to slay the long-armed giants. Sancho tells him that they are not giants but windmills that rotate in the wind, turning a millstone, which is a circular stone used to grind grain. Don Quixote does not believe Sancho. Instead, Quixote believes that a magician changed the windmills into giants to hurt him. Quixote goes on to joust with one but is knocked off his horse by it. This interaction gave rise to the term *"tilting at windmills,"* which means to attack imaginary enemies.

During the American Revolution in the United States, a windmill was built on Cape Cod to pump seawater for making salt. On August 29, 1854, the inventor **Daniel Halladay** (1826 to 1916) patented the first commercially viable windmill worldwide, which he built in his Connecticut machine shop. The differences between his and previous windmills were that he added a tail fin (**wind vane**) to allow his windmill automatically to change direction to face the wind, and his windmill maintained a constant speed by changing the **pitch** (steepness of slope) of its sails without human oversight. Halladay's windmills were sold by the thousands. They were used to pump water and grind grain on farms and ranches and to help open up the western United States to rail transport. Trains required water for their steam engines, and windmills were used to pump water along rail routes for this purpose. Between 1854 and 1970, over 6 million windmills were installed in the United States.

In July 1887, **Professor James Blyth** (1839 to 1906), an electrical engineer at Anderson's College, Glasgow, built and operated the world's first wind turbine. The turbine used cloth sails to turn a rotor. It was used to produce electricity that charged an energy storage device, and the stored electricity was used to light up his cottage.

Near the same time, between 1887 and 1888, **Charles F. Brush** (1849 to 1929) of Cleveland, Ohio, built a much larger wind turbine with a unique design. It was 17 meters in diameter, had 144 rotor blades, and was made of cedar wood. The blades were mounted on an 18-meter tower and generated a peak of 12 kilowatts of electricity that he used to charge 12 batteries, which in turn powered 100 light bulbs, 3 arc lamps, and several motors in his mansion. The wind turbine and battery bank operated for 20 years.

In 1891, **Paul la Cour** of Denmark invented a wind turbine that he used to electrolyze water to produce hydrogen gas. The hydrogen was used to power several hydrogen gas lamps.

Subsequently, between 1900 and 1940, hundreds of thousands of small wind turbines, with nameplate capacities of 5 to 25 kilowatts, were installed to produce electricity in rural areas of the United States. These areas lacked access to the electric grid.

Large-scale commercial wind turbine installations began in California. Between 1981 and 1986, the state installed 15,000 wind turbines in three locations – **Altamont Pass, Tehachapi Pass**, and **San Gorgonio Pass**. The three farms had a combined nameplate capacity exceeding 1 gigawatt. Today, Altamont Pass is still the largest wind farm in the world in terms of number of turbines (almost 5,000), but no longer in terms of nameplate capacity. From 1990 to 2000, the center of wind farm development moved to Europe. During that period, 10 gigawatts were installed in Europe, compared with 2.2 gigawatts in the United States.

In 1991, the first offshore wind turbine, with nameplate capacity of 225 kilowatts, was installed in water 25 meters deep offshore of Sweden. Denmark installed the first offshore wind farm in 1992. It consisted of eleven 450-kilowatt turbines in 2 to 4 meters of water, 3 kilometers from shore.

Since the year 2000, onshore and offshore wind turbine development has taken off worldwide.

11.2   Types of Wind Turbine

Wind turbines that spin vertically around a horizontal axis are called **horizontal-axis wind turbines**. They are by far the most common type of turbine today. Wind turbines that provide grid electricity are usually mounted on tall towers, with a hub height of 80 to 150 meters. The **hub height** is the height above the ground or water surface where the horizontal axis, or **rotor**, of a turbine is located. Some horizontal-axis turbines have low hub heights for use in backyards and in locations where wind turbine heights or nameplate capacities are restricted.

Almost all horizontal-axis wind turbines today are **three-blade turbines**, and all face the wind (thus are **upwind turbines**). In the past, some horizontal-axis turbines faced away from the wind (thus were **downwind turbines**). The advantage of a downwind turbine is that the wind naturally rotates the turbine horizontally to the exact opposite direction of the wind. The horizontal rotation of a horizontal-axis wind turbine around its vertical axis is called **yaw control**. Yaw control with downwind turbines is automatic, so does not require mechanical equipment. With upwind turbines, however, complex mechanical equipment is needed to rotate the turbine blades continuously to face the changing wind direction. Wind tails, such as those used on the Halladay windmill, can help with yaw control only for small turbines.

The main problem with downwind turbines, though, is that they do not receive power when each blade passes by the tower, since the tower itself blocks the wind. Also, because the wind speed is first high, then zero, then high again as each blade passes the tower, the blades are subject to enormous **wind shear** (variation of wind speed with distance). Wind shear increases wear and tear on and failure of the blades. Because of these two problems (power loss and wear), no downwind turbines are being built today.

Another type of horizontal-axis turbine that was previously used but is no longer manufactured is the **two-blade turbine**. Advantages of the two-blade over the three-blade turbine are that the fewer blades a turbine has, the less expensive the turbine is to build and the less each blade's turbulence causes wear, tear, and efficiency loss on the next blade. The reduced turbulence in a two-blade system also causes the blades to spin faster, decreasing generator costs. For these reasons, a two-blade turbine would seem to be less costly than a three-blade turbine. However, three-blade turbines absorb horizontal and vertical wind shear more evenly, thus are more efficient than and require fewer repairs and less down time than do two-blade turbines. Finally, because of their lower spin rate and more even distribution of wind shear, three-blade turbines are quieter. For those reasons, only three-blade turbines are manufactured on a large scale today.

Wind turbines that spin horizontally around a vertical axis are called **vertical-axis wind turbines**. These turbines are usually short so can be used in urban regions or regions with height restrictions. An early vertical-axis turbine was the **Darrieus turbine**, which had two or more curved blades attached at each end to the top and bottom of a vertical rotating shaft. This turbine looked like an eggbeater and was named after the French aeronautical engineer, **Georges Jean Marie Darrieus** (1888 to 1979), who was granted a patent on it in 1927. Its main problem was that its blades fatigued easily. Today, many variations of vertical-axis turbine designs exist.

Advantages of vertical- over horizontal-axis turbines are that the former do not need to rotate to track the wind, they are easier to maintain because their gearbox is closer to the ground, they cost less to install because their towers are shorter, and they can be grouped more closely to each other in a wind farm. In addition, they can be deployed in places with legal height restrictions, such as near airports and in some urban areas, or where tall turbines are a disadvantage, such as in military bases within combat zones.

The main disadvantage of vertical-axis (and short horizontal-axis) turbines is that because their heights are limited, they are exposed to slower winds than are tall horizontal-axis turbines. Since instantaneous wind power is proportional to the cube of the instantaneous wind speed, this means a lot less power for vertical-axis turbines, increasing their cost per unit energy significantly. A second issue is that vertical-axis turbine blades can fatigue easily owing to the variation in forces acting on them during each rotation and because winds near the ground are more turbulent than are winds aloft. This can cause greater wear and tear and repair requirements for vertical-axis turbines than for tall horizontal-axis turbines.

11.3   Wind Turbine Parts

A horizontal-axis, upwind wind turbine consists of a tower, blades, rotor, nacelle, wind vane, and anemometer.[204] The **nacelle** houses several additional components, including the shafts, gearbox, generator, controller, braking system, pitch system, yaw drive, and yaw motor.

The **tower** holds up all the other components of the turbine and is generally hollow inside, except for wires and a ladder. The nacelle sits at the top of the tower. The rotor and blades are connected to the nacelle at one end. The wind vane and anemometer are attached to the other end of the nacelle.

The kinetic energy in the wind rotates the **blades**, which are connected to the **rotor**. The spinning rotor turns a low-speed shaft, which sits inside the nacelle, about 3 to 20 revolutions per minute for large turbines and up to 400 revolutions per minute for residential turbines. A gearbox connects the low-speed shaft to a **high-speed shaft**, which turns at 750 to 3,600 revolutions per minute. This is a sufficient spin rate for a **generator** to convert rotational mechanical energy into electricity.

Many wind turbines today, called **direct-drive** or **gearless turbines**, operate without a gearbox. Instead, the low-speed shaft is connected directly to the generator, which is larger and heavier than is the generator in a geared turbine.

With both geared and gearless turbines, the generators produce 50- or 60-hertz AC electricity (depending on country) although some generators produce DC electricity. A **heat exchanger** keeps the generator cool.

An **anemometer**, which sits outside the back (downwind side) of the nacelle, measures wind speed and communicates it to the **controller**, which is inside the nacelle. When the wind speed first increases above a low threshold, the controller starts the rotor up from rest. The controller also shuts the rotor off when the wind speed increases beyond a high threshold. The **pitch system** changes the steepness of the slope of (or pitches) the blades. The reason is to minimize the blades' direct contact with the wind in order to control the rotor speed or to stop the rotor from spinning when the wind speed exceeds a limit or falls below a minimum wind speed needed for operation. The **brake** stops the rotor in an emergency as a backup to the pitch system or as a "parking brake" during maintenance. The **wind vane**, which is near the anemometer outside the nacelle, measures wind direction and communicates the information to the yaw drive. The **yaw drive** rotates the turbine horizontally around a vertical axis to face the wind. The **yaw motor** provides power for the yaw drive.

The primary materials needed for wind turbines include pre-stressed concrete (for towers), steel (for towers, nacelles, and rotors), copper (for generator coils and electricity conduction), aluminum (for nacelles), wood epoxy (for rotor blades), glass-fiber-reinforced plastic (for rotor blades), carbon-filament reinforced plastic (for rotor blades), and neodymium (for permanent magnets in generators).

The large-scale growth of wind power will not be constrained by limits in the quantities of these materials. The major components of concrete – gravel, sand, and limestone – are abundant, and concrete can be recycled and reused. The world does have somewhat limited reserves of economically recoverable iron ore (on the order of 100 to 200 years at current production rates), but the steel used to make towers, nacelles, and rotors for wind turbines is 100 percent recyclable. The production of millions of wind turbines would consume less than 10 percent of the world's low-cost copper reserves. Other conductors of electricity could also be used instead of copper.

Most permanent magnets are made of a neodymium–iron–boron alloy. This magnet was developed in 1982 and is among the strongest permanent magnets available commercially. Although **neodymium** is abundant in the Earth's crust, it cannot be found isolated in nature. Instead, it appears, along with other heavy metals, called **rare-earth metals**, in ores, such as **monazite** and **bastnasite**, which are mineral group names. These ores contain small amounts of all rare-earth metals. Rare-earth metals are a group of 17 elements, including neodymium, in the periodic table, that tend to be found together. They all appear lustrous and silvery-white so are difficult to distinguish from each other. Although they are abundant in nature, rare earths are concentrated enough for mining in only a few geographically dispersed locations and are rarely found in economically viable ore deposits.

Wind turbines with permanent magnet generators require about 202 kilograms of neodymium oxide per megawatt of wind nameplate capacity.[205] With an estimated 13.9 terawatts nameplate capacity of wind needed to provide 44 percent of the world's all-purpose end-use power demand in 2050 with wind, the amount of neodymium needed among all wind turbines in 2050 is about 10.2 percent of the world's known reserve base, by far most of which is in China.

An alternative to neodymium for permanent magnets is iron nitride. This compound contains both iron and nitrogen, thus no rare-earth metals. The magnetic energy from iron nitride is reported to be over twice the maximum of a magnet containing neodymium.[71] Since iron nitride contains common elements, it can readily be mass produced and is now being commercialized.[206]

11.4 Wind Turbine Mechanics

A wind turbine blade rotates when the wind flows around it. A blade is like an airfoil. It is round (convex) on the top and flatter or more concave on the bottom. As such, the distance from the front tip to the back tip of the blade is greater on top than on bottom. Wind hitting the airfoil splits. Some of the air flows over the top and the rest, under the bottom. Because the top distance is greater, air flowing over the top travels faster than air flowing under the bottom in an effort for the two air parcels to meet each other at the back tip. Because air flowing over the top must travel further and faster, it spreads out more, causing air pressure above the foil to drop relative to air pressure under the foil. The lower pressure on top and higher pressure on bottom of the airfoil creates an upward **lift force** perpendicular to the airflow.

The lift force causes the blade to move in the direction of the lift force, just as the lift force acts upward over an airplane wing to keep the airplane afloat against the force of gravity. Because a wind turbine is constrained to rotate in a circle, the lift force accelerates the spin of the turbine in the clockwise direction so long as the more rounded top of the blade faces the clockwise direction. There is no advantage or disadvantage to turbines spinning clockwise versus counterclockwise, but all commercial turbines since 1978 have been built to spin clockwise. Previous to that, wind turbines and windmills mostly spun counterclockwise.

The **angle of attack** is the angle between the airflow and the airfoil. If the front tip and back tip of the airfoil are parallel to the wind, the angle of attack is zero degrees. If the front tip is then lifted so that the wind slightly impinges on the underside of the airfoil, the angle of attack is positive. Angles of attack of wind turbines are usually less than 10 degrees.

In order to impart a lift force on the airfoil, the wind must give up some kinetic energy, converting it to mechanical energy. The removal of kinetic energy slows down the wind past the turbine.

The other force the wind imparts on a wind turbine is the **drag force**, which consists of two components. First, viscous friction on the surface of the airfoil slows down the wind, removing more kinetic energy from it. That energy from the wind is converted into an equal and opposite drag force of the wind, pushing the airfoil slightly in the direction of the wind. Second, because the wind speed downstream of the turbine is slower than the wind speed upstream, the air pressure downstream is lower than that upstream. This results in a pressure force acting from the higher pressure upstream to the lower pressure downstream part of the airfoil. This force is also a drag force that is added to the first drag force in the direction of the wind.

Wind turbines operate best with high lift-to-drag ratios. Thus, they are built to maximize lift and minimize drag. This can be accomplished by increasing the angle of attack up to a point. For example, increasing the angle of attack from 0 to 5 degrees increases the lift-to-drag ratio from about 50 to 123 for one particular turbine. However, increasing the angle of attack further to 10 degrees decreases the ratio down to about 98.[207] Higher angles of attack decrease the ratio further.

The lift-to-drag ratio decreases at high angles of attack because turbulence builds up on the top of the airfoil with high angles of attack. This causes the air flowing over the top to no longer stick to the airfoil. As a consequence, air pressure above the top of the airfoil increases, reducing lift and slowing the spinning blade. This slowdown is referred to as **stall**. A benefit of stall is that, when wind speeds are above a threshold, the wind turbine controller and pitch system can increase the angle of attack to increase stall until the blades slow down sufficiently or stop spinning. This method of reducing or stopping power output by a turbine is called **active stall control**.

Related to active stall control is pitch control. **Pitch control** is the computer-controlled change in the angle of attack of wind turbine blades to maximize power output for all wind speeds. With pitch control, if wind power output exceeds the maximum nameplate capacity of the generator, but the wind speed is still below the maximum allowable wind speed set by the manufacturer of the turbine, the controller and pitch system can reduce the angle of attack. This reduces the lift to drag ratio, reducing output without stopping the turbine from spinning or producing power. Reducing the angle of attack in this manner is specifically called **feathering** the turbine's blades.

Finally, a method of limiting the power output of a wind turbine to the maximum that the generator can handle without involving moving parts or electronics is called **passive stall control**. With passive stall control, blades are designed so that they twist with increasing distance from the rotor so as to induce stall automatically when the wind speed exceeds the speed that generates the maximum allowable power. The turbine will continue to spin but not any faster than the speed that gives the maximum allowable power. The main problem with passive stall control is that it is difficult to design a turbine that ensures stall occurs under all conditions, so a safety margin is needed. As a result, passive stall control turbines do not operate under optimal conditions, and they are limited primarily to smaller turbines.[208]

11.5 Wind Turbine Generators

Wind turbine generators convert the rotational energy of a wind turbine's high-speed shaft into electricity. The more the wind rotates the blade, the more the generator can produce electricity, up to the nameplate capacity of the generator.

The three main categories of generators for electricity production are dynamos, AC synchronous generators, and AC asynchronous generators. Dynamos produce DC electricity and are often attached to small wind turbines. In an AC synchronous generator, the high-speed shaft operates at a fixed speed. Most generators worldwide are synchronous generators used to produce electricity in coal, natural gas, nuclear, and other power plants that generate electricity at a constant rate. In an AC asynchronous generator, the high-speed shaft operates at a variable speed. Wind turbines use AC asynchronous generators, owing to the variable nature of wind energy production.

Generators for wind turbines operate based on electromagnetic induction, and thus are **induction generators**. Induction generators can also act as motors by operating in reverse, turning electricity into rotating motion. In fact, an induction generator often acts as a motor to start up a wind turbine. Once the wind itself begins rotating the turbine blades, the motor returns to producing electricity as a generator. All wind turbine generators produce 50 hertz frequency (in Europe and most countries) or 60 hertz frequency (United States and other countries) three-phase AC power.

Two important induction generators are wound field generators and permanent magnet generators.

A two-pole **wound field generator** consists of two coiled wires mounted on opposite sides (180 degrees apart) of a rotating structure called a **generator's rotor**, not to be confused with a wind turbine's rotor. A wind turbine's rotor is connected to a low-speed shaft, which is connected to a gearbox, which is connected to a high-speed shaft, which is connected to the generator's rotor. When the wind is blowing, the high-speed, rotating shaft turns the generator's rotor at 750 to 3,600 rotations per minute. The generator's rotor is surrounded by a stationary structure, called a **stator**, that has three sets of coiled wires spaced 120 degrees apart. The rotor first creates a magnetic field around its two sets of coiled wires when a DC current passes through the wires. This current must be created by an outside power source, such as from the power grid. When the rotor and its attached coils and their magnetic fields start spinning inside of the stationary stator, the stator coils see the magnetic fields of the spinning rotor coils approach them and then move away from them quickly. The change in magnetic flux seen by the stator coils induces a three-phase AC current in the coils. The faster the generator's rotor spins, the greater the voltage generated across the coils. The AC current is then sent by wire to its destination.

A four-pole wound field generator is similar, except that it has four coiled wires mounted 90 degrees apart on the generator's rotor. A four-pole generator needs half as many rotations per minute from the high-speed shaft as does a two-pole generator to obtain the same output frequency of 50 or 60 hertz.

A **permanent magnet generator** is similar. But in this case, permanent magnets, instead of coils, are connected to the generator's rotor. The magnets are separated by equal distance. As the rotor spins at high speed, the stator coils see a changing magnetic field from the moving magnets on the rotor. The changing magnetic field induces an alternating current in the stator coils. The AC current is then sent to its destination. Unlike with a wound field generator, a permanent magnet generator does not need an external DC power source to initiate a magnetic field. This is useful particularly if the wind farm is located far from an electric power grid. Reducing the external DC power requirement also reduces the need for batteries and capacitors. In addition, magnets in a permanent magnet generator weigh less than the copper wires in a wound field generator that they replace.

On the flip side, controlling voltage is difficult with permanent magnets. Permanent magnets also do not operate well under high temperature and require more cooling than without magnets. Finally, most permanent magnets used today contain neodymium, which is subject to price fluctuations and shortages as it is mined in only a few places worldwide. Some concern also exists as to whether enough neodymium exists for permanent magnet generators to be used on a large scale in wind turbines. Aside from the fact that other types of asynchronous generators, including wound field generators, can be used instead of permanent magnet generators, the answer is that enough neodymium does exist (Section 11.3). In addition, iron nitride magnets, which contain more common elements than neodymium magnets, are now being commercialized as an alternative.

11.6 Power in the Wind and Wind Turbine Power Output

Like solar PV panels, individual wind turbines generate electricity variably over time. In other words, wind turbine output fluctuates up and down over short timescales (sometimes seconds) and over long timescales (hours, days, weeks, and months). In this section, wind and wind turbine characteristics are defined.

11.6.1 Wind Speed Frequency Distributions

Determining a wind turbine's power output over a period of time requires knowing how often different wind speeds occur. A frequency (probability) distribution of wind speed gives the percentage of all wind speeds measured over a month or year that are of a given speed. A simple frequency distribution, for example, is that 40 percent of all wind speeds during a year are less than 6 meters per second and 60 percent are greater than 6 meters per second. Real frequency distributions, though, consider many more wind speed intervals.

Most wind speed frequency distributions can be characterized as a Rayleigh distribution. A **Rayleigh frequency distribution** is a probability distribution of wind speed that looks similar to a bell curve but stretched toward higher wind speed. It is characterized by an average wind speed and gradually decreasing probabilities of slower and faster wind speeds. Good wind energy locations are generally those with mean wind speeds of 7 meters per second or faster at hub height. The Rayleigh frequency distribution is a specialized case of the **Weibull frequency distribution** of wind speed. Some wind speed frequency distributions are better described as Weibull distributions.[209]

11.6.2 Betz Limit

The wind passing through the swept area of a wind turbine's blades at a given instant contains a certain amount of power, called the **power in the wind**. Wind turbines can extract only a portion of this power. The maximum percentage of power in the wind that can be extracted by a wind turbine, regardless of its design, is 59.3 percent. This is called **Betz's law**. The German physicist **Albert Betz** (1885 to 1968) derived this law in 1919.

The basic idea is that, because a wind turbine extracts kinetic energy from the wind, the wind on the downwind side of the turbine has a lower speed and air pressure than on the upstream side of the turbine. The pressure difference causes air passing through the turbine to spread out, expanding in volume. If a turbine extracted all the kinetic energy in the wind, air would stop behind the turbine, preventing any more wind from passing through the turbine's blades. If the wind speed downwind of the turbine equaled the wind speed upwind of the turbine, then no kinetic energy would be extracted. Betz's law states that the wind speed downwind of a turbine can be no less than one-third that upstream of the turbine. The resulting power extraction at this minimum ratio is 59.3 percent. This is the Betz limit. It means that no wind turbine can extract more than 59.3 percent of the power in the wind that it is exposed to.

The best wind turbines today extract 45 to 47 percent of the power in the wind; thus they reach 76 to 79 percent of the Betz limit. A turbine's extraction efficiency depends largely on the number of blade rotations per minute. Blades that spin too slowly allow too much wind to pass. Blades that spin too quickly create turbulence that affects other blades, reducing lift and extraction efficiency.

11.6.3 Wind Turbine Power Curve

A wind turbine **power curve** is a graph of the instantaneous power output of a wind turbine as a function of wind speed. The maximum instantaneous power output of the turbine on the power curve at any wind speed is the nameplate capacity, or **rated power**, of the wind turbine. A wind turbine's nameplate capacity is limited by the nameplate capacity of the generator in the nacelle of the turbine. Wind turbines produce power as a function of wind speed. Between zero and the **cut-in** wind speed (generally 2 to 3.5 meters per second), a turbine produces no power because the power generated is so low at those speeds that it would be uneconomical to produce. Above the cut-in wind speed, the instantaneous power output increases roughly proportionally to the cube of the wind speed.

A wind turbine's **rated wind speed** is the wind speed at which the turbine reaches its maximum power output (rated power). The power output stays at the rated power for all wind speeds above the rated wind speed, up to a limit, owing to pitch control or passive stall control. However, wind turbines operating under realistic conditions have difficulty maintaining exactly constant power with increasing wind speed because of the time delay associated with pitch control and because of the approximation of turbine blade design in the case of passive stall control.

With increasing wind speed above the rated wind speed, the power output remains roughly constant until the wind reaches the **cut-out wind speed**. At that speed, the power output is decreased to zero. The shutoff is accomplished through pitch control or active stall control and is needed to prevent damage to the turbine. Most turbines, even when shut off, are designed to survive wind speeds up to a **destruction wind speed**, of 50 meters per second for up to 10 minutes.

> Transition highlight
> A class of turbines has been designed to withstand sustained 10-minute wind speeds of up to 57.5 meters per second.[210] These **typhoon-class** wind turbines are placed primarily offshore in locations prone to hurricanes or typhoons. Sustained 1-minute wind speeds in a Category 4 hurricane are, for example, 58.6 to 69.3 meters per second.

11.6.4   Wind Turbine Energy Output and Capacity Factor

How much energy and power do wind turbines actually extract during the year?

The **capacity factor** of a wind turbine is the annually averaged power produced by a turbine divided by the rated power (nameplate capacity) of the turbine. Alternatively, it is the energy produced per year divided by the maximum possible energy produced per year by the turbine.

If a wind turbine theoretically ran for a full year at its rated power, its annual average power output would equal its rated power (nameplate capacity), and its capacity factor would equal 1. However, the capacity factor of a turbine is always less than 1 because the annual average near-surface wind speed anywhere in the world is always less than the rated wind speed of a turbine. Real capacity factors range from 10 percent for old turbines at locations with low wind speeds to 56 percent for modern turbines at some high-wind-speed offshore locations. For example, the 2018 capacity factor of the Hywind floating wind farm (five 6-megawatt turbines located 29 kilometers offshore of Peterhead, Scotland) was 56 percent.

Transition highlight

The annual average capacity factor of wind turbines in the United States increased from 25.2 percent for those installed between 1998 and 2001 to 41.4 percent for those installed between 2014 and 2019.[211]

11.6.5 Factors Reducing Wind Turbine Gross Annual Energy Output

A wind turbine's energy output is lost owing primarily to four factors: transmission and distribution losses, downtime losses, curtailment losses, and array losses.

11.6.5.1 Transmission and Distribution Losses

Like with electricity from all sources, wind electricity traveling through transmission and distribution lines suffers line losses. Given that many good land-based sites for wind energy development are far from population centers, wind electricity must often be transmitted long distance. If the distance is greater than 600 kilometers, high-voltage direct current (HVDC) transmission lines should be used. For shorter distance, high-voltage alternating current (HVAC) lines should be used. Most offshore wind will be located within 200 kilometers of a coastline, and most of the world's population lives near the coast. As such, most offshore wind transmission will be with HVAC lines.

11.6.5.2 Downtime Losses

A wind turbine's energy output is also reduced when the turbine is down for regularly scheduled maintenance, equipment failure, or refurbishment of the turbine.

An analysis of repair data from 1,500 onshore wind turbines over about 10 years found that the average downtime for repairs was 1.6 percent (6 days) of the year. Minor failures (such as failure of the electrical system, electrical controls, the hydraulic system for pitch control, and yaw system), which represented 75 percent of the problems, caused only about 5 percent of the downtime. Major failures (such as failure of the rotor blades, rotor hub, drive train, gearbox, and generator), which represented the rest of the problems, caused about 95 percent of the downtime.[212]

Offshore wind turbines generally require more repair-related downtime than do onshore turbines because harsh weather conditions, including high waves and strong winds, prevent access for several days per month during stormy months. In addition, merely scheduling a boat ride to an offshore wind turbine can take additional time compared with driving to an onshore turbine. Third, the harsher weather conditions and faster wind speeds offshore increase wear and tear compared with onshore wind turbines. On the other hand, because of the lack of terrain and cooler ocean surface than land surface, offshore turbines are exposed to less turbulence, and thus have less wear and tear due to turbulence than do onshore turbines. Offshore turbines are down for scheduled maintenance for 4 to 6 days per year[213] and unscheduled repairs for 4 to 8 days per year for a total of 8 to 14 days per year (2.2 to 3.8 percent of the year).

In comparison, the average coal plant and combined cycle natural gas plant in the Eastern United States were down 10.6 percent and 3 percent, respectively, of the days between January and June 2015 for unscheduled maintenance.[214] Coal plants are down another 6 percent of the year for scheduled maintenance.

A difference, though, between outages of **centralized power plants** (coal, nuclear, natural gas) and outages of **distributed power plants** (wind, solar, wave) is that, when individual solar panels or wind turbines are down, only a small fraction of electricity production is affected. When a centralized plant is down, a large fraction of the grid is affected. When more than one large, centralized plant is offline at the same time, an entire grid can be affected.

11.6.5.3 Curtailment Losses

A third factor that reduces a wind turbine's output is curtailment. **Curtailment**, or **shedding**, is the deliberate reduction in output of an electricity generator below what it could otherwise produce. Grid operators may order curtailment under contract with a large electricity producer when electricity supply exceeds demand. Curtailment is avoided if excess electricity is stored in electricity storage or used on site to produce heat, cold, or hydrogen, which are either stored or used immediately.

11.6.5.4 Array Losses

Finally, competition for the same kinetic energy in the wind among wind turbines close to each other in a wind farm reduces wind turbine output compared with when the turbines are far away from each other. Losses due to competition among turbines are commonly called **array losses**.

If wind powers 37.1 percent of the world's all-purpose end-use energy in 2050, competition among wind turbines for limited kinetic energy may reduce annually averaged wind power output by about 6.7 percent. Reductions in densely packed farms or multiple farms close to each other are greater than in less densely packed farms or farms spread far apart from each other.

11.6.5.5 Overall Loss

The overall percentage loss in wind energy output due to the four processes just discussed is estimated as follows. Transmission and distribution losses are expected to decline from a world average of 8.3 percent in 2014 to about 6.2 percent by 2050.[190] Downtime for onshore and offshore wind turbines should decline slightly from today's losses to about 1.5 percent by 2050. Curtailment of wind, which is high in some countries today, should go to near zero in a 100 percent WWS world because wind electricity will be used for heat, cold, and hydrogen production in addition to for normal electricity. Finally, array losses in a 100 percent WWS world in 2050 are expected to increase to about 6.7 percent. The overall loss in 2050 under these conditions is thus estimated to be a mean of about 14 percent.

11.7   Wind Turbine Footprint and Spacing Areas

Two types of land or water areas associated with wind farms are their footprint area and spacing area.

11.7.1   Footprint Area

The **footprint area** of a wind farm is the topsoil or water area touched by bases of all the wind turbine towers in the farm. It does not include the areas of the turbine bases under the ground or underwater because a wind farm's footprint area represents land above the ground that can be used for other purposes. The footprint area does include the areas of any permanent roads (those covered with pavement) installed because of the wind farm, but it does not include areas of unpaved roads, which can readily be reverted back to their natural conditions. Transmission lines between the wind farm and grid that are underground also do not count toward footprint. The land areas of aboveground transmission line pads touching the ground due solely to the wind farm do count toward footprint. However, the areas of transmission tower pads are trivially small, as indicated by the fact that vegetation can grow under transmission tower lattices.

11.7.2   Spacing Area

The **spacing area** of a wind farm is the land or water area between wind turbines and beyond the edge of a farm required to (1) prevent one turbine's blades from touching another's, (2) prevent turbines from falling onto one another or onto nearby structures, (3) minimize wear and tear on a downstream turbine resulting from the turbulent wake of an upstream turbine, and (4) minimize array losses (competition among all turbines for limited available kinetic energy in the wind). The spacing area between onshore turbines can be used for multiple purposes, including for agriculture, cattle grazing, ranching, solar arrays, forests, or open space. Water between offshore turbines is similarly open water. Because footprint areas in a wind farm are small, wind farms leave 98 to 99.5 percent of land or water undisturbed or usable for other purposes.

Spacing area is an important parameter because it affects a wind farm's installed power density and output power density. **Installed power density** is the nameplate capacity (megawatts) per square kilometer of spacing area in a wind farm. **Output power density** is the average power output (megawatts) per square kilometer of spacing area. The larger the spacing area, the smaller the installed and output power densities. A wind farm's capacity factor is simply the ratio of the farm's output power density to its installed power density. Alternatively, the farm's capacity factor is its average power output divided by its nameplate capacity. As such, the capacity factor is independent of spacing area.

No unique way exists to define spacing area. At one extreme, it could be defined as the surface area of the Earth. This would give trivially small installed and output power densities. At the other extreme, it could be defined as the circular area on the ground around one wind turbine's tower, with the circle's radius equal to the turbine tip height, multiplied by the number of wind turbines in the wind farm. The **tip height** of a turbine equals its hub height plus one blade radius. The tip height is used to define the minimum spacing area of a single turbine because that is the distance a turbine will extend to if it falls to the ground in any direction. Laws in many locations require structures to lie beyond the tip height of a turbine.

Although there is no correct method to determine spacing area, methods that assign large, fixed rectangles, circles, or polygons to each turbine erroneously include areas that are not part of a wind farm. For example, they erroneously count, as part of a wind farm, space outside of a wind farm's boundary, space between clusters of turbines within a farm, and overlapping space that results when each turbine is assigned a large rectangle, for example.

A way to define spacing that overcomes these three issues is as follows. The methodology involves defining and combining three areas. The areas assume that a wind farm consists of one or more clusters of individual wind turbines. The first area is simply a circular area around each turbine, where the radius of each circle is the tip height of the turbine. This distance is chosen because it is the maximum physical horizontal distance a turbine can extend to if it falls to the ground. This distance also ensures that, if a wind turbine is on the edge of a wind farm, no space beyond the tip height will be counted as part of the wind farm spacing. Thus, the edge of a farm extends horizontally to no more than one tip height from each tower at the edge of the farm.

The second area is the spacing area of each cluster of turbines in the wind farm. The cluster area is obtained by tracing a line around the outside edges of the circles around each wind turbine on the outer edge of the cluster. This cluster area includes the circular areas of all wind turbines within the cluster. A cluster of turbines within a wind farm is separated from another cluster when the distance from a turbine tower at the edge of the first cluster to the closest neighboring turbine tower exceeds three tip heights. The area in between clusters is not included as part of the spacing area of the wind farm.

The third and final area is the total spacing area of the wind farm. This is simply the sum of the spacing areas of all the clusters within the farm.

The method just described avoids counting areas outside of cluster boundaries as part of a wind farm, avoids counting areas between clusters as part of a wind farm, and avoids counting overlapping areas, previously assigned to individual turbines, as part of a wind farm. It also accounts for actual distances between wind turbines. This method also ensures that the addition of a wind turbine to the farm increases the spacing area required by the farm, as it should in reality.

A study of 16 onshore and 7 offshore wind farms among 13 countries across five continents using this method found that installed power densities ranged from 7.2 megawatts per square kilometer for offshore European farms to 19.8 megawatts per square kilometer for onshore European farms and to 20.5 megawatts per square kilometer for onshore farms across four other continents.[215] In sum, the spacing area of a wind farm is important for determining the farm's installed and output power densities.

11.8 Global and Land Wind Resources

Large regions of fast onshore winds worldwide include the Great Plains of the United States and Canada; northern Europe; parts of Russia; the Gobi and Sahara Deserts; much of the Australian desert areas; New Zealand; parts of South Africa; parts of Peru; and southern South America. Windy offshore regions include the North Sea, the east and west coasts of North America, offshore of Australia, and offshore of the east coast of Asia, among other locations.

In order to determine total energy available for wind turbines worldwide or over land, it is necessary to account for competition among turbines for limited kinetic energy in the wind. When one wind turbine converts energy in the wind to mechanical energy to spin a turbine's blades, less kinetic energy is available for other wind turbines in the wind farm and in the world. As more wind turbines become operational, each turbine is able to extract less and less energy. At some point, the addition of one more wind turbine worldwide results in no additional energy generation. At that point, the annual average power extracted by the existing turbines is called the **saturation wind power potential**.[135] The saturation wind power potential is important because it gives the upper limit to how much average power is available from wind turbines installed worldwide (land plus ocean) or over land alone at a given hub height.

As the cumulative nameplate capacity of wind turbines increases over land plus ocean worldwide, the total extractable power among all wind turbines increases, but with diminishing returns. Above 3,000 terawatts of nameplate capacity installed evenly worldwide, for example, no additional power can be extracted from the wind at 100 meters hub height. The resulting worldwide limit to extractable power (not nameplate capacity) is about 253 terawatts. Over land outside of Antarctica, the limit is about 72 terawatts.[135]

> **Transition highlight**
> The world's near-surface winds contain far more power than is needed for all humanity. For example, the world needs only about 4 terawatts of annual average wind power output in 2050 for wind to provide 45 percent of the world's end-use power demand after all energy sectors have been electrified.[5] As such, 18 and 63 times the extractable wind power over land and worldwide, respectively, are available compared with the power needed to meet this demand. Thus, there is no wind resource barrier to obtaining even 100 percent of the world's 2050 all-purpose electric power from wind.

Finally, when wind farms are far apart from each other, their output power can increase by up to a factor of 4.6 for the same nameplate capacity summed over all farms, compared with if all wind farms are adjacent to each other. This is due to the reduced competition for available kinetic energy among wind turbines when wind farms are far apart versus close together.

11.9 Wind Turbine Impacts on Climate, Hurricanes, and Birds

This section examines the impacts of wind turbines on global and local climate, hurricanes, and birds. Briefly, wind turbines cool the globe on average owing to their reduction in water vapor, a greenhouse gas. Second, lots of offshore wind turbines can help to dissipate a hurricane from the outside in, thereby reducing hurricane wind speeds and storm surge. Third, although wind turbines kill birds, they kill far fewer birds than fossil-fuel generators, buildings, and cats.

11.9.1 Wind Turbine Impacts on Climate

Whereas wind turbines can cause a local warming of the ground downstream of a wind farm, they cause a net cooling of the Earth in the global average. The reasons are as follows.

Wind turbines extract kinetic energy from the wind, reducing wind speeds downstream of a wind farm. A reduction in wind speed reduces evaporation of liquid water from soil or a water body. Since evaporation is a cooling process, a reduction in evaporation warms the ground or the ocean or a lake surface. Some studies have found an increase in ground temperature near a wind farm.[216]

However, because wind turbines reduce evaporation and thus reduce water vapor in the air, they also reduce condensation of vapor to form clouds, reducing cloudiness. Because condensation releases heat to the air, less condensation cools the air. As such, those two processes caused by wind turbines – less evaporation at the surface and less condensation in the air – cancel each other out.

However, water vapor is a greenhouse gas that traps heat radiation emitted by the surface of the Earth, in the lower atmosphere. Thus, less water vapor means that more heat radiation can now escape to the upper atmosphere or outer space. This escape of energy from the lower atmosphere caused by wind turbines through this mechanism causes a net global cooling. However, such cooling is measurable only with a large number of turbines.

> **Transition highlight**
> Suppose 2.5 million 5-megawatt turbines are distributed over 139 countries to provide a projected 37.1 percent of the world's all-purpose end-use energy in 2050 after all energy sectors have been electrified. In this case, wind turbines reduce globally averaged water vapor throughout the atmosphere by 0.29 kilograms per square meter of ground, and near-surface air temperature by about 0.03 degrees Celsius. This temperature decrease represents about 3 percent of global warming through 2020. Thus, installing this number of wind turbines by 2050 may reduce about 3 percent of today's global warming.[136]

In sum, wind turbines may serve an additional benefit beyond replacing the electricity from polluting fossil fuels. They may also directly cool the surface, thereby reducing global warming beyond that caused by their replacing fossil-fuel energy generators.

11.9.2 Wind Turbine Impacts on Hurricanes

Because wind turbines extract kinetic energy from normal winds, wind turbines can extract kinetic energy from hurricane winds as well. However, if hurricane winds are too strong, they will topple an individual wind turbine. On the other hand, thousands of turbines near each other may slow a hurricane's winds sufficiently to prevent damage to the turbines. Also, a hurricane contains so much kinetic energy that thousands of turbines are needed to measurably reduce the damage from a hurricane.

About half of a hurricane's damage is due to high winds, and the other half is due to storm surge. **Storm surge** is the rise in water levels generated by a storm above and beyond water levels caused by normal tides under normal conditions. Storm surge in a hurricane is enhanced by three factors. One is that warm ocean temperatures due to global warming have caused water to expand and ice to melt, increasing the background sea level height. Second, the low surface air pressure in a hurricane raises the ocean sea level further compared with the sea level height under normal atmospheric pressure. Third, the strong counterclockwise flow in the Northern Hemisphere and clockwise flow in the Southern Hemisphere of winds around a hurricane's core over long distances create tall waves that lift water to great heights when the waves hit land.

Wind turbines in the past have been designed to withstand only 50 meters per second wind speeds. Some turbines now are built to withstand 57.5 meters per second wind speeds. Both of these wind speeds correspond to Category 3 hurricane wind speeds on the Saffir–Simpson scale. Yet hurricanes can reach Category 4 (58.6 to 69.3 meters per second) or Category 5 (greater than 69.3 meters per second) status. Thus, an important question to address is whether many wind turbines collectively can keep hurricane wind speeds below the destruction wind speed of a turbine while extracting sufficient power from the hurricane to dissipate it.

Because hurricane paths are not predictable, offshore wind farms will always be built primarily to generate electric power year-round in order to pay for themselves. However, if enough offshore farms are built, some farms can be placed strategically in front of a city to maximize protection of the city as well, thus serving a dual benefit.

It is impossible to know how a real wind farm might diminish a hurricane, because once a hurricane has passed the farm, it is not possible to replicate the hurricane in the absence of the farm. However, such an experiment can be carried out with a computer model that predicts hurricane formation and accounts for extraction of kinetic energy by wind turbines. In such a case, two hurricane simulations are needed – one with turbines and one without. Here, analyses of turbines in the presence of two hurricanes – Katrina and Sandy – are discussed.

Hurricane Katrina occurred from August 23 to 31, 2005. It damaged New Orleans, Louisiana, and caused 1,836 deaths. Peak wind speeds offshore were 78.2 meters per second, making it a Category 5 hurricane offshore. During Hurricane Katrina, storm surge caused flooding in New Orleans, which is below sea level. The flooding was exacerbated by the breach of many levies built to prevent flooding.

Hurricane Sandy formed October 22, 2012, and dissipated November 2, 2012. It hit the U.S. east coast between Washington D.C. and New York. Its peak measured wind speed was 35.8 meters per second. It was one of the widest hurricanes to hit the United States and one of the few to hit the northeast coast. Its damage was due primarily to flooding of the low elevation coast caused by storm surge.

Based on computer model simulations, large arrays of offshore wind turbines within 100 kilometers of the coast could have reduced peak wind speeds and storm surge significantly in both hurricanes. In the case of Katrina, large wind farms (totaling 78,000 7.58-megawatt wind turbines) to the southeast of New Orleans could have avoided the damage. In the case of Sandy, large wind farms (totaling 112,000 7.58-megawatt turbines) between Washington D.C. and New York City could have ameliorated much damage.

For Katrina, the farms could have reduced peak wind speeds by about 36 percent and storm surge by 6 to 71 percent, depending on location along the coast. For Sandy, wind farms could have reduced peak wind speeds by 36 percent and storm surge by 12 to 21 percent.[217]

Wind turbines reduce hurricane wind speeds for the following reason. Wind turbines first reduce the outer rotational winds of the hurricane, then subsequently increase its central pressure and decrease its peak wind speeds in the eye wall (near the center) of the hurricane. As such, wind turbines dissipate a hurricane from the outside in. The reason can be elucidated further as follows.

Wind turbines are exposed first to the hurricane's outer rotational winds, which are slower than eye-wall winds. The reduction in outer rotational wind speeds by wind turbines decreases wave heights there because wave heights are somewhat proportional to wind speeds. Since waves are a source of friction, decreasing wave height reduces surface friction.

The angle at which surface winds moving around a hurricane's eye wall converge toward the eye wall depends on friction. The less friction, the more circular the flow. The more friction, the more the winds converge inward toward the eyewall. Thus, the decrease in friction due to wind turbines decreases wind convergence toward the eye wall, causing the winds to flow more in a circle around the hurricane's eye wall.

The resulting decrease in wind moving toward the eye wall decreases the upward spiraling of air around the eye wall and the outward flow of air aloft away from the eye wall. Because fast outflow aloft relative to inflow at the surface drops a hurricane's surface pressure, strengthening the hurricane, a decrease in winds spiraling upward and flowing away aloft conversely increases surface air pressure, weakening the hurricane.

The increase in surface air pressure due to wind turbines then reduces the horizontal pressure difference between the center of the hurricane and the outside of the hurricane. The reduction in pressure difference slows hurricane winds. Slower winds, in turn, reduce wave heights further, reducing convergence to the center, reducing the spiraled lifting of air and outflow aloft, increasing central pressure further in a positive feedback. In this way, slowing the outer rotational winds of a hurricane with wind turbines dissipates the hurricane from the outside in.

Because wind turbine arrays dissipate a hurricane from the outside in, peak hurricane wind speeds reaching the turbines never reach the destruction wind speed of a turbine. By reducing hurricane wind speeds, wind turbines also reduce storm surge.

Although the simulations discussed included an enormous number of wind turbines, any smaller number of turbines has a proportionally smaller benefit and thus can still help. The United States may need about 53,000 offshore 5-megawatt turbines to provide 7.1 percent of its 2050 all-purpose end-use energy from offshore wind.[5] Dividing this number into wind farms of 100 to 300 turbines, placing those farms in separate clusters (where each farm is separated by distance to minimize competition for available kinetic energy), and placing each cluster of wind farms strategically in front of major cities at risk of a hurricane may reduce future hurricane damage.

Because offshore wind farms will be built to generate year-round electricity but may also reduce damage due to hurricanes, they are cost-effective in comparison with other techniques of reducing hurricane damage, such as building sea walls. Whereas turbines pay for themselves from the sale of the electricity they produce, sea walls have no other function than to reduce storm surge. Sea walls also do not reduce hurricane winds, which wind turbines do.

11.9.3   Wind Turbine Impacts on Birds and Bats

Birds and bats can fatally collide with the spinning blades of a wind turbine. Some bats also die from the drop in pressure behind the wind turbine, which causes their organs and blood vessels to expand to equalize the pressure. Of particular concern is the collision of **raptors** (birds of prey) with wind turbines. Raptors include eagles, ospreys, kites, hawks, buzzards, harriers, vultures, falcons, caracaras, and owls.

Estimates of avian mortalities, based on 2017 U.S. wind electricity production, range from 76,000 to 990,000 per year. These seem like large numbers until other sources of avian death are compared.

In the United States, cats are the number one threat to birds, killing an estimated 2.4 billion per year.[218] Windows of buildings kill about a billion birds per year. Transmission and distribution lines kill 8 million to 57 million birds per year and communication towers another 7 million per year. All of these are much larger sources of bird deaths than are wind turbines. The number of birds that died from wind turbines in the United States in 2017 ranged from 0.002 percent to 0.028 percent of all other causes of bird death.

Coal and natural gas electric power plants also kill more birds per unit energy and in total than do wind turbines.[219] As such, transitioning from coal or gas to wind reduces bird kills. Coal and natural gas used for electric power kill birds in three major ways. One is through the destruction in bird habitat due to invasive mining, such as mountaintop removal in the case of coal, and conventional drilling and fracking in the case of natural gas. The second is through air and water pollution, acid rain, and climate change caused by fossil-fuel power plants. The same air pollution from fossil plants that kills birds contributes to the 7 million premature human deaths worldwide each year. The third way that coal and natural gas kill birds is through power plant structures and equipment, which birds collide with or are electrocuted by.

Even nuclear power kills a similar number of birds to wind turbines, per kilowatt-hour of electricity produced.[219] Nuclear causes bird fatalities in two major ways. One is from contaminated ponds associated with uranium mining and milling. Abandoned open-pit uranium mines also form hazardous lakes. Second, birds collide with nuclear power plant buildings and equipment.

Raptor deaths comprise between 5 and 15 percent of all bird deaths. Bat deaths per unit energy are about 50 percent higher than are all bird deaths.[220,221] Many bat deaths occur near wind turbines in the Northeastern and upper Midwestern United States. A non-wind-turbine major source of bat death since 2006 has been white nose syndrome, which has decimated populations of bats in North America.

Wind turbines today are safer for birds and bats than were those built in the 1980s and 1990s. Early wind turbine towers often had lattices that birds could perch on, increasing their proximity to the turbine blades. Also, spin rates were not previously controlled, resulting in higher tip speeds at a given wind speed. The removal of lattices and the controlling of the spin rate at high wind speeds have contributed to safer (but not completely safe) turbines. Because many birds fly low, taller turbines also permit birds to fly under turbine blades. In addition, radar systems are now being used to alert wind farm operators of approaching flocks of birds, including raptors. Some endangered birds, such as condors, are fitted with a global positioning satellite transmitter. Wind farm operators can slow down or stop wind turbines if they are alerted that a condor is approaching. Devices that emit high-frequency ultrasonic sounds are now being tested to deter bats from entering a wind farm. Another way of reducing bird mortalities is to site wind turbines out of migratory paths of birds.

Transition highlight

An encouraging study found that painting just one of a wind turbine's three blades black (and keeping the rest white or gray) reduced bird mortalities by 72 percent compared with bird deaths from neighboring unpainted turbines.[222] Raptor fatalities were reduced the most. The reduced deaths were due to the increased contrast, and thus visibility, caused by painting a single blade black.

12 STEPS IN DEVELOPING 100 PERCENT WWS ROADMAPS

So far, this book has examined the main technologies needed for a 100 percent clean, renewable energy and storage system. Virtually all of these technologies exist today, and none is a miracle technology. This chapter focuses on combining the technologies together in countries, states, cities, and towns to provide end-point roadmaps for a transition. Such roadmaps provide targets for meeting all-purpose power demand with 100 percent WWS in the annual average by some year, often 2050, but ideally sooner, such as 2035. Chapter 13 discusses methods of matching power demand continuously (rather than in the annual average) with WWS supply, storage, and demand response. Roadmaps and grid stability analyses are helpful for giving policymakers, utilities, and the public confidence that a transition will not cause grid failures, particularly during extreme weather events.

12.1 Projecting End-Use Energy Demand

In a 100 percent WWS world, all energy sectors are electrified or powered with direct heat, where the electricity and heat are provided by WWS. Some electricity is used for hydrogen, primarily for use in fuel cells for transportation; but also for some industrial processes, such as steel production; and for electricity and heat production from fuel cells in some microgrids. Some heat is provided by district heating. Some electricity and heat are stored in electricity, heat, cold, and hydrogen storage. Energy efficiency and energy reduction measures are adopted to reduce demand. Grid operators also shift the times of peak demand with demand response to help keep the grid stable.

The first step in developing a roadmap to transition the energy infrastructure of a country, province, state, city, or town to 100 percent WWS is to project current annually averaged end-use energy demand across all energy sectors to a future year in a business-as-usual case.

A **business-as-usual case** is a future scenario describing how energy demand may change over time in a country if technologies and policies do not change much. It assumes moderate economic growth, some population growth, the use of renewable energy but growing only at historic rates, the use of modest energy efficiency measures, and modest reductions in energy use between the present and the future.

End-use energy is energy directly used by a consumer. It is the energy embodied in electricity, natural gas, gasoline, diesel, kerosene, and jet fuel that people use directly. It equals primary energy minus the energy lost in converting primary energy to end-use energy, including the energy lost during transmission and distribution and waste heat. **Primary energy** is the energy naturally embodied in chemical bonds in raw fuels, such as coal, oil, natural gas, biomass, uranium, and renewable (hydroelectric, solar, wind) electricity, before the fuel has been subjected to any conversion process.

The conversion from primary energy to end-use energy differs for different energy sectors and types of fuels. In the electricity sector, for example, end-use energy equals primary energy minus the energy lost during the generation, transmission, and distribution of electricity. For instance, when coal is burned to produce electricity, only about one-third of the energy embodied in the coal is converted to electricity. The rest is waste heat. Further, some of the electricity produced is lost during transmission and distribution. The end-use electricity in this case is the electricity that consumers use in the end, not the primary energy that was contained in the coal nor the energy in the electricity at the power plant itself.

In another example, solar electricity produced by a PV panel is primary energy. Some of that electricity is lost during transmission and distribution. The solar electricity that actually reaches a consumer after transmission and distribution losses is end-use energy. Similarly, electricity produced at a hydropower plant is primary energy. The electricity remaining after transmission and distribution losses is end-use energy.

In the transportation sector, the energy embodied in crude oil is primary energy. Converting crude oil to end-use products, including gasoline, diesel, kerosene, refinery gas, and jet fuel, involves little loss of the primary energy in crude oil, so the end-use energy in these cases is close to the primary energy in crude oil.

Natural gas used for heating and cooking is similar to the natural gas recovered from a well, so primary energy and end-use energy are similar. On the other hand, when natural gas is burned for electricity, only a portion of the natural gas is converted to electricity and some of that electricity is lost during transmission and distribution. As such, the end-use energy in the delivered electricity is much less than the primary energy in the natural gas creating the electricity.

Transition highlight

In 2018, the annually averaged, end-use power demand (energy consumption, in terawatt-hours per year divided by the number of hours per year) among all energy sectors in 145 countries representing 99.7 percent of world emissions was about 13.1 terawatts, or trillion watts. Of this, 21.3 percent was electricity demand. By 2050, this demand may grow in a business-as-usual case to 20.4 terawatts (by 56 percent) if no large-scale transition to WWS occurs.[5] The growth is due to a population increase, partly mitigated by lower energy use per person resulting from some modest shifts from coal to natural gas, biofuels, bioenergy, some WWS, and some energy efficiency.

12.2 Powering Future Energy With WWS

The second step in roadmap development is to transition 2050 business-as-usual end-use energy for each fuel type in each energy sector to electricity, some hydrogen created from electricity, and some heat. The electricity and heat are produced by WWS. Such a transition also involves implementing energy efficiency measures.

The energy sectors to transition are the residential, commercial, transportation, industrial, agriculture/forestry/fishing, and military sectors.

Energy in the **residential sector** includes electricity and heat consumed by households, but not for transportation.

Energy in the **commercial sector** includes electricity and heat consumed by commercial and public buildings, but not for transportation.

Energy in the **industrial sector** includes energy consumed by industry. Industry includes businesses that manufacture iron, steel, cement, chemicals, petrochemicals, iron-free metals, nonmetallic minerals, transport equipment, machinery, food, tobacco, paper, pulp, wood, wood products, textiles, and leather. Industry also includes printing, construction, and mining businesses.

Energy in the **transportation sector** includes energy consumed for any type of transport by road, rail, pipeline, air, and water, and for the use of highways by agricultural and construction machines. For pipelines, the energy required is for the support and operation of the pipelines, including for moving natural gas and oil through pipes. The transportation category excludes fuel used for non-highway use of agricultural machines, used for fishing vessels, and delivered to international ships, since those are included under the agriculture/forestry/fishing category.

Energy in the **agriculture/forestry/fishing sector** includes energy consumed by agriculture, hunting, forestry, or fishing. For agriculture and forestry, it includes consumption of energy for electricity and heat and for vehicles aside from agricultural machines using the highway. For fishing, it includes energy for inland, coastal, and deep-sea fishing, including fuels delivered to ships of all flags (including for international shipping) that have refueled in the country and energy used by the fishing industry.

Finally, energy in the **military sector** includes fuel used by the military for all transport (ships, aircraft, tanks, on-road, and nonroad transport) and bases (forward operating bases, home bases), regardless of whether the fuel is used by the country or another country.

Ninety-five percent of the technologies needed to transition all sectors are available today. The main technologies still missing are those for most air and marine transport and for some industrial processes.

The main WWS electricity generation technologies already available include onshore and offshore wind turbines, concentrated solar power plants, geothermal electricity plants, solar photovoltaics on rooftops and in power plants, tidal and ocean current power devices, wave power devices, and hydropower plants.

Vehicles used for transportation in a WWS world include battery-electric vehicles and hydrogen fuel-cell vehicles, where the hydrogen is produced with an electrolyzer that runs on WWS electricity (green hydrogen). Whereas battery-electric and hydrogen fuel-cell vehicles are available now for most types of land transport, they are still being developed for air and marine transport.

Battery-electric vehicles will dominate two- and three-wheel transport; short- and long-distance light-duty transport; and most truck, construction machine, and agricultural equipment transport. Battery-electric vehicles will also dominate short- and moderate-distance trains, short-distance boats and ships (ferries, speedboats, tugboats, dredgers), short-distance military equipment, and aircraft traveling less than 1,500 kilometers.

Of all commercial aircraft flight distances traveled worldwide, about 85.2 percent by number and 53.9 percent by distance are short-haul flights (less than 3 hours in duration, with a mean distance of 783 kilometers).[61] As such, all but 15 percent of aircraft flights worldwide by number will be electrified with batteries. Some short-distance trains will run on overhead-wire electricity.

Hydrogen fuel cells coupled with electric motors will dominate long-distance, heavy-duty trucks; long-distance trains; long-distance ships; long-haul aircraft; and long-distance military equipment.

Ground-, air-, and water-source electric heat pumps will provide building air heating and cooling. Heat pumps will also heat domestic water. In some cases, an electric resistance element will be added for low temperatures. In other cases, rooftop solar hot water heaters may preheat the water for the heat pump. In some tropical countries, only rooftop solar hot water heaters are needed. Clothes dryers will be electric heat pump dryers. Cook stoves will be electric induction. Lawnmowers, leaf blowers, chainsaws, and other machines will run on electricity.

Electric arc furnaces, induction furnaces, resistance furnaces, and dielectric heaters will provide high temperatures for industrial processes. Heat pumps and CSP steam will provide low-temperature heat for industry. In sum, all fossil-fuel and bioenergy combustion for energy will be replaced with WWS electricity and heat.

12.3 Changes in Energy Needed upon a Transition to WWS

The third step in developing a roadmap is to calculate the reduction in annual-average, end-use energy demand that arises from transitioning to 100 percent WWS electricity and heat for all purposes. The reductions occur for five main reasons:

(1) the higher efficiency of electricity and electrolytic hydrogen over combustion for transportation;

(2) the higher efficiency of electric heat pumps over combustion for air and water heating in buildings;

(3) the higher efficiency of electricity over combustion for high-temperature industrial heat;

(4) eliminating the energy needed to mine, transport, and process fossil fuels, biofuels, bioenergy, and uranium; and

(5) improving end-use energy efficiency and reducing energy use beyond what will occur under business-as-usual.

These reductions are discussed in turn.

12.3.1 Efficiency of Electricity and Electrolytic Hydrogen over Combustion for Transportation

First, replacing end-use energy supplied by fossil fuels (natural gas, gasoline, diesel, kerosene, jet fuel) and biofuels for transportation with that supplied by WWS electricity and green hydrogen eliminates waste heat of combustion. This is because electricity and electrolytic hydrogen have higher energy output-to-work-input ratios than do fossil fuels and biofuels. This factor is embodied in the **electricity-to-fuel ratio,** which is the energy required for an electric or hydrogen fuel-cell machine to perform the same work (move the same distance) as a business-as-usual machine running on fossil fuels or biofuels. This ratio is calculated here for battery-electric vehicles and hydrogen fuel-cell vehicles versus **internal combustion engine** vehicles.

12.3.1.1 Efficiency of Battery-Electric Vehicles over Fossil-Fuel Vehicles

An example of the greater efficiency of electricity over combustion arises with battery-electric cars in comparison with internal combustion engine cars. Only 17 to 20 percent of the end-use energy embodied in gasoline, for example, is used to move a gasoline passenger car. This is the **tank-to-wheel efficiency** of the car. The rest of the energy (80 to 83 percent) is waste heat.

A battery-electric passenger car, on the other hand, converts 64 to 89 percent (plug-to-wheel efficiency) of electricity at the plug (before charging the car) into motion, and the rest is waste heat.

The tank-to-wheel efficiency of a fossil-fuel car divided by the plug-to wheel efficiency of a battery-electric car is the fraction of a fossil-fuel car's end-use energy consumption that is needed to move an equivalent battery-electric car the same distance with electricity. This electricity-to-fuel ratio ranges from 0.19 to 0.31. In other words, a fossil-fuel car requires 3.2 to 5.3 times the energy in gasoline that a battery-electric car needs in electricity at the plug (before charging) to drive the same distance. By 2050, the average electricity-to-fuel ratio is expected to be closer to 0.19 owing to improvements in vehicle charging efficiencies and battery efficiencies, and the greater use of permanent magnet motors.

12.3.1.2 Efficiency of Hydrogen Fuel-Cell Vehicles over Fossil-Fuel Vehicles

A portion of future transportation will use hydrogen fuel cells. Hydrogen fuel-cell passenger cars are less efficient than are battery-electric passenger cars but are still more efficient than are internal combustion engine cars.

Hydrogen fuel cells convert hydrogen to DC electricity. The DC electricity is then converted with an inverter to AC electricity, which is used in an AC motor to produce rotation to turn wheels or a propeller.

The overall plug-to-wheel efficiency of a hydrogen fuel-cell passenger car accounts for multiple energy losses. These include losses in the electrolyzer used to produce hydrogen, in the compressor used to compress hydrogen, in the fuel cell within the car, in the DC-to-AC inverter in the car, and in the AC motor in the car. The efficiency also accounts for hydrogen leaks. The overall efficiency ranges from 23 to 43 percent. Thus, 23 to 43 percent of the electricity used to produce hydrogen for a passenger car actually moves the vehicle.

Dividing the low and high tank-to-wheel efficiencies of a fossil-fuel passenger car by the high and low plug-to-wheel efficiency of a hydrogen fuel-cell passenger car, respectively, gives the electricity-to-fuel ratio of a hydrogen fuel-cell car replacing a gasoline car as 0.4 to 0.87. Given that fuel cells are improving more rapidly than are gasoline cars, the ratio 0.4 is more appropriate for 2050.

Since the plug-to-wheel efficiency of a hydrogen fuel-cell passenger car is only 26 to 67 percent that of a battery-electric car, a hydrogen fuel-cell passenger car needs proportionately more wind turbines or solar panels to run than does a battery-electric passenger car to go the same distance. Thus, most short- and moderate-distance transportation in a 100 percent WWS world will be battery-electric transport. On the other hand, for heavier and longer-distance vehicles, such as long-distance aircraft, ships, trains, trucks, and military vehicles, hydrogen fuel-cell vehicles are more efficient than battery-electric vehicles.

12.3.2   Efficiency of Electric Heat Pumps for Buildings

Air-source heat pumps move heat from one place to another rather than create new heat. As a result, they are much more efficient than are fuel-burning heaters or electric resistance heaters for low-temperature heat in buildings. Air-source heat pumps have a coefficient of performance of 3.2 to 4.5, whereas ground-source heat pumps have a coefficient of performance of 4.2 to 5.2.[94] Thus, only 1 joule (unit of energy) of electricity is needed to move 3.2 to 5.2 joules of hot or cold air with an electric heat pump. This compares with a coefficient of performance of 0.97 for electric resistance heaters and 0.8 for typical natural-gas-powered boilers. In those cases, 1 joule of electricity creates only 0.97 or 0.8 joules of heat, respectively. As such, a heat pump reduces energy demand substantially compared with an electric resistance heater or a natural gas boiler.

For example, the electricity-to-fuel ratio of an electric heat pump powered by WWS replacing a natural gas boiler in the residential sector is 0.2. This is calculated assuming that the average coefficient of performance of all (ground-, air-, and water-source) heat pumps is 4. It also assumes that the average coefficient of performance of natural gas heaters is 0.803, which itself assumes that 98 percent of combustion heaters have a coefficient of 0.8 and 2 percent have a coefficient of 0.95. Dividing 0.803 by 4 gives 0.2.

The use of electric heat pumps for cooling does not change the electricity-to-fuel ratio for air conditioning because an electric air conditioner already acts like a heat pump – it removes heat from a room and transfers it to the outside air. As such, an air conditioner has a similar coefficient of performance to a heat pump. It just does not run in reverse to produce heat. Refrigerators, like air conditioners, also act like heat pumps but only for cooling, not for heating.

12.3.3 Efficiency of Electricity over Combustion for High-Temperature Heat

Replacing fossil fuels and biomass for high- medium-, and low-temperature industrial heat with WWS electricity also reduces energy requirements. In a WWS world, oil, gas, coal, biofuel, and waste combustion heating for high- and medium-temperature heating processes (above 100 degrees Celsius) will be replaced with a mixture of electric resistance furnaces, arc furnaces, induction furnaces, dielectric heaters, and electron beam heaters. Heat pumps and CSP steam will replace those fuels for low-temperature heating processes (below 100 degrees Celsius).

Most natural gas furnaces used for producing high- and medium- temperature heat are about 80 percent efficient (coefficient of performance of 0.8). In other words, about 80 percent of the energy in the chemical bonds of the natural gas is converted to useful heat. The rest of the energy is lost either as waste heat or incompletely combusted natural gas that escapes as exhaust. Some new natural gas boilers that recapture waste heat and latent heat of condensation have an efficiency of up to 107 percent. The efficiency can be higher than 100 percent since the efficiency above 100 percent is due to capture of the latent heat. Assuming 80 percent of conventional boilers have an efficiency of 80 percent and the rest have an efficiency of 107 percent gives a mean efficiency of business-as-usual fuels for industry of 85.4 percent.

On the other hand, an electric resistance furnace for producing high temperatures is about 97 percent efficient. Thus, about 97 percent of the electricity going into an electric resistance furnace gets converted to useful heat for industry. The rest is waste heat that is lost due to conduction of heat out of the furnace without the heat being used. The efficiency of an average heat pump is about 400 percent (since the coefficient of performance is 4). Thus, heat pumps move four units of heat for every one unit of electricity they consume.

In sum, the electricity-to-fuel ratio for WWS replacing business-as-usual fuels for high- and medium-temperature industrial heating is

0.88. That for low-temperature industrial heating is 0.21. Assuming 85 percent of heating is for high-temperature heat and 15 percent is for low-temperature heat (an estimate for the industrial sector) gives a mean electricity-to-fuel ratio for WWS replacing business-as-usual electricity for industry as 0.78.

12.3.4 Eliminating Energy to Mine, Transport, and Process Conventional Fuels

Fourth, producing all energy with WWS eliminates the need to mine, transport, and process fossil fuels, bioenergy, and uranium. Worldwide, about 11 to 13 percent of all energy is used for these purposes. A WWS energy economy eliminates the need for such energy. Instead, wind comes right to the wind turbine and sunlight comes right to the solar panel, so no energy is needed to mine wind or sunlight.

Energy for the mining and transport of fuels is consumed in the industrial and transportation sectors. In the industrial sector, the energy consumed is for fossil-fuel and uranium mining operations and for petroleum refining. In the transportation sector, energy is needed to push natural gas and oil through pipes and to move oil tankers, coal trains, gasoline and diesel trucks, trucks carrying biomass crops, and trucks and barges carrying liquid biofuels. About 2 percent of oil, 50 percent of coal, and 80 percent of natural gas in the transportation sector are used just to transport fossil fuels, bioenergy, and uranium.[223] This need for energy is eliminated upon a transition to 100 percent WWS.

12.3.5 Increasing Energy Efficiency and Reducing Energy Use

Finally, increasing energy efficiency and reducing energy use beyond what will occur in a business-as-usual scenario reduce end-use demand further. Most business-as-usual projections of energy demand between today and 2050 are based on moderate economic growth and account for some end-use energy efficiency improvements and reductions in energy use. However, such improvements are modest. Additional policy-driven energy efficiency measures and incentives can reduce end-use energy demand further.

Most of the additional improvements will be due to increasing energy efficiency in the residential and commercial sectors. For example, additional efficiency can be obtained through better insulation of pipes and weatherizing homes. It can also be obtained through more efficient light bulbs and appliances and through the faster implementation of energy-efficient technologies. Improvements in vehicle design and in lightweight materials will also help.

12.3.6 Overall Reduction in End-Use Demand

When conventional energy is converted to WWS energy, annual-average end-use energy demand decreases substantially due to the five reasons just discussed. For example, among 145 countries representing 99.7 percent of all world fossil-fuel carbon dioxide emissions, a transition in 2050 reduces end-use demand by 56.4 percent, from 20.4 to 8.9 terawatts.[5] Of this reduction, 21.8 percentage points are due to the efficiency of WWS over business-as-usual transportation; 13.3 percentage points are due to the efficiency of heat pumps for building heating; 3.4 percentage points are due to the efficiency of WWS electricity for industrial heat; 11.3 percentage points are due to eliminating energy in the mining, transporting, and refining of fossil fuels, bioenergy, and uranium; and 6.6 percentage points are due to end-use energy efficiency improvements and reduced energy use beyond those in the business-as-usual case.

In sum, transitioning from fossil fuels, bioenergy, and uranium to 100 percent WWS has the potential to reduce energy demand worldwide significantly by reducing waste heat, eliminating unnecessary energy in finding and processing of fuels, and improving energy efficiency.

12.4 Performing a Resource Analysis

The next step in developing a 100 percent WWS roadmap is to perform a resource analysis for each WWS energy generating technology. The purpose is to determine the maximum renewable energy resource available to meet demand in each country, state, city, or town. Resource analyses can be performed through data analysis, computer modeling, or both.

Data include local data or remotely sensed data. Local data include, for example, continuous measurements of wind speed and sunlight at a weather station. Remotely sensed data include mostly satellite products. Satellite products are derived from measurements combined with model predictions. For example, satellite-derived wind speeds near the surface of the Earth are obtained by combining data from a satellite instrument, such as a microwave radiometer, a microwave scatterometer, or an ultraviolet laser, with results from a computer model.

Although local data are considered the most accurate, they are available only at a few locations. Satellite products are less accurate, particularly in the presence of clouds, but satellite products are available over most of the world and at high resolution. Two major problems with both local data and satellite products are (1) they cannot be used to predict WWS energy resources in the future and (2) they cannot be used to predict wind resources in the presence versus absence of wind turbines.

Three-dimensional atmospheric models, on the other hand, can cover high spatial and time resolutions, predict the future, account for competition among wind turbines for limited energy in the wind, and estimate wind and solar resource worldwide. However, model predictions are generally less accurate than are either local data or remotely sensed satellite products.

Analyses of local data, satellite products, and computer model results suggest that enough solar energy and wind energy exist, independently, to power the entire world many times over for all purposes. The world solar resource is much larger than is the world wind resource. CSP resources are lower than are solar PV resources because CSP is limited to locations with significant sunlight. No other WWS technology can power the whole world on its own, but all can play a role.

Another important component of a resource analysis is to determine the land or rooftop area that can feasibly be used to harvest renewable energy. A quantification of the rooftop area available for solar PV is needed in each country, province, state, city, and town. Rooftop areas include areas over buildings, parking canopies, parking structures, and road canopies. Rooftop areas suitable for PV exclude areas facing north in the Northern Hemisphere or south in the Southern Hemisphere. They also exclude areas continuously shaded and areas between rows of panels needed for walking and for avoiding shading of one row by another.

Land available for onshore wind and utility PV must also be determined. In considering available land, it is necessary to determine not only which land is exposed to fast winds or lots of sunlight, but also which land is not excluded by law or practical constraint. Legal exclusions arise due to zoning ordinances that prevent development in sensitive areas or areas close to homes or other buildings.[224] Practical exclusions arise due to topography. For example, building wind turbines near the top of the Himalayas is not practical.

12.5 Selecting a Mix of WWS Energy Generators to Meet Demand

The next step in developing a 100 percent WWS roadmap is to quantify a mix, for each region of interest, of WWS electricity and heat generators that can supply the end-use all-purpose energy in the region in the annual average. The penetration of each WWS electricity generator in each city, state, region, or country is limited by the following constraints:

(1) each type of renewable generator cannot draw more energy than is available based on the renewable resource in the region;
(2) the land area required by all WWS generators should be no more than a few percent of the total land area of the region;
(3) no new conventional hydropower dams should be installed if possible, but existing ones can be improved;
(4) wind and solar, which are complementary in nature, should both be used in similar proportions to the extent possible;
(5) wind power should be installed, where possible, in cold regions;
(6) the average cost among generators, storage, and transmission/distribution should be kept low; and
(7) a technology's use should be minimized in a region if its capacity factor is very low there.

With these basic guidelines, it is possible to provide a first estimate of the WWS resources needed to power a town, city, state, region, or country for all energy purposes in the annual average. First estimates must then be refined, as described in Chapter 13, in order to ensure that energy demand is matched with supply continuously.

12.6 Estimating Avoided Energy, Air Pollution, and Climate Costs

Transitioning to 100 percent WWS will reduce the private and social costs of energy. The **private (business) cost of energy** is the marketplace cost of energy. It does not account for health or climate costs.

The **social (economic) cost of energy** is the total cost of energy to society. It is the private cost of energy plus all externality costs associated with energy. An **externality cost** is a cost not captured in the market. The most relevant externality costs related to energy are health costs, climate costs, and nonhealth and nonclimate environmental costs. In this section, these costs are discussed.

12.6.1 Private Energy Cost

The private (nonexternality) cost of energy includes the costs of electricity generation and storage, heat generation and storage, cold generation and storage, hydrogen generation and storage, transmission and distribution, and machines and appliances that use electricity, heat, or cold.

Two metrics commonly used to calculate electricity generating costs are the cost per unit energy generated (cost per kilowatt-hour of electricity produced) and the annual cost of energy (cost of energy per year). The annual cost of energy is simply the cost per unit energy multiplied by the energy consumed per year. A 100 percent WWS system may reduce annually averaged energy demand worldwide, compared with a business-as-usual system, by about 56.4 percent in 2050.[5] As such, the same cost per unit energy between a business-as-usual and WWS system translates to a 56.4 percent lower annual energy cost to consumers when WWS is used.

Here, the method of calculating the cost per unit energy, also called the levelized cost of electricity, of an electric power generator is first described. The **levelized cost of electricity** is the value, in today's dollars, of the total cost of electricity over the life of a generating plant, divided by today's value of the total energy generated by the plant over its life. The levelized cost of electricity is the average price that an electricity generator must be paid before taxes for it to break even over its lifetime. It accounts for the electricity generator's upfront capital cost, its annual operation and maintenance cost, its annual fuel cost (if any), its lifetime, and the decommissioning cost at the end of its lifetime. It also accounts for the **discount rate**. For renewable energy generators (wind, solar, geothermal, hydro, tidal, and wave), the fuel cost is zero. For hydrogen, the fuel cost is the cost of water used to produce the hydrogen by electrolysis.

There are two types of discount rate – a private discount rate and a social discount rate.

A **private discount rate** is the interest rate that banks charge builders and consumers for taking out loans. Such loans may be used to pay for the construction of a power plant or to build a house. The private discount rate is also the **opportunity cost of capital**. In other words, it is the rate of return that can be obtained by investing capital in a market. Private discount rates are appropriate only for short-term public projects that, dollar-for-dollar, crowd out private investment.[225,226] The private discount rate in 2020 in the United States was between 3 and 6 percent.

A **social discount rate** is the discount rate used in a **social cost analysis,** which is a cost analysis that include externality costs, such as health or climate costs. The **social cost** of an investment is the investment's private cost plus its externality costs. A social discount rate is used primarily when the costs and benefits of a project occur at different times and over more than one generation. Such projects are called **intergenerational projects.**

Social discount rates are smaller than private discount rates, because society, as a whole, cares more about the welfare of distant future generations than does the average consumer or investor, who is generally concerned with near-term impacts during his or her lifetime. As a result, social discount rates appropriately weigh the present value of future impacts higher than do private discount rates. The (incorrect) use of a high private discount rate in the evaluation of long-term climate-change mitigation undervalues future social benefits and thus biases present-day investments away from efforts that provide long-term benefits to society. In order to properly evaluate long-term costs and benefits from the perspective of society, a social discount rate must be used. The consensus among experts of the mean social discount rate that should be used is 2 percent.[225,227,228]

A study of 145 countries found that a transition from business-as-usual energy to WWS energy may reduce the 2050 levelized cost of energy by about 11.1 percent. Because WWS also reduces all-purpose end-use energy requirements by 56.4 percent, WWS may reduce the annual energy cost by 62.7 percent, from $17.8 to $6.6 trillion per year.[5]

The lower private costs of energy due to WWS are being borne out in the real world. In 2021, for example, onshore wind and utility PV were the least expensive sources of new electricity in the United States[161] and Europe.[229] They were near 50 percent the cost of electricity from new natural gas combined cycle electric power plants in both cases.

12.6.2 Avoided Health Costs from Air Pollution

Transitioning homes, cities, states, provinces, and countries to WWS reduces air pollution health problems immediately. The combustion of fossil fuels, biofuels, and biomass releases air pollutant gases and particles. In the presence of ultraviolet rays of sunlight, some of the gases cook to form ozone and aerosol particles. Particles smaller than 2.5 micrometers in diameter penetrate deeper into people's lungs than do larger particles. Particles, mostly small ones, are the most dangerous component of air pollution, causing up to 90 percent of air pollution deaths and illnesses. Ozone causes much of the rest.

Today, about 7 million people die each year from air pollution health problems arising from fossil-fuel, biofuel, and biomass burning. Because emission control technologies are expected to improve between today and 2050, air pollution deaths and illnesses from a business-as-usual infrastructure may decline from close to 7 million per year in 2016 to about 5.3 million per year in 2050.[5] Reductions may occur in almost all world regions despite higher populations in all regions. The main exception is Africa, where population growth may be so high that it outpaces technological improvement, resulting in more air pollution deaths and illnesses in 2050 than today in a business-as-usual case.

The cost and emotional damage of over 5 million deaths and hundreds of millions more illnesses per year worldwide in 2050 are enormous. Air pollution costs are due to hospitalizations, emergency room visits, lost workdays, lost school days, higher workman's compensation premiums, higher insurance premiums, higher taxes, and loss of companionship. Costs are also due to agricultural losses, visibility losses, and structure damage arising from air pollution.

Conversely, eliminating such air pollution damage results in a better quality of life and lower costs. The cost reduction is usually estimated by first determining a parameter called the value of statistical life. The **value of statistical life** is a widely used metric determined by economists to assign a cost to a reduction in the risk of dying. It is the value of reducing one statistical death in a population. For example, if the average person in a city of 100,000 is willing to pay $80 to reduce her or his death risk by 1/100,000, the statistical value of reducing one mortality is $8 million. The value of statistical life is also determined from how much more employers pay their workers who have a higher risk of dying on the job.

The cost savings from reducing a single air pollution mortality is estimated as the value of statistical life. This value varies by country and changes each year. The cost savings due to reducing air pollution illness aside from death are then estimated as a percentage (usually around 15 percent) of the cost savings from eliminating all air pollution deaths. The cost savings due to reducing all nonhealth environmental damage (which does not include global warming, since that cost is determined separately) is estimated as a percentage (usually around 10 percent) of the cost savings from eliminating all air pollution death and illness.

With these assumptions in mind, a worldwide transition to 100 percent WWS may avoid about $33.6 trillion per year in health costs in 2050. Such cost savings arise from preventing over 5 million deaths, hundreds of millions more illnesses, and nonhealth environmental damage, aside from climate change damage, each year. This is the worldwide WWS air pollution social cost savings.

12.6.3 Avoided Climate Change Damage Costs

The damage that greenhouse gas emissions cause to the global economy through their impacts on climate is referred to as the **social cost of carbon**. Units of the social cost of carbon are U.S. dollars per metric tonne of CO_2-equivalent emissions. Climate change damage costs include costs arising from higher sea levels (coastal infrastructure losses), reduced crop yields for certain crops, more intense hurricanes, more droughts and floods, more wildfires, more air pollution, more migration due to famine, more heat stress and heat stroke, more disease of certain types, more fishery and coral reef losses, and greater energy requirements for air conditioning, among other impacts. Only a portion of these costs is offset by higher yields of some crops due to global warming and lower air heating requirements in some locations.

The social cost of carbon emissions is likely to increase over time as carbon dioxide and other warming agents accumulate in the atmosphere and temperatures continue to rise. The accumulation will accelerate as more people rise out of poverty in developing countries and consume energy faster than their predecessors did. As such, the social cost of carbon is tied to the gross domestic product of a country. Based on data from several studies, an estimate of the social cost of carbon in 2010 was $250 ($125 to $600) per tonne of CO_2-equivalent emissions. Accounting for the growth in damage between 2010 and 2050 gives a projected social cost of carbon of $500 ($280 to $1,060) per tonne of CO_2-equivalent emissions in 2050. Consequently, the climate damage to the world in 2050 due to fossil fuels and bioenergy in a business-as-usual case is estimated to be a mean of about $31.8 trillion per year.[5]

12.6.4 Summary of Avoided Energy, Health, and Climate Damage Costs

The total social cost of business-as-usual energy is the sum of the private cost, the air pollution cost, and the climate cost of energy. Worldwide (among 145 countries), those costs in a business-as-usual case are approximately $17.8 trillion, $33.6 trillion, and $31.8 trillion per year, respectively, for a total of $83.2 trillion per year.[5] Thus, health and climate costs of emissions exceed energy costs by a lot, so eliminating emissions reduces social costs enormously.

Using WWS results in zero emissions of air pollutants and climate-affecting pollutants from energy. It also reduces private energy costs from $17.8 trillion to $6.6 trillion per year, or by 62.7 percent. The $6.6 trillion per year cost is the annual energy cost of a worldwide *Green New Deal*. The **Green New Deal** is the name given to a policy proposal that calls for, among other goals, a country to transition to 100 percent clean, renewable energy for all purposes as fast as possible. WWS also reduces social energy costs from $83.2 trillion to $6.6 trillion per year, or by 92 percent. The lower private energy cost is due to the 56.4 percent lower annual energy requirements and the 11.1 percent lower cost per unit energy.[5] The lower WWS annual social cost is due to the lower WWS annual energy cost together with the elimination of health and climate costs associated with business-as-usual fuels.

13 KEEPING THE GRID STABLE WITH 100 PERCENT WWS

One of the greatest concerns facing the implementation of a worldwide 100 percent clean, renewable energy and storage system is whether electricity, heat, cold, and hydrogen will be available when they are needed. In other words, can a 100 percent WWS grid avoid blackouts?

The electric power grid in a 100 percent WWS world will be very different from that today. Today, electricity comprises about 20 percent of all end-use energy (or 40 percent of primary energy). In a 100 percent WWS world, electricity will comprise close to 100 percent of all end-use energy, which itself will equal primary energy less transmission and distribution losses. The rest of end-use energy will come from geothermal or solar heat. The sectors that will be electrified (transport, buildings, industry, agriculture/forestry/fishing, and the military) will use more energy-efficient technologies than their fossil-fuel counterparts. Such technologies include battery-electric vehicles, hydrogen fuel-cell vehicles, and heat pumps, among others. The reduction in energy use due to the use of more efficient technologies will reduce overall energy demand substantially. Demand will also decrease because no more energy will be used to mine, transport, or process fossil fuels, bioenergy, or uranium for energy. End-use energy efficiency will increase, and policies will encourage less energy use. A future electric grid will also be coupled with electricity, heat, cold, and hydrogen storage. Finally, a future grid will have more long-distance electrical transmission instead of fossil-fuel pipelines.

Thus, the main challenge in a future grid will be to match electricity, heat, cold, and hydrogen demand with 100 percent WWS electricity and heat supply plus storage while using demand response. This chapter discusses meeting such demand both on the short timescale (seconds to minutes) and long timescale (months to seasons to years).

13.1 Variable versus Intermittent Resources

Winds, waves, and sunlight produce electricity that varies over short timescales (seconds to minutes) and over long timescales (months to seasons to years) owing to continuous changes in the weather and climate. Thus, these energy sources are **variable** resources. Another word commonly used to describe a variable resource is an **intermittent** resource. However, all energy resources, even those not affected by the weather, are intermittent owing to the fact they are shut down during scheduled and unscheduled maintenance. Variable resources are those affected by both the weather and maintenance.

One concern with the use of variable WWS resources is whether they can provide electricity, heating, and cooling whenever such energy is needed. Any electricity system must respond to changes in electricity demand over periods of seconds, minutes, hours, seasons, and years, and accommodate unanticipated changes in the availability of electricity generation. It is not possible to control the weather; thus, a sudden change in demand often cannot be met by a variable WWS resource unless storage, demand response, or long-distance transmission, or all three, are coupled with electricity generation.

The concern about matching demand with supply, though, applies to all energy sources, not just to the variable ones. For example, because geothermal, tidal, coal, and nuclear resources can provide relatively constant (**baseload**) supplies of electricity during the day, they rarely match energy demand, which varies continuously and significantly. In addition, all baseload electricity generators are down for maintenance anywhere from 3 to 40 percent of all hours of a year. As a result, gap-filling resources, such as natural gas, hydropower, pumped hydropower, and now batteries, are currently used to meet peaks in demand with grids dominated by baseload resources.

Even nuclear power, which has been designed in some countries to ramp its electricity production up and down slowly, does not match demand. In France, for example, nuclear power ramps to 100 percent of its power in 20 to 100 minutes. But this is 40 to 400 times slower than the ramp-up rate of hydropower or pumped hydropower storage (100 percent in 15 to 30 seconds), infinitely slower than that of a battery, and 4 to 20 times slower than that of an open cycle natural gas turbine (100 percent in 5 minutes).[32]

13.1.1 Risk of Grid Failure with the Current Grid

Although concern exists about stability of the current fossil-fuel-dominated grid, the grid generally works, increasing the desire of grid operators to maintain the status quo. But, with increasingly extreme heatwaves and cold spells caused in part by global warming, the fragility of the current grid is being exposed. Because renewables have been added to the grid in most places, some people have blamed renewables for blackouts that have occurred. Close inspection, though, indicates that grid failures in those cases were caused mostly by problems with the grid itself or with fossil-fuel unreliability, not with renewable electricity variability.

Two examples are the August 14 to 15, 2020, summer blackout in California and the February 14 to 18, 2021, winter blackout in Texas. Close to 50 percent of California's electricity and 23 percent of Texas's electricity was from WWS in 2020. Despite many people blaming the growth of solar and wind in California for the blackouts, the heads of the California Public Utility Commission, California Independent System Operator, and California Energy Commission confirmed that *"renewable energy did not cause the rotating outages"*.[230] Instead, a variety of factors, including an unexpected unavailability of imports from across the west, led to the blackouts.

In the case of Texas, low temperatures caused natural gas, coal, nuclear, and wind electricity generators to fail, with natural gas being the largest source of electricity and failure.[231] Some frozen wind turbines were shut down because none had de-icing equipment. A simple solution is to add de-icing equipment to wind turbines in Texas. Wind turbines with de-icing equipment have operated successfully in cold weather in Iceland, Norway, Sweden, Alaska, Canada, and Russia.

13.1.2 Countries and Regions with 100 Percent WWS and a Stable Grid

Blackouts can occur with a 100 percent WWS grid if the grid is not planned properly. However, proper planning and management can prevent these. Also, a 100 percent renewable grid dominated by hydropower risks little chance of a blackout because hydropower is useful for both baseload and peaking power. Hydropower can readily fill in gaps in wind and solar supply.

> **Transition highlight**
> Impressively, by 2021, at least 10 countries had obtained 97.5 to 100 percent of all the electricity they produced from WWS resources. These include Iceland, Norway, Costa Rica, Albania, Paraguay, Bhutan, Namibia, Nepal, Ethiopia, and the Democratic Republic of the Congo (Figure 13.1). The dominant renewable in all 10 countries was hydropower. Two other countries (Kenya and Tajikistan) produced more than 90 percent of their own electricity from WWS. In Kenya, the primary source of electricity was geothermal; in Tajikistan, it was hydropower. In addition, Scotland produced from WWS more than 90 percent of the electricity that it consumed. Most of that came from wind. Some of the countries with high WWS penetration (Paraguay, Bhutan, and Nepal) produced more electricity than they consumed and exported the difference. For example, Paraguay exported to Argentina and Brazil. Bhutan and Nepal both exported to India.

Only two countries shown in Figure 13.1 generate most of their electricity with something other than hydropower.

In 2020, 92 percent of Kenya's electricity production was from WWS, with 44 percent from geothermal. In July 2021, Kenya announced that it will double its geothermal nameplate capacity by 2030. This addition alone ensures that Kenya will produce more than 100 percent of its electricity with WWS.

In 2021, 91.2 percent of Scotland's electricity consumption was from WWS, with 71 percent from wind. During the first 6 months of 2019, wind provided twice the electricity needed to power all the homes in Scotland.[235]

In addition, in many world regions, WWS has supplied nearly or at 100 percent of all electricity consumed for limited periods.

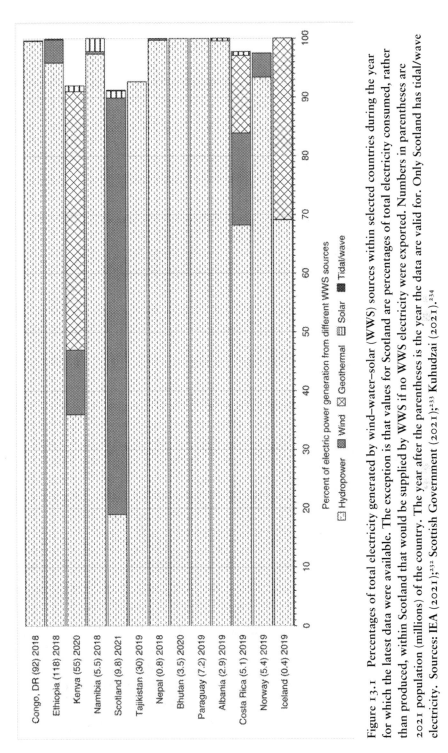

Figure 13.1 Percentages of total electricity generated by wind–water–solar (WWS) sources within selected countries during the year for which the latest data were available. The exception is that values for Scotland are percentages of total electricity consumed, rather than produced, within Scotland that would be supplied by WWS if no WWS electricity were exported. Numbers in parentheses are 2021 population (millions) of the country. The year after the parentheses is the year the data are valid for. Only Scotland has tidal/wave electricity. Sources: IEA (2021);[232] Scottish Government (2021);[233] Kuhudzai (2021).[234]

For example, during the first 6 months of 2019, the Coquimbo region of Chile (population 742,000) produced 99.7 percent of its electricity from WWS, with 74.9, 20.8, and 4 percent coming from wind, solar PV, and hydropower, respectively.[236]

On May 5, 2019, renewables alone provided 96 percent of California's electricity from 3 to 3:30 p.m. During the same day, 91 percent of electricity in the state was provided by renewables from 12:30 to 5:30 p.m. Similarly, on April 24, 2021, WWS supplied 88.4 percent of California's electricity at 3:25 p.m. The mix of resources was 55.9 percent solar, 24.8 percent wind, 4.9 percent geothermal, 0.73 percent small hydropower, and 2.12 percent large hydropower. From 11:00 a.m. to 5:00 p.m. on the same day, WWS supplied over 80 percent of California's electricity. California's move to 100 percent WWS is being assisted by the installation of batteries. As of September 2021, California had installed 1.44 gigawatts of 4-hour batteries.

In June 2021, the isolated town of Onslow in Australia provided 100 percent of its electricity for 80 minutes from a microgrid containing rooftop and utility PV plus battery storage. For a 6.5-day period ending December 29, 2021, South Australia supplied an average of over 100 percent of its demand with wind (64.4 percent), rooftop PV (29.5 percent), and utility PV (6.2 percent).[237]

Progress is also being made in populous countries. In 2020, renewables overtook both coal and nuclear to become the second-largest source of United States electricity after natural gas. The breakdown was 40 percent natural gas, 21 percent renewables, 20 percent nuclear, and 19 percent coal.[238] However, electricity is only 20 percent of end-use energy, and other sectors have not transitioned to the same degree.

13.1.3 Effects of Climate Change on the Reliability of Conventional and WWS Resources

Weather and climate affect both conventional and WWS energy resources. However, the increased frequency of heat waves due to global warming is a greater danger to the current energy infrastructure than to a WWS infrastructure. The main reason is that coal, natural gas, oil, biomass, and nuclear electric power plants depend on cooling water, usually from a river, lake, or the ocean, to operate safely. The effectiveness of the cooling system of such power plants is constrained by the laws of physics, regulations, and access to cooling water.[239] For example, if cooling water temperature rises too high, or if river or lake water levels drop too much, power plants that require cooling must be shut down; otherwise, they risk damage to marine life or themselves.

During an August 2018 heatwave in Europe, for example, high water temperatures caused Sweden to shut down two nuclear reactors; France to shut four reactors; and Germany, Switzerland, and Finland to reduce power from several reactors. Germany also shut a coal plant because of high water temperatures, shut another because of low water levels, and reduced output from two natural gas plants. Similarly, during a July 2019 heat wave, France and Germany each shut a nuclear reactor, and France reduced output from six more reactors.

High water temperatures (from heat) and low water levels (from drought) are not the only climate-related causes of power plant shutdowns. Others include wildfires, lightning, hurricanes, and typhoons. Between the 1990s and the 2010s, for example, the frequency of climate-related nuclear plant outages due to all these factors increased by a factor of 8.[240] Such outages are expected to increase in the decades to come.

The only WWS generator that has issues related to cooling water is CSP. However, CSP plants are more and more cooled by air instead of water, since most are located in deserts, where water is scarce. Air-cooling reduces the efficiency of CSP output slightly, but not materially, compared with water-cooling.

Global warming can impact hydropower output quite a bit, both for the better and worse. Water supplies for hydropower depend on rainfall, and global warming is changing weather patterns, causing drought in some places and flooding in others. In the global and annual average, though, evaporation must equal precipitation. In other words, the same amount of water that goes into the air must come out. As the Earth warms on average, more water evaporates from the oceans and soils. This means more rainfall plus snowfall must occur in the global average. Thus, although global warming increases droughts in some locations, it increases rainfall to a greater extent in other locations. As such, although hydropower resources will decline in some regions as the Earth warms, hydropower resources will likely increase in the global average.

The seasonal variation and yearly uncertainty in hydropower output, though, can cause problems, particularly if hydropower is the main source of electricity in a country. For example, although Tajikistan usually supplies far more hydropower electricity than the country needs (exporting the difference), it has, until recently, had to require its citizens to reduce electricity use during winters because of low reservoir water levels. Tajikistan addressed this problem by building additional hydropower plants with reservoirs.

The increased frequency of drought in some locations due to global warming is increasing the concern about water shortages affecting hydropower output and drinking water supplies. The western United States, for example, which has a lot of hydropower, risks losing a portion of hydropower output due to extended and more frequent drought. However, the western United States and California in particular import much of their hydropower from Canada. So, the loss of local hydropower can be replaced by increasing imported hydropower. In addition, the west has enormous solar and wind resources. As such, during a drought, more sunlight is available so additional solar electricity production substitutes in for the reduced hydropower availability.

Another impact of high temperatures is on solar PV output. High air temperatures reduce the output of solar PV panels, but they do not cause the panels to shut down. On the other hand, high temperatures often occur when the sky is cloud-free, so more sunlight is available to PV panels during hot periods.

The effects of climate change on wind energy output are more complex. Wind energy output depends on wind speeds 30 to 200 meters above the ground or ocean surface. Buildings and vegetation slow down the wind, as does air pollution. For example, measurements between 1974 and 1994 from 79 stations in Southeast China indicated that something caused wind speeds to decrease by 24 percent, resulting in a 60 percent decrease in wind power.[241] This **disappearing wind syndrome** was attributed to the construction of new buildings and the growth of vegetation near meteorological stations. However, heavy air pollution may have also played a role, as it can reduce near-surface wind speeds by up to 8 percent.[242]

The reason that air pollution reduces wind speeds is as follows. Some air pollutants absorb sunlight, heating the air but also preventing the light from getting to the surface, cooling the ground. Remaining air pollutants reflect sunlight, also cooling the ground relative to the air. Thus, all air pollutants cool the ground relative to the air. Warmer air above a cooler ground makes it more difficult for air or winds near the ground to be lifted or for air or winds high above the ground to descend to the surface. Since horizontal winds speeds high above the ground are always faster than near the ground (wind speeds at the ground itself are always zero), it is now more difficult for fast horizontal winds high up to be transported lower down to the height of wind turbines. Thus, air pollution decreases wind speeds available to wind turbines. The consequence is that, as air pollution is cleaned up during a transition to 100 percent WWS, wind power output will increase.

However, wind speeds are also affected by changes in climate. For example, global warming may be increasing wind energy resources in northern Europe, the southern U.S. Great Plains, and Brazil, but decreasing wind resources in southern Europe and the western United States.[243]

In sum, global warming has shifted some locations and magnitudes of wind resources, and air pollution has decreased wind speeds at the height of modern wind turbines. As such, transitioning to 100 percent WWS will undo some of those changes.

Global warming is expected to have little impact on geothermal electricity or tidal electricity since geothermal electricity depends on temperatures far below the Earth's surface, and tidal electricity depends on the gravitational attraction between the moon and the Earth. Because waves are driven by offshore winds, global warming is likely to affect wave power similarly to how it affects offshore wind power.

Finally, higher temperatures due to global warming will increase summer air conditioning demand and decrease winter heating demand. This will shift more electricity demand from winter to summer. Most all air conditioning and heating in a WWS world will be powered by electric heat pumps.

13.2   Methods of Meeting Energy Demand Continuously

Given the minute-by-minute, daily, seasonal, and yearly variations in both energy demand and WWS supply, it is critical to develop methods to ensure the reliable matching of demand with supply on all these timescales. In a 100 percent WWS world, all energy produced and consumed will be in the form of electricity or heat, with electricity dominating, but with heat (for both buildings and industry) being a substantial share of energy consumed. Thus, a good portion of electricity must be converted to heat, either with electric heat pumps or high-temperature industrial machines. A portion of the electricity will also be used for cooling. Another portion of electricity will be needed to produce hydrogen to be used for transportation, for some industrial nonheating processes, for some electricity and heat production in remote microgrids, and for ammonia production.

There are several steps in designing an integrated energy system in which electricity, heat, cold, and hydrogen demand are matched with 100 percent WWS electricity and heat supply plus storage, demand response, and transmission. These are discussed below. The overall strategy, though, is divided into two categories of steps. The first step is to create a WWS infrastructure. The second step is to match demand with supply, storage, and demand response within the infrastructure. These steps are summarized in the next two sections.

13.2.1 Steps in Creating an Infrastructure for Providing All Energy with WWS

The first category of steps is to create a WWS infrastructure. This category involves taking the following actions:

A. Transition transportation to run on batteries and hydrogen fuel cells, where the hydrogen is produced from electricity.
B. Transition all building air and water heat so that it is provided with electric heat pumps, direct solar and geothermal heat, and/or district heat. Also, ensure that district heat is produced by electric heat pumps, direct solar heat, and/or direct geothermal heat. Use district heat immediately or store it in water tanks or underground. The types of underground storage include boreholes, water pits, and aquifers.
C. Transition air conditioning in buildings to run on the same heat pumps as those used for air heating in buildings, or to run on a district cooling loop that is part of a fourth-generation or higher district heating system. The district cooling system should include cold storage in water tanks and/or ice.
D. Electrify high-, medium-, and low-temperature heat generation needed for industry with electric arc furnaces, induction furnaces, resistance furnaces, dielectric heaters, electron beam heaters, CSP steam, and/or heat pump steam.
E. Transition all remaining combustion appliances, machines, and processes to electric ones. These include moving to electric induction cookers, electric fireplaces, electric leaf blowers, electric lawnmowers, and electric chainsaws, for example.

F. Build wind farms in cold regions because wind output and heat demand both increase with lower temperatures, and the wind electricity can help meet the heat demand.

G. Interconnect, through the transmission system, geographically dispersed wind, wave, and solar resources to turn some of their variable supply into steadier supply while reducing overall transmission requirements.

H. Build both wind and solar together, since wind and solar are complementary in nature.

I. Increase energy efficiency in buildings by using LED lights, installing energy-efficient appliances, reducing air leaks in buildings, and increasing insulation in walls and windows.

J. Improve passive heating and cooling in buildings by using thermal mass, ventilated facades, window blinds and films, and night ventilation.

K. Provide all electricity with onshore and offshore wind, solar PV on rooftops and in power plants, CSP, geothermal power, tidal power, wave power, and hydroelectric power.

L. Store WWS electricity in CSP storage, pumped hydropower storage, stationary batteries, vehicle batteries, flywheels, compressed air storage, and/or gravitational storage with solid masses.

M. Build more short- and long-distance high-voltage AC transmission, long-distance high-voltage DC transmission, and AC distribution lines to interconnect WWS supply and demand centers.

With respect to Step F, building wind farms in cold regions increases the reliability of the overall grid. The reason is that wind energy production in cold regions correlates strongly with heat demand in buildings. This result has been obtained from both data analysis[244] and computer modeling.[245] The correlation is strongest when averaging over large, cold regions, such as Canada, Europe, Russia, and the United States.[245] Even on an hourly timescale, wind energy output is well matched with the demand for heat in buildings. Wind turbine electricity in cold regions is best used to run ground-, water-, or air-source heat pumps to provide heat for buildings.

With respect to Step G, interconnecting geographically dispersed wind, wave, or solar farms to a common transmission grid smooths out electricity supply significantly.[246] In fact, interconnecting wind farms over regions as small as a few hundred kilometers in size can eliminate hours of no power summed over all farms. For example, in a

550 by 700 square-kilometer region of the Great Plains of the United States, low wind speeds resulted in no output from an isolated wind farm 7.6 percent of the hours during a year between noon and 3 p.m. However, with three interconnected wind farms spread out over the same region, the number of no-power hours decreased to 2.6 percent. With eight wind farms spread out over the same region, the number decreased all the way to zero percent. In other words, interconnecting geographically dispersed wind farms eliminates times of the year when no wind power output occurs over a region.[248]

Interconnecting wind farms also means that wind, aggregated geographically, acts more like a single wind farm with steadier, closer-to-baseload, power production. In fact, when 19 geographically dispersed wind sites in the Midwest, over a region 850 by 850 square kilometers, were hypothetically interconnected, between 33 and 47 percent of yearly averaged wind power was calculated to be available at the same baseload reliability as a coal-fired power plant.[248] The benefits of interconnecting increased with more and more interconnected sites, although with diminishing returns.

Transition highlight

One more benefit of interconnecting geographically dispersed wind farms is that it reduces transmission requirements with little power loss. For example, connecting 19 wind farms to a common point and then connecting that point to a far-away city reduces the need for long-distance transmission by, for example, 20 percent with only a 1.6 percent loss of energy.[248]

Co-locating a wind and a wave farm has similar benefits to interconnecting geographically dispersed wind farms. Namely, it dampens overall power variations[249] and reduces transmission requirements.[250] Thus, for example, combining wind and wave power at the same location results in lower variability and fewer hours of no power than a wind farm or wave farm acting alone.

With respect to Step H, it is important for grid stability to build both wind and solar in the same region. The reason is that wind and solar are complementary in nature. During the day, when the wind is not blowing, the sun is often shining, and vice versa. This occurs because, when surface air pressure is high (such as under a high-pressure system), winds are generally slow. At the same time, descending air in a high-pressure system evaporates clouds, increasing sunlight to the surface. Conversely, when surface air pressure is low (such as under a low-pressure system), winds are generally fast. At the same time, rising air in a low-pressure system increases cloudiness, reducing sunlight to the surface.

The anticorrelation between wind and solar is remarkably consistent worldwide.[245] The implication of this is that co-locating wind and solar farms reduces the variability of the overall power output of wind alone or solar alone. Thus, both should be built together (either co-located or built in the same region) where possible to smooth out the overall power supply.

13.2.2   Steps for Matching Demand with Supply, Storage, and Demand Response

The second category of steps for keeping the grid stable involves matching instantaneous electricity, heat, cold, and hydrogen demand with WWS electricity and heat supply, storage, and demand response.

Demand response is the use of financial incentives to shift the time of a flexible demand to a future time when more energy is available to satisfy the demand. **Flexible demands** are demands for electricity, heat, cold, or hydrogen that can be shifted forward in time during a day or over several days by demand response.

For instance, charging an electric vehicle is a flexible demand because the time of charging can be shifted to a nonpeak time of grid electricity use through a financial incentive, such as a lower nighttime electricity rate. In fact, any electricity demand that can be shifted in time is a flexible demand. Examples are electricity used for (1) a wastewater treatment plant, (2) charging a stationary battery, (3) pumping water to an upper reservoir in a pumped hydropower system, (4) storing energy in a flywheel, (5) heating water in a hot water tank with a heat pump, (6) producing ice for later use, or (7) producing and storing hydrogen for later use.

Similarly, electricity demand for high-temperature industrial heat is a flexible demand if the heating can be scheduled flexibly to take place at different times of day. For example, a utility may establish an agreement with an industrial manufacturing plant or a wastewater treatment plant for the plant to use electricity during only certain hours of the day in exchange for a better electricity rate. In this way, the utility can shift the time of electricity use to a time when more WWS electricity supply is available.

Similarly, the demand for electricity to charge battery-electric vehicles is a flexible demand because such vehicles are generally charged at night, and it is not critical which nighttime hours the electricity is supplied so long as the full power is provided sometime during the night. In this case, a utility can use a smart meter to provide electricity for battery-electric vehicle charging when WWS is available and to stop charging when it is not. Utility customers would sign up their vehicles under a plan by which the utility controlled the time it supplied power for charging vehicles.

Inflexible demands (loads), on the other hand, are electricity, heat, cold, or hydrogen demands that cannot be shifted in time. Examples are electricity for refrigerators, lighting, and electric induction cooktops.

Whereas the electricity stored in an onboard battery used to drive an electric vehicle or to power an electric leaf blower is an inflexible demand, the electricity used to charge the batteries in both cases is a flexible demand. Similarly, whereas low-temperature heat and cold demands for building heating and cooling are inflexible demands, the electricity or direct heat used to provide heat or cold for the storage media – water tanks, borehole fields, water pits, and aquifers – are flexible demands. Finally, whereas the use of hydrogen in a hydrogen fuel-cell vehicle is an inflexible demand, the electricity used to produce hydrogen for the vehicle is a flexible demand.

Incentives for demand response can be either a lower price for electricity during times of low demand or a direct payment to the customer in exchange for the customer agreeing not to use electricity during the time of peak demand. The demand that is shifted can be an electricity, heat, cold, or hydrogen demand.

The detailed steps for meeting demand with supply, storage, and demand response are as follows.

A. Determine the annual end-use energy demand in each energy sector. Size WWS electricity and heat generators to meet the annual average demand.

B. Estimate the additional generation, storage, and demand response needed to meet demand continuously over time rather than just annually. Do this by first quantifying (1) electric and direct heat demands for low-temperature heating that must be met; (2) electric demands for cooling and refrigeration that must be met; (3) electric demands for producing, compressing, and storing hydrogen for fuel cells that must be met; (4) all other electric demands (including industrial heat demands) that must be met. Of these demands, identify which ones are inflexible, and which are flexible and thus subject to demand response.

C. Once the generation, storage, and demand response system is designed, carry out the following steps to match instantaneous demand with supply.

D. When the current WWS electricity supply exceeds the current inflexible electricity or heat demand, use the electricity supply to satisfy the demand.

E. When the current WWS solar plus geothermal heat supply exceeds the current inflexible heat demand, use the heat supply to satisfy the demand.

F. Use the remaining instantaneous WWS electricity or heat supply to satisfy as much existing flexible electric or heat demand as possible.

G. Use any excess electricity remaining after that to fill electricity, heat, cold, or hydrogen storage. Select which type of storage to fill first.

H. Use any excess geothermal or solar heat after satisfying inflexible and flexible heat demand to fill heat storage.

I. When the current inflexible plus flexible electricity demand exceeds the current WWS electricity supply from the grid, use electricity storage to fill in the gap in supply. Use the electricity to supply inflexible demands first.

J. If electricity storage is depleted and flexible demand remains, use demand response to shift the flexible demand to a future hour.

K. If inflexible plus flexible heat demand in buildings exceeds current supply, use stored heat. If stored heat is exhausted, shift flexible heat demand to a future hour with demand response.

L. If inflexible plus flexible cold demand in buildings exceeds current supply, use stored cold. If stored cold is exhausted, shift flexible cold demand to a future hour with demand response.

M. Provide hydrogen demand with current hydrogen supply, where the current hydrogen is produced by current electricity. If current hydrogen supply is exhausted, use stored hydrogen.

N. If the measures above are insufficient to match energy demand with supply continuously, do one or more of the following: (1) increase the nameplate capacity of some electricity generators; (2) increase electricity, heat, cold, or hydrogen storage nameplate capacity; (3) increase the nameplate capacity of the transmission system to allow more renewables that are far away, thus subject to different weather patterns, to help fill in gaps in supply; (4) use electricity stored in vehicle batteries to help fill in gaps in supply on the grid; and/or (5) integrate weather forecasts into electricity planning to improve the efficiency of the grid.

In the next sections, some of the steps required for matching demand with supply, storage, and demand response are discussed in more detail.

13.2.3   Specify WWS Generator Requirements to Meet Continuous Demand

The first step in the process of matching demand with supply, storage, and demand response is to estimate future (e.g., 2050) all-sector end-use annual energy demand, then to propose a mix of WWS electricity and heat generators to match those annual-average demands. Chapter 12 discussed the methodology for carrying out this step.

Generators sufficient for meeting demand in the annual average are not sufficient for meeting demand every minute. Demand varies continuously, as does electricity supply from most WWS resources. Although WWS generators provide more than enough electricity during some minutes of a year, they provide less electricity than needed during others. As such, additional methods must be used to match demand with supply. One method is to oversize the nameplate capacity of electricity and heat generators. A second is to use storage. A third is to use demand response.

Oversizing the nameplate capacities of electricity generators increases the average power supply to the grid during the year. With a higher average supply of electricity, more peaks in demand are met before storage is even considered.

However, to guarantee that demand will be met *every* minute of a year without any storage or demand response, it may be necessary to oversize the nameplate capacity of WWS generation by a factor of 5 to 10 or more. This will result in an exorbitant cost of energy, particularly because the WWS generators will produce 5 to 10 times more electricity than used, in the annual average. As such, 80 to 90 percent of the excess electricity will be wasted. When the electricity is wasted, the cost per kilowatt-hour produced goes up, in this case, by a factor of 5 to 10.

13.2.4 Specify Storage and Demand Response Requirements to Meet Continuous Demand

No future proposal calls solely for oversizing of the nameplate capacity of generators to meet peaks in demand. Future proposals call for combining some additional generation with storage and demand response. To determine the ideal amount of storage and demand response needed to meet continuous demand, it is important first to characterize the demand. Specifically, total annual average demand should be separated into

(1) electricity and heat demands needed for low-temperature air and water heating,
(2) electricity demands needed for cooling and refrigeration,
(3) electricity demands needed to produce, compress, and store hydrogen
(4) all other electricity demands (including industrial heat demands).

Each of these demands should then be divided into flexible and inflexible demands.

Flexible demands include most electric and heat demand that can be used to fill cold storage and low-temperature heat storage, all electricity used to produce hydrogen (since all hydrogen can be stored), and all remaining electric and heat demands subject to demand response. For example, electricity needed to heat an electric heat-pump hot water tank is a flexible demand because the heat can be produced from electricity any time before the hot water is needed. Similarly, electricity for charging an electric car is a flexible demand because it can be used to charge the battery any time before the car is needed for driving.

Inflexible demands include electricity for lights and computers, for example. However, electricity needed immediately to produce hot or cold air when insufficient hot or cold storage is available is also an inflexible demand. Similarly, electricity needed immediately to charge an electric car battery when the driver has stopped at a charging station during a long road trip is an inflexible demand.

Inflexible demands must be met immediately. However, if the cost of an inflexible demand is high enough, many individuals will forgo the demand to save money. As such, no demand is truly inflexible. Demands are inflexible only up to a certain cost. Beyond that cost, demands can be shifted in time with demand response.

Electricity storage is needed to meet inflexible electricity demand when current electricity generation is exhausted. Whereas heat, cold, and hydrogen storage are relatively inexpensive, lithium-ion battery storage is still expensive as of 2022. On the other hand, iron–air batteries for stationary electricity storage may have the potential to decrease battery storage costs by a factor of 5 to 10.[36] Other forms of electricity storage (CSP storage, pumped hydropower storage) are already less expensive than lithium-ion batteries but are limited to certain geographic regions or face other hurdles that limit the pace at which they will be adopted.

When the cost of stationary battery electricity storage is no longer a barrier, battery storage will be expanded rapidly enough, not only to fill short-term gaps in electricity supply, but also to store large amounts of electricity that can be used for days or weeks if the current WWS electricity supply drops to zero for that length of time. Fortunately, storing battery electricity for days, weeks, or months does not require a battery that provides hundreds or thousands of hours of storage at the peak discharge rate. To the contrary, long-duration electricity storage can be obtained simply by concatenating existing-technology batteries that have 4, 8, or 12 hours of storage.[251]

For example, suppose a battery stores up to 4 hours of electricity when the electricity is discharging at the peak discharge rate of 10 kilowatts. This means that the maximum storage capacity of the battery is 40 kilowatt-hours. Suppose also that 10 kilowatts of power must be supplied to meet demands on the grid for 100 hours (4.17 days) straight because no WWS generation is available on those days. The storage capacity needed in this case is then 1,000 kilowatt-hours. One way to provide the necessary storage is to invent a new 100-hour battery that has a peak discharge rate of 10 kilowatts. An easier way, though, is just to concatenate together 25 existing-technology 4-hour batteries, each with a peak discharge rate of 10 kilowatts. Once one 4-hour battery is depleted, the second is used, and so on.

There are two main advantages of the second option. The first is that 4-hour batteries already exist, whereas 100-hour batteries do not. The second and more important advantage is that having 25 of the 4-hour batteries with a peak discharge rate of 10 kilowatts each allows the peak discharge rate from storage to equal 250 kilowatts over a 4-hour period. Such a peak discharge rate may be needed if demands on the grid are that high. The 100-hour battery, on the other hand, can never provide more than 10 kilowatts of power, so cannot even home-charge many electric cars at their peak charge rate (20 kilowatts). Thus, having a short storage time maximizes the flexibility of batteries both to meet peaks in demand (kilowatts) and to store electricity for long periods (kilowatt-hours). In summary, concatenating 25 10-kilowatt batteries, each with 4 hours of storage, allows for either 100 hours of storage at the peak discharge rate of 10 kilowatts or 4 hours of storage at the peak discharge rate of 250 kilowatts, or anything in between.

Hydropower and batteries are similar in that hydropower can provide electricity at a high discharge rate for a short period or at a low discharge rate for a long period. The main disadvantage of hydropower (unless it is pumped hydropower) is that it recharges only naturally. Pumped hydropower, on the other hand, can be charged whenever excess electricity is available. Also, hydropower is limited by how many new dams can be built and how much water is available behind each dam.

Most other electricity storage types, such as gravitational storage, compressed air storage, and CSP with storage, operate over time-scales of minutes to days to a couple of weeks (if the discharge rate is slow).

Heat and cold can be stored in boreholes, water pits, and aquifers over periods of minutes to seasons. The cost of such storage is relatively low, so such storage can be and has been used season-ally. Hydrogen storage tanks can also store hydrogen for minutes to seasons.

13.2.5 Procedures for Matching Demand with Supply in Cases of Over and Under Generation

Once electricity and heat generator sizes, storage sizes, and demand response characteristics are defined, a procedure is needed to ensure instantaneous energy demand matches instantaneous supply plus storage plus demand response continuously. Procedures are needed for two main situations. One is when instantaneous generation exceeds demand. The other is when instantaneous demand exceeds generation. Procedures for each case are discussed next.

13.2.5.1 Steps When Instantaneous WWS Electricity or Heat Supply Exceeds Instantaneous Demand

This section discusses the steps taken when current (instantaneous) WWS electricity or heat supply exceeds the current electricity or heat total demand. Total demand consists of flexible and inflexible demands. Whereas flexible demands may be shifted forward in time with demand response, inflexible demands must be met immediately.

If WWS instantaneous electricity or heat supply exceeds instantaneous inflexible electricity or heat demand, then the supply is used to satisfy that demand. The excess WWS is then used to satisfy as much current flexible electric or heat demand as possible.

If any excess electricity exists after inflexible and current flexible demands are met, the excess electricity is sent to fill electricity, heat, cold, or hydrogen storage. Electricity storage is filled first. Excess CSP energy goes to CSP electricity storage. Remaining instantaneous CSP electricity and excess electricity from other sources are then used to fill battery storage, followed by pumped hydropower storage, flywheel storage, gravitational mass storage, compressed air energy storage, cold water storage, ice storage, hot water tank storage, and underground thermal energy storage. Remaining electricity is used to produce hydrogen. Any residual after that is shed.

Heat and cold storage are filled by using excess electricity to run a heat pump, which moves heat or cold from the air, water, or ground to a thermal storage medium (water or soil usually). Hydrogen storage is filled by using electricity to produce hydrogen with an electrolyzer, then to compress the hydrogen before it is put into a storage tank.

If any excess direct geothermal or solar heat exists after it is used to satisfy inflexible and flexible heat demands, it is then used to fill either district heat storage (water tank and underground heat storage) or heat storage in small hot water tanks in buildings.

13.2.5.2 Steps When Instantaneous Demand Exceeds Instantaneous WWS Electricity or Heat Supply

When current inflexible plus flexible electricity demand exceeds current WWS electricity supply from the grid, the first step is to use electricity storage (CSP, battery, pumped hydro, hydropower, flywheel, gravitational mass, and compressed air energy storage) to fill in the gap in supply. The electricity is used to supply the inflexible demand first, followed by the flexible demand.

If electricity storage becomes depleted and flexible demand persists, demand response is used to shift the flexible demand to a future minute or hour.

If the inflexible plus flexible heat demand subject to storage exceeds WWS direct heat supply, then stored district heat (in water and soil) is used to satisfy district heat demands subject to storage. Stored heat in building water tanks is used to satisfy building water heat demands. If stored heat becomes exhausted, then most of any remaining low-temperature air or water heat demand becomes an inflexible demand that must be met immediately with electricity. The rest becomes a flexible heat demand that can be met with current electricity or shifted forward in time with demand response.

Similarly, if the inflexible plus flexible cold demand subject to storage exceeds the cold energy available in ice or water storage, most of the excess cold demand becomes an inflexible demand that must be met immediately with electricity. The rest becomes a flexible cold demand that can be met with current electricity or shifted forward in time with demand response.

Finally, if the current hydrogen demand depletes hydrogen storage, the remaining hydrogen demand becomes an inflexible electrical demand that must be met immediately with current electricity supply.

13.2.6 Measures Needed When Instantaneous Demand Cannot Be Met with Current Supply or Storage

If the measures just described are insufficient for meeting power demand with supply continuously, the system must be modified to ensure reliability. The standard for reliability is a **loss of load expectation** of no more than 1 day (24 hours) in 10 years. Loss of load arises through the failure to match instantaneous demand with instantaneous supply or storage. It can arise because of a severe lull in the supply of both wind and sunlight over a long period, scheduled or unscheduled maintenance of many energy generators at the same time, an unexpected heavy demand due to a heat wave or a cold spell, transmission line congestion, or a transmission line outage.

If supply does not match demand consistently, it will be necessary to do one or more of the following: (1) increase the nameplate capacity of the electric power generators, (2) increase the electricity, heat, cold, or hydrogen nameplate storage capacity available, (3) increase the nameplate capacity of the transmission system to allow more renewables that are far away and subject to different weather patterns to help fill in gaps in supply, (4) use electricity stored in vehicle batteries to help fill in gaps in supply on the grid, and/or (5) integrate weather forecasts into system operations to improve efficiency of the grid. Some of these measures also help to increase the efficiency of the grid, even when supply matches demand continuously. These techniques are discussed next.

13.2.6.1 Oversizing Wind, Water, and Solar Generation to Help Meet Demand

Oversizing the nameplate capacity of WWS installations can help to increase the number of times during a year that available WWS electricity supply exceeds inflexible demand. This solution, therefore, reduces the number of outages that occur during a year. Although oversizing helps to meet hours of peak demand, it also results in excess supply during many other hours. Ideally, the excess supply is put into electricity storage or used to produce heat, cold, and/or hydrogen, which are either stored or used immediately. Using excess WWS in these ways avoids WWS curtailment, reducing system cost.

13.2.6.2 Oversizing Storage to Help Meet Peaks in Demand

Oversizing storage is a second way to improve matching power demand with supply and storage on the grid. Because wind, solar, and wave power are variable WWS resources, they often overproduce electricity. Existing storage may not be enough to hold the excess electricity. Purchasing more storage can help to allow the excess electricity to be used during additional hours when WWS electricity is not available. The same applies to heat, cold, and hydrogen storage.

13.2.6.3 Increasing Transmission Nameplate Capacity to Help Meet Demand

Interconnecting geographically dispersed variable renewable resources such as wind, solar, and wave power, smooths the aggregate power supply from these generators. Similarly, interconnecting wind and solar, which are complementary in nature, smooths out overall supply. As such, an additional solution to keeping the grid stable is to import far-away variable renewable electricity through an upgraded transmission system. The long-distance transmission lines should be HVDC lines to reduce line losses, thus reducing system costs. Although each country will ideally produce its own energy, interconnecting renewables among nearby countries reduces overall costs as well.[252] The reason is that some countries have better wind and hydropower resources than do others, whereas others have better solar resources. Interconnecting nearby countries to take advantage of the plentiful resources in each country reduces times that no power is available and allows storage to be replenished more regularly. Interconnecting also allows aggregate transmission requirements to be reduced with little loss in annual energy output.

13.2.6.4 Helping to Balance Demand with Vehicle-to-Grid

Another method to help match demand with supply is to store electric energy in battery-electric vehicles, then to withdraw such energy when it is needed for the grid. This concept is referred to as **vehicle-to-grid**.[253] With vehicle-to-grid, a utility operator enters into a contract with an individual battery-electric vehicle owner to allow electricity transfers back to the grid any time during an agreed-upon period in exchange for a lower electricity price.

Vehicle-to-grid is expected to wear down batteries faster by increasing the number of charge and discharge cycles the battery goes through during a year. However, only a small percentage of vehicles need participate in a vehicle-to-grid plan to help meet demand with supply. One study suggests only 3.2 percent of vehicles in the United States are needed to smooth out U.S. electricity demand when 50 percent of demand is supplied by wind.[254] Through computer modeling, other studies[255,256] further found that using vehicle-to-grid would help to match demand with supply and storage hourly in the eastern United States and in the Aland Islands (Baltic Sea), respectively. An alternative to vehicle-to-grid is simply to avoid charging battery-electric vehicles when electric power is in short supply and charge them when too much WWS electricity is available. Such an action is a form of demand response.

13.2.6.5 Using Weather Forecasts to Plan for and Reduce Backup Requirements

Forecasting the weather (winds, sunlight, waves, tides, and precipitation) gives grid operators more time to plan ahead for a backup energy supply. Good forecast accuracy can also allow reserves to be shut down more frequently, increasing the overall grid efficiency.[257] Forecasting is done usually with a numerical weather prediction model, the best of which can produce minute-by-minute predictions 1 to 4 days in advance with good accuracy. Forecasting is also done using statistical analyses of local measurements. The use of forecasting reduces uncertainty and makes planning more dependable, thus reducing the impacts of variability, reducing costs.

13.2.7 Ancillary Services: Load Following, Regulation, Reserves, and Voltage Control

An electric power system needs to meet two major requirements. One is to match current demand with current generation plus storage. The second is to manage power flows between and among individual transmission facilities. This section discusses both issues in light of practical grid operations.

Meeting instantaneous demand is difficult because demand and supply continuously vary in time. Demand can vary over seconds to minutes owing to the random turning on and off of millions of individual inflexible demands. Demand can also vary seasonally because of changes in heating and cooling demand during the year. WWS electricity and heat production also varies minutely to seasonally.

Ancillary services are services performed by equipment and people that are, according to the U.S. Federal Energy Regulatory Commission (FERC), *"necessary to support the transmission of electric power from seller to purchaser given the obligations of control areas and transmitting utilities within those control areas to maintain reliable operations of the interconnected transmission system"*.[258]

The main ancillary services include load following, regulation, spinning reserves, supplemental reserves, replacement control, and voltage control. These are discussed, in turn.

13.2.7.1 Load Following

Load following is the use of power generators or storage to follow the rate of change (megawatts per minute) of the demand for electricity, averaged over a 5 to 15-minute period. Load following can also be accomplished by shifting part of the demand forward in time (demand response).

> **Transition highlight**
>
> Load following can be accomplished by rapidly reducing the power consumption of large industrial customers. Some industrial demands that can be reduced quickly include air liquefaction; induction and ladle metallurgy; water pumping with variable speed drive; and production by electrolysis of aluminum, chlor-alkali, potassium hydroxide, magnesium, sodium chlorate, and copper.[258] Indeed, as the U.S. National Research Council states,[259]
>
> *"The ability of industry to cut peak electric loads is a motivator for utilities to incentivize demand response (shifting loads to off-peak periods) in industry... In combination with peak-load pricing for electricity, energy efficiency and demand response can be a lucrative enterprise for industrial customers."*

Load following can similarly be accomplished by reducing power consumption in residences and commercial buildings. For example, demand response can help push the time of battery charging for electric vehicles to late at night. It can also push the time of heating water in a hot water tank or of running a dishwasher, washing machine, or dryer to a time of low overall electricity demand on the grid. Similarly, demand response can push the time of running a wastewater treatment plant or hydrogen production by electrolysis to a time of low demand.

Load following generators or storage devices can have a relatively modest ramp rate. Thus, in a WWS world, hydropower, a portion of CSP output, and a portion of solar PV output may be used to follow demand, as can CSP storage or any electricity storage. In fact, any WWS generator can load follow, since if a generator projects that it will have electricity available over a coming period (such as due to wind or solar forecasting), it can compete in a real-time electricity market to sell the electricity for load following.

13.2.7.2 Regulation

Regulation is the automatic (through electronic grid controls) use of pre-contracted power generators, storage, or demand reduction to fill in small gaps between the actual demand and the demand that results from load following. Over a 5- to 15-minute period, the actual demand varies on sub-minute timescales because of hundreds to millions of people turning demands on or off. Load following meets only the average change in demand over a period, not the minute-by-minute change in demand. Minute-by-minute regulation can be met with generators, storage, and demand reduction methods that have ramp rates 5 to 10 times the minimum ramp rates required for load following.

In a WWS world, regulation will be performed primarily with hydropower, pumped hydropower storage, batteries, flywheel storage, compressed air storage, and gravitational storage with solid masses. The first two ramp up to 100 percent power within 15 to 30 seconds; the rest, in less than a second. The best electricity producers for regulation are generally fast-responding storage technologies that can repeatedly operate through many cycles without degrading. Batteries, for example, respond quickly (within milliseconds), have output that can be controlled precisely, and can operate through many cycles. As such, they are ideal for regulation.

When intermittent WWS generators, such as wind, solar, or wave power, are aggregated together over a large geographic region, their overall variability decreases. This reduces the regulation requirement for, and thus the cost of, wind, solar, and wave power.[258] In other words, aggregation makes it easier for battery storage to meet the difference between instantaneous fluctuation in demand and WWS generation.

Because of the fast response time required for regulation services, such services are not obtained from market signals, which respond too slowly. Instead, generators, storage, and demand reduction used for regulation are contracted ahead of time.

Frequency Regulation

Frequency regulation helps to maintain the AC frequency on the grid. The AC frequency on the grid is required to be close to 60 hertz in the United States and some other countries, and 50 hertz in Europe and remaining countries. The deviation allowed from this mean frequency in the United States is from 59.7 to 60.3 hertz. In most of Europe, it is from 49.8 to 50.2 hertz. In some places in Eastern Europe, it is broader, from 47 to 53 hertz. Frequency is maintained within a narrow range because too high or too low of a frequency can destroy electrical equipment on the grid and end-use electrical appliances and devices.

When the frequency on the grid falls above the range allowed, averaged over a minute, grid operators reduce the frequency down into the permissible range. They do this by reducing electricity added to the grid from generation or storage or by increasing demand on the grid. In other words, a frequency that is too high can be reduced by slightly reducing the quantity of electricity added to the grid. Conversely, adding slightly more electricity to the grid can push a low frequency into the normal range. The increase or decrease of generation, storage, or demand to regulate frequency on the grid is called **frequency regulation**.

Transition highlight
Traditionally, grid operators have increased frequency by increasing generation from gas, coal, oil, nuclear, or hydroelectric plants. In a WWS world, operators will increase frequency by increasing hydroelectric, geothermal, wind, and solar PV output as well as output from CSP storage, batteries, pumped hydropower storage, flywheels, compressed air storage, and gravitational storage with solid masses. Conversely, grid operators will reduce frequencies that are too high by reducing generation and storage or turning on artificial demands.

In the case of wind turbines, an operator can increase generation by running some turbines in partial load mode such that, when a decrease in grid frequency occurs, the operator increases turbine output by varying blade pitch. Alternatively, the operator may control short releases of electricity from wind turbines to the grid.[260] Similarly, some solar PV plants can be run at less than full output such that, when a decrease in grid frequency occurs, output is increased. Inverters can also be optimized to provide frequency control to the grid.[261]

Spinning, Supplemental, and Replacement Reserves and Voltage Control

Three ancillary services – spinning reserves, supplemental reserves, and replacement reserves – are used to supply demand in the event that a large power generator or transmission line goes down.

Spinning reserves are electricity generators or storage media, online and connected to the grid, that can increase output immediately in response to a major generator or transmission outage. Spinning reserves are required to reach full output within 10 minutes.

Supplemental reserves are the same as spinning reserves, but they are not required to respond immediately. They can be offline but, when turned online, they must reach full output within 10 minutes.

Replacement reserves are the same as supplemental reserves, but they must reach full output within 30 minutes. They are also used to replenish spinning and supplemental reserves to their precontingency status.

Voltage control is the injection or absorption of reactive power to maintain transmission-system voltages within required ranges. Such control is needed on the timescale of seconds. Whereas **real power** is the energy per unit time used to run a motor or a heat pump, **reactive power** is imaginary power that does not do any useful work but simply moves back and forth within power lines. It is a byproduct of a circuit that has capacitors or inductors. Reactive power not only smooths voltage on the transmission grid by supplying or absorbing it, but it also provides needed power on a timescale of seconds to avoid blackouts. Transformers, motors, and generators all need reactive power to produce a magnetic flux.

13.3 Studies on Meeting Demand with 100 Percent WWS

Several countries now provide 100 percent of the electricity that they produce with WWS (Figure 13.1). However, electricity is only 20 percent of total end-use energy. In all countries, all energy must be provided by electricity and heat from WWS. Because no country has an all-sector 100 percent WWS energy system, computer modeling of grid stability with a 100 percent WWS system is needed to optimize its buildout. In this section, past modeling of 100 percent systems is first described. Then, results from a study for meeting all-sector energy demand with WWS supply, storage, and demand response for most of the world is discussed.

13.3.1 Previous Studies of Matching Demand with or near 100 Percent WWS

Many studies support the potential of matching power demand with 100 or near 100 percent WWS supply in one or more energy sectors.

Whereas several of these studies examine matching demand with supply in the annual average, dozens of others treat matching demand with supply at a time resolution ranging from 30 seconds to a few hours. Both types of studies are summarized below.

In 1975, Sorensen[262] examined the supply of wind and solar in Denmark and suggested that enough in combination might be available to supply all building heat, transportation, industrial, and electricity demand in the annual average by 2050. In 1976 Amory Lovins wrote an article called *Energy Strategy: The Road Not Taken*,[263] describing how the United States might be able to transition entirely to renewable energy and efficiency, a transition he called the *soft energy path*. In 1996, Sorensen then estimated[264] the annual average quantity of biomass, biofuels, and biogas energy together with WWS (wind, hydro, and solar) energy that might be needed to provide all world energy.

In 2001, Jacobson and Masters[265] estimated the number of wind turbines needed, in the annual average, to replace 60 percent of coal in the United States to satisfy the 1997 Kyoto Protocol.

Subsequently, in 2005, Jacobson et al.[266] modeled the health and climate benefits of transitioning 100 percent of the U.S. on-road vehicle fleet to hydrogen fuel-cell vehicles, where the hydrogen was produced by either electrolysis (with the electricity from wind), steam reforming of natural gas, or coal gasification. The result was that wind-electrolysis was the cleanest among the options for producing hydrogen for transportation.

Also in 2005, Archer and Jacobson[267] used data from 7,753 surface stations and 446 stations measuring vertical profiles of wind worldwide to map and quantify the land and coastal wind energy potential for every inhabited continent worldwide. They concluded that enough wind energy is available in high-wind-speed locations (where the mean annual wind speed exceeds 6.9 meters per second at an 80-meter hub height) to power the world for all purposes several times over.

In 2005 and 2007, respectively, Czisch[268] and Czisch and Geibel[269] developed a least-cost optimization model that simulated the electric power grid over Europe, North Africa, and the Middle East divided into 19 regions interconnected by HVDC transmission. They found that a renewable electricity supply of wind, solar PV, CSP, bio-energy, and hydroelectric power could keep the grid stable at low cost for 1 year.

In 2006, Lund[270] examined optimized mixes of wind, solar PV, and wave energy that meet 100 percent of electricity demand in Denmark. He concluded that 50 percent wind, 20 percent PV, and 30 percent wave may be optimal.

In 2009, Hoste et al.[271] found that by combining baseload geothermal, intermittent wind, intermittent solar, and gap-filling hydroelectric, and by assuming an expanded transmission network, California might be able to meet 100 percent of its hour-by-hour electricity demand during April and July 2020. Thus, bundling intermittent wind and solar with baseload geothermal and gap-filling hydropower may increase the ability of all resources to meet demand.

Soon after, Jacobson[133] evaluated the effects of 9 electricity sources and 12 options for producing energy for vehicles on 11 impact categories (climate, air pollution, land use, water supply, grid reliability, wildlife, energy security, and more). The study concluded that WWS resources (onshore and offshore wind, solar PV, CSP, geothermal, hydroelectric, tidal, and wave) may be the best in terms of minimizing such impacts and should be used to advance solutions to global warming, air pollution, and energy security.

Later in 2009, Jacobson and Delucchi[272] then estimated the WWS resources needed to provide 100 percent of the world's 2030 end-use power demand for all purposes. The main barriers were stated to be social and political, not technical or economic. Material requirements presented some challenges but not limitations. The study considered grid stability and land requirements. Results were obtained without nuclear power, fossil fuels with carbon capture, direct air capture, biomass, biofuels, or biogas. As such, the study was the first to propose a worldwide all-sector WWS energy system that eliminated energy-related air pollution, climate-relevant emissions, and energy insecurity simultaneously.

In 2011, Jacobson and Delucchi then developed more detailed roadmaps for both the world and the United States to meet annually averaged power demand with WWS supply.[78,274] Their group subsequently developed roadmaps of the same type for New York,[275] California,[276] Washington State,[277] the 50 United States,[145,251] 139 countries,[223] 143 countries,[278] 145 countries,[5] 53 towns and cities in North America,[279] and 74 metropolitan areas worldwide.[280]

In 2009, Lund and Mathiesen[273] modeled the hour-by-hour matching of electricity demand with wind, wave, solar, and biomass supply in Denmark for the year 2050 and determined that such a system can remain reliable.

In 2010, Mason et al.[281] modeled the electricity system in New Zealand assuming 22 to 25 percent wind, 12 to 14 percent geothermal, 1 percent biomass, 53 to 60 percent baseload hydro, 0 to 12 percent added hydro for filling gaps in supply, and assumptions about demand response. They found that matching demand with supply with this system (near 100 percent WWS) was feasible.

In 2011, Hart and Jacobson[257] modeled the California electric power grid under 2005 to 2006 conditions as if the grid were transitioned to renewables. They found that matching power demand with 99 percent of electricity supplied from noncarbon sources was possible, even without considering demand response or large-scale storage. Gaps in supply were filled primarily by hydropower.

Between 2011 and 2022, dozens of studies among 19 independent groups and 117 scientists supported the hypothesis that 100 percent renewables is possible. They found that matching 100 percent or near 100 percent time-dependent power demand with time-dependent supply and storage in one or more energy sectors is feasible at low cost.[5,136,245,278,282–325] Of note is a study from the U.S. National Renewable Energy Laboratory that found that a 100 percent renewable and stable U.S. electricity grid might cost less than 5 cents per kilowatt-hour, which is less than the cost of electricity from a new natural gas plant.[325]

Two reviews[326,327] of several 100 percent renewables studies similarly found that the methods used in the studies were rigorous. Another review discusses the growth over time in the number of 100 percent renewable energy papers.[328]

13.3.2 Matching Demand with WWS Supply, Storage, and Demand Response in 24 World Regions

Most studies mentioned discuss meeting demand in specific parts of the world. An important question is whether demand can be met among all energy sectors in all regions of the world with 100 percent WWS together with storage and demand response. This section summarizes results from one modeling study of matching demand with 100 percent WWS supply and storage in 145 countries divided into 24 world regions.[5] The regions include a mix of 9 large multi-country regions (Africa, Central America, Central Asia, China region, Europe,

India region, the Middle East, South America, and Southeast Asia) and 15 individual countries or pairs of countries (Australia, Canada, Cuba, Haiti–Dominican Republic, Israel, Iceland, Jamaica, Japan, Mauritius, New Zealand, the Philippines, Russia–Georgia, South Korea, Taiwan, and the United States). The 145 countries represent more than 99.7 percent of global fossil-fuel carbon dioxide emissions.

Electricity grids were assumed to be interconnected perfectly within each country and among all countries in each region. Winds and solar fields and building heat and cold demands were predicted in each country every 30 seconds for 3 years with a global weather-prediction model. Such a model predicts the weather instantaneously everywhere worldwide. Modeled wind speeds and solar output were then used to estimate electric power generation from a set of onshore and offshore wind turbines, wave devices, solar PV panels on rooftops, utility-scale solar PV panels, CSP plants, and solar thermal heat collectors in each country every 30 seconds for the 3 years. The time-dependent electricity production was used as input into a second model, called a grid model, that matches power demand with supply, storage, and demand response in each region or country.

Additional inputs into the grid model include geothermal and hydropower electricity generation; geothermal heat generation; electricity, heat, cold, and hydrogen storage sizes; transmission losses; time-dependent electricity, heat, cold, and hydrogen energy demands; and characteristics of demand response.

Ten types of storage were treated. Two were for heat storage in buildings (hot water tank storage and underground thermal energy storage), two were for cold storage (cold water tank storage, and ice storage), four were for electric power storage (batteries, pumped hydropower storage, conventional hydropower storage, and storage associated with CSP plants), and two were for vehicle transport (batteries and hydrogen storage).

Demand response was treated by allowing flexible demands, if they could not be met immediately, to be shifted forward every 30 seconds for up to a maximum of 8 hours. If a flexible demand was not satisfied by then, it was converted into an inflexible demand that had to be satisfied immediately. Heat pumps provided all building air-cooling and air and water heating. Some building heating and cooling was subject to district heating.

In each of the 24 world regions, the model was run forward in time following the proposed priorities for matching demand discussed earlier. Simulations were run from 2050 to 2052. Multiple successful simulations were run for each region, but only the lowest-cost solution is presented.

Energy supply matched total energy demand (end-use energy demand plus changes in storage plus losses plus shedding) every 30 seconds for all 3 years in all 24 world regions encompassing all 145 countries.

Energy social costs are energy private costs (costs of energy alone) plus health and climate damage costs due to energy. In a 2050 100 percent WWS world, WWS energy private costs equal WWS energy social costs because WWS eliminates all health and climate costs associated with energy. This is because WWS generation, storage, and transmission result in zero emissions. Further, in 2050, WWS technology production and decommissioning will be free of energy-related emissions. The health and climate costs of zero emissions are zero.

The private energy cost per unit energy includes the costs of new electricity and heat generation, short-distance transmission, long-distance transmission, distribution, heat storage, cold storage, electricity storage, hydrogen production/compression/storage, and heat pumps for filling district heating.

Since a 100 percent WWS system relies a lot on electricity storage, one big question is, will battery prices drop sufficiently to make a transition to WWS seamless? If battery prices do drop, then that eliminates the major roadblock to a rapid transition, which is the concern about avoiding blackouts on the grid due to low availability of WWS electricity.

In 2021, lithium-ion battery cell and pack prices were an average of $101 and $132 per kilowatt-hour of storage, respectively.[329] A battery cell is an individual container that stores electricity in the form of chemical energy. For example, the cylindrical battery that goes into a flashlight is a single cell. A battery pack is a package that contains battery cells, software, and a heating and cooling system. Battery packs are installed in an electric car to deliver electricity to the car, or on the wall of a home to deliver electricity to the home. Battery packs (including cells) cost about 30 percent more than the battery cells alone. Based on the 2021 battery cost data and a trend in battery prices,[330] the 2035 battery cell and pack costs are projected to be less than $25 and $33 per kilowatt-hour of storage, respectively. The 145-country study discussed here[5] assumed a battery pack cost of $60 per kilowatt-hour of storage. As such, if battery costs do indeed drop to $33 per kilowatt-hour of storage, then a transition to 100 percent WWS should be even faster, more feasible, and more cost-effective than discussed here.

From the study, the 2050 WWS mean private cost per unit energy among all 24 regions is approximated to be 8.5 cents per kilowatt-hour of all energy consumed, in 2020 U.S. dollars. Costs for individual regions range from 6.5 cents per kilowatt-hour (Canada) to 12.6 cents per kilowatt-hour (South Korea). The largest portion of cost is generation (which includes the costs of capital, operation, maintenance, and decommissioning), followed by transmission and distribution, battery electricity storage, hydrogen production, and thermal energy storage, respectively.

The overall **upfront capital cost** of transitioning the 145 countries while keeping the grid stable, to be spent between 2020 and 2050, may be about $61.5 trillion, in 2020 U.S. dollars. This is the approximate capital cost of a worldwide *Green New Deal*. Individual regional ranges are $2.8 billion for Iceland to $13.3 trillion for China. The U.S. cost is about $6.7 trillion. The cost for Europe is about $5.9 trillion.

A more relevant metric is the **annual energy cost**. This is determined by multiplying the cost per unit energy by the energy consumed per year, for each the WWS and business-as-usual cases. Since a WWS infrastructure consumes 56.4 percent less energy per year in 2050 than does a business-as-usual infrastructure,[5] the annual energy cost must be much lower with WWS.

In fact, transitioning from business-as-usual to WWS decreases the annual private energy cost by 62.7 percent, from $17.8 to $6.6 trillion per year, among all 145 countries. Transitioning also reduces the annual social energy cost by 92 percent (from $83.2 to $6.6 trillion per year). This larger reduction in the annual social cost versus private cost occurs because WWS eliminates health and climate costs, which are included in the social cost per unit energy but not in the private cost per unit energy.

The large annual saving in private energy cost resulting from moving to WWS means that the capital cost of a WWS system can be paid back quickly. For example, averaged over 145 countries, the annual cost savings result in a capital cost payback time of 5.5 years. The even larger annual saving in social energy cost results in a capital cost payback time of less than a year. In sum, the capital cost of WWS pays for itself with energy, health, and climate cost savings rapidly, and even the amount paid back is through energy sales rather than subsidy. Whereas private energy cost savings can be seen by individual consumers, social energy cost savings are seen by a country as a whole.

> ### Transition highlight
> In island countries, such as in the Caribbean, the actual price paid for business-as-usual electricity is currently a mean of about 33 cents per kilowatt-hour. This price reflects, among other factors, the cost of transporting fuels to the islands and price hikes due to frequent supply shortages. Such high costs should not occur when WWS electricity, which is produced locally, is combined with storage.
>
> The current business-as-usual energy cost per unit energy on the islands of Haiti–Dominican Republic, Cuba, and Jamaica (33 cents per kilowatt-hour) is 2.8 to 3.4 times the estimated costs of 100 percent WWS in those countries (8.5, 12.2, and 9.6 cents per kilowatt-hour, respectively). In addition, energy requirements in those three regions with WWS are 59, 43, and 53 percent lower, respectively, than are those in the business-as-usual case, so the annual private energy cost people will pay with WWS will be one-fifth to one-eighth that which they pay currently.

The 2050 WWS cost of energy per unit energy is relatively low for big regions, such as Canada, Russia, Africa, China, Europe, and the United States, and for small countries with good WWS resources, such as Iceland and New Zealand. The large geographical areas of the big regions allow for the effective aggregation of wind and solar power, resulting in less intermittency of the combined resources. These regions also have a good balance of solar and wind, which are complementary in nature seasonally. They also have substantial existing hydropower resources that can provide peaking power. Iceland has a lot of hydropower, geothermal electricity and heat, and wind.

Costs are highest in small countries with high population densities, such as Cuba, Israel, Mauritius, South Korea, and Taiwan. These countries have little room for onshore wind or utility solar PV, which are the least-expensive forms of WWS electricity generation. Nevertheless, the WWS private energy cost per year in all five of these regions is still 39 to 55 percent lower than with business-as-usual, indicating that a transition to WWS reduces energy costs even under the least favorable circumstances.[5]

13.4 Estimating Footprint and Spacing Areas of WWS Generators

One question about an all-renewable energy infrastructure is, how much land area does it require? This question arises primarily because wind turbines require space between them to operate efficiently, and solar panels require surface area to collect sunlight.

To answer this question, it is important first to differentiate between land footprint and spacing. Footprint (Section 11.7.1) is the physical area on the top surface of soil or water needed for each energy device. It does not include areas of underground structures. Spacing is the area between some devices, such as wind turbines, wave devices, and tidal turbines, needed to minimize interference of the wake of one turbine with the flow of wind to downwind turbines. Spacing area can be used for multiple purposes, including rangeland, farm land, ranching land, forest land, solar panels, open space, and open water.

Footprint is the most relevant metric for determining land use because that is the land area that cannot be used for any purpose aside from energy generation. Spacing area can be used for non-energy purposes and for just growing natural vegetation.

In a 100 percent WWS system, new land footprint arises only for new solar PV plants, CSP plants, onshore wind turbines, geothermal plants, solar thermal plants, and hydroelectric plants. Offshore wind, wave, and tidal generators are in the water, so they do not take up new land, and rooftop PV does not take up new land. New land spacing area arises only for new onshore wind turbines.

The footprint area of a wind farm is relatively trivial. It is primarily the area of the cement around the bases of all the wind turbine towers in the farm that are visible above the ground. Cement underground is not part of the footprint since it doesn't prevent vegetation from growing on the soil above it. Paved roads associated with wind farms are also considered footprint, but unpaved roads are not since such roads can return to their natural state over time.

The total new land area for footprint required to power 145 countries with 100 percent WWS is about 0.17 percent of the 145-country land area.[5] Almost all of that is for utility PV and CSP.

The spacing area for onshore wind to power these countries is about 0.36 percent of the 145-country land area. Since WWS uses no fuel to run, WWS has no footprint associated with mining fuels to run. However, both WWS and business-as-usual energy infrastructures require one-time mining for raw materials for new equipment.

Together, the new land footprint and spacing areas for 100 percent WWS across all energy sectors sum to 0.53 percent of the 145-country land area. This is equivalent to about 1.52 times the land area of California land area for virtually all world energy. Most of this is multi-purpose spacing land. In fact, solar PV panels can be installed on some of the space between wind turbines.

In comparison, about 37.4 percent of the world's land was used for agriculture in 2016, and 2.5 percent was urban land in 2010.[331] Also, the land area required for the fossil-fuel infrastructure in the United States alone is about 1.3 percent of the United States land area.[8] Thus, replacing fossil fuels with 100 percent WWS should reduce land requirements substantially.

13.5   Job Creation and Loss Due to a Transition

A transition to WWS results in both the loss and production of jobs. Construction and operation jobs are lost in the fossil-fuel, bio-energy, and nuclear industries. Such jobs are mainly in the areas of mining, transporting, and processing of fuels and building and running power plants. Transitioning also reduces jobs in building internal combustion engines, natural gas water and air heaters, natural gas stoves, natural gas turbines, coal plants, pipelines, fueling stations, natural gas storage facilities, and refineries.

However, a transition creates jobs in the manufacturing and installation of solar PV panels, CSP plants, wind turbines, geothermal plants, tidal devices, and wave devices. It also creates jobs in the electricity, heat, cold, and hydrogen storage industries. For example, jobs are needed to produce batteries, pumped hydropower storage, compressed air storage, gravitational storage, CSP storage, borehole storage, water pit storage, aquifer storage, ice storage, water tank storage, and hydrogen storage. More transmission lines are needed in a WWS world, so employment increases in that sector as well. Jobs are also needed to install battery charging stations and to build electric vehicles, electric heat pumps, electric induction cooktops, electric lawnmowers, electric arc furnaces, electric induction furnaces, and electric resistance furnaces. Electric vehicles that are needed include on-road passenger vehicles, nonroad vehicles, trucks, buses, trains, ships, aircraft, agricultural machines, construction machines, and military vehicles. Jobs are also needed to weatherize homes, make homes more energy efficient, and retrofit homes with electric appliances.

Construction and operation jobs include direct, indirect, and induced jobs. **Direct jobs** are jobs for project development, onsite construction, onsite operation, and onsite maintenance of the electricity generating facility. **Indirect jobs** are revenue and supply chain jobs. Such jobs include suppliers of construction materials and components; analysts and attorneys who assess project feasibility and negotiate agreements; workers at banks financing the project; equipment manufacturers; and manufacturers of blades and replacement parts. The number of indirect manufacturing jobs is included in the number of construction jobs. **Induced jobs** result from the reinvestment and spending of earnings from direct and indirect jobs. They include jobs resulting from increased business at local restaurants, hotels, and retail stores, and for childcare providers, for example.

Construction jobs (direct, indirect, and induced) are full-time jobs (40 hours per week) but temporary since they last only the duration of the construction project. If each WWS construction project lasts 1 year but the entire WWS infrastructure is built over 30 years, then the number of long-term, full-time (permanent) construction jobs is simply the number of temporary (1-year) construction jobs divided by 30 years. Knowing the number of permanent construction jobs is important for comparing with the number of permanent operation jobs.

Operation jobs are considered permanent in that they last as long as the energy or storage facility lasts. They are needed to manage, operate, and maintain an energy generation or storage facility. In a 100 percent WWS system, permanent jobs are effectively indefinite because, once a plant is decommissioned, another one must be built to replace it. The new plant requires additional construction jobs as well.

Transition highlight

Although a transition to WWS may reduce the world workforce by almost 1 percent owing to job losses in the fossil-fuel, biofuel, and nuclear industries, it will more than make up for those losses with job gains in WWS generation, storage, and transmission. In fact, a transition may increase the number of long-term, full-time jobs created over lost by about 28.4 million among 145 countries. This is the sum of 25.4 million permanent construction jobs and 30.2 million operation jobs produced minus the number of jobs lost (27.2 million).

In sum, a worldwide transition to 100 percent WWS while keeping the grid stable is expected to save consumers money by reducing energy, health, and climate costs; minimize land use; and create many more long-term, full-time jobs than lost. As such, little downside to a transition is anticipated.

14 TIMELINE AND POLICIES NEEDED TO TRANSITION

The solution to global warming, air pollution, and energy security requires not only a technical and economic roadmap but also popular support, political will, and a rapid rollout of the solution. In fact, the main barriers to transitioning to 100 percent clean, renewable energy are neither technical nor economic; instead, they are social and political. People need to be confident that a solution is possible, to understand what changes they can make in their own lives to solve the problems, to make such changes, and to support policymakers who pass laws speeding a transition. Policymakers, themselves, need to understand the urgency and thus take bold steps. One of the most important factors leading to a change is education about what is needed and how fast it is needed. This chapter discusses the necessary timeline for a transition, obstacles in the way of meeting the timeline, actions that individuals can take, and policies that are needed to overcome the obstacles to meet the 100 percent goal on time.

14.1 Transition Timeline

A critical step in implementing a transition to 100 percent WWS is to develop a transition timeline. The perfect timeline is one in which a 100 percent transition occurs in all energy sectors immediately. That won't happen. Jacobson and Delucchi[272] postulated that transitioning all world energy by 2030 is technically and economically feasible but noted that, for social and political reasons, a completion date two decades later, such as 2050, may be more practical. Since then, most 100 percent WWS roadmaps developed[136,144,278] have proposed an 80 percent transition by 2030 and 100 percent by no later than 2050. Because technologies are advancing and WWS costs are dropping more quickly than previously thought, newer studies[5,251] propose an 80 percent transition by 2030 and 100 percent ideally by 2035 but certainly no later than 2050. For electricity and buildings, a transition by 2030 is possible. For industry and transportation, a transition by 2035 is more feasible. This is particularly the case because some technologies, such as long-distance hydrogen fuel-cell aircraft, may not be commercially available until near 2030.

Although new WWS infrastructure will be installed upon the natural retirement of the current infrastructure, policies are needed to force the remaining existing infrastructure to retire early to allow the conversion to WWS at the rapid pace necessary to solve air pollution, climate, and energy security problems.

14.1.1   Transition Timelines for Individual Technologies

Some technologies can transition faster than others. Below is a proposed timeline for transitioning technologies in different sectors, either naturally or through aggressive policies.

Development of super grids and smart grids: As soon as possible, countries should expand the transmission-and-distribution systems to supply geographically dispersed WWS energy to demand centers. The grid should be managed with modern **smart grid** digital communication technology that detects and reacts to local changes in supply and use by invoking storage, demand response, geographically dispersed generation, and backup WWS power generation.

Power plants: by 2022, no more construction of new coal, nuclear, natural gas, oil, or biomass fired power plants should occur; all new power plants built should be WWS.

Heating, drying, and cooking in the residential and commercial sectors: by 2022, all new devices, machines, and equipment for heating, drying, and cooking should be powered by electricity, direct heat, and/or district heating.

Marine vessels: by 2022 to 2030, all new boats (speed boats, yachts, barges, dredgers, towing vessels, fishing boats, patrol vessels, ferries, military boats, etc.) and ships (container ships, cruise ships, military ships, etc.) should be electrified and/or use electrolytic hydrogen, and all new and existing port operations should be electrified. This should be feasible for ports and ships because ports are centralized, and few ships are built each year. Policies are needed to incentivize the early retirement of fossil-fuel-powered ships that do not naturally retire before 2030.

Industrial heat: by 2023, all new high-, mid- and low-temperature industrial process technologies should be electric technologies, such as arc furnaces, induction furnaces, resistance furnaces, dielectric heaters, electron beam heaters, and heat pumps.

Rail and bus transport: by 2023, all new buses should be electric and new trains should be electric or hydrogen fuel-cell.

Nonroad vehicles: by 2023, all new nonroad land-based vehicles (agricultural machines, construction vehicles, military vehicles) should be electric.

Heavy-duty truck transport: by 2022 to 2030, all new heavy-duty trucks and buses should be battery-electric or hydrogen fuel-cell.

Light-duty on-road transport: by 2023, all new light-duty on-road vehicles should be battery-electric.

Short-haul aircraft: by 2024, all new small, short-range aircraft should be battery-electric.

Long-haul aircraft: by 2027 to 2035, all remaining new aircraft should be hydrogen fuel-cell or battery-electric.

Although much new WWS infrastructure can be installed upon the natural retirement of conventional infrastructure, new policies are needed to force remaining existing infrastructure to retire early in order to allow the complete conversion to WWS to occur on time.

Some suggest that retiring fossil plants early will create **stranded assets**, incurring a large cost. However, averaged worldwide, the annual social cost of a WWS energy system is less than 10 percent that of a fossil system. As such, replacing a fossil plant before the end of its life saves substantial damage and money. Transitioning also increases the number of jobs available. Society thus benefits from stopping the operation of existing fossil-fuel plants and replacing them with new WWS plants as soon as possible.

During the transition, fossil fuels, bioenergy, and existing WWS technologies are needed to produce the new WWS infrastructure. However, if no WWS infrastructure or machines were produced, much of the conventional energy would still be used to produce new business-as-usual infrastructure and machines. Further, as the fraction of WWS energy increases, conventional energy generation used to produce WWS infrastructure decreases, ultimately to zero. At that point, all new WWS infrastructure will be produced with WWS energy. In sum, the time-dependent transition to WWS infrastructure may result in a temporary increase in emissions before such emissions are eliminated.

14.1.2   How the Proposed Timeline May Impact Carbon Dioxide Levels and Temperatures

The WWS strategy proposed here is to transition the world to 80 percent WWS by 2030 and 100 percent by no later than 2050, but ideally by 2035, while eliminating non-energy emissions at the same pace.

An important question then is, how long will carbon dioxide levels take to recover once carbon dioxide emissions have been halted? Carbon dioxide levels in the air in 1750, before the Industrial Revolution, were about 276 parts per million. In 2021, they were up to a peak of almost 420 parts per million. Global computer simulations suggest that eliminating 80 percent of all human-created carbon dioxide emissions by 2030 and 100 percent by 2050 may reduce carbon dioxide levels in the air down to about 350 parts per million by 2100,[8] which is a reasonably safe level, albeit higher than the 1750 level. On the other hand, holding constant or increasing carbon dioxide emissions going forward will increase carbon dioxide levels in the air substantially compared with today.

How would globally averaged near-surface air temperatures respond if we eliminate emissions as proposed? Global air temperature rise is roughly proportional to cumulative carbon dioxide emissions, regardless of the timing of those emissions. As such, the more the cumulative emissions, the greater the temperature rise. The less the cumulative emissions, the lower the rise.

Between 1850 and 2019, the world emitted an estimated 2,390 billion tonnes of carbon dioxide from fossil-fuel combustion, cement manufacturing, and land use change.[7] This has resulted in the average global surface temperature between 2011 and 2020 being about 1.09 degrees Celsius warmer than between 1850 and 1900.

To limit global warming to no more than 1.5 degrees Celsius above the 1850 to 1900 mean temperature (with a 50 percent probability), we can allow no more than another 500 billion tonnes of carbon dioxide emissions from all human sources after 2020. To limit warming to no more than 2 degrees Celsius, we can allow no more than 850 billion tonnes of carbon dioxide emissions.[7]

In 2020, global carbon dioxide emissions from fossil-fuel burning and cement production were about 34.1 billion tonnes per year and those from land use change were about 5.9 billion tonnes per year, for a total of 40 billion tonnes per year. 2020 was a pandemic year, and total emissions were about 7 percent lower than in 2019.

If carbon dioxide emissions stay constant at 2020 levels going forward, enough carbon dioxide will accumulate in the air for the Earth to warm by 1.5 degrees Celsius relative to 1850 to 1900 levels by 2032 and 2 degrees Celsius by 2041.

On the other hand, if emissions of carbon dioxide from all energy and non-energy sources decrease 80 percent from 2020 levels by 2030 and 100 percent by 2050, additional cumulative emissions of only 340 billion tonnes of carbon dioxide will go into the air by 2050. This will avoid 1.5 degrees Celsius global warming since 1850. This is the strategy proposed here: to transition the world to 80 percent WWS by 2030 and to 100 percent by no later than 2050, but ideally by 2035, and eliminate non-energy emissions at the same pace. To accomplish this goal, aggressive policies are needed.

14.2 Obstacles to Overcome for a Transition

While technically and economically feasible, a transition of the entire world's energy infrastructure to clean, renewable energy in all sectors is a daunting task that faces significant social and political hurdles. Below, several of these challenges are discussed.

14.2.1 Vested Interests in the Current Energy Infrastructure

Possibly the greatest challenge to overcome in transitioning the world to 100 percent clean, renewable energy is the challenge of repelling vested interests. Fossil-fuel companies have accumulated enormous wealth and political influence since the start of the Industrial Revolution. As a result, they have been able to implement legislation that has given them an entrenched financial benefit in many countries through tax code breaks, direct grants, and subsidies. In addition, most of the existing energy infrastructure is a fossil-fuel infrastructure, and most people's day-to-day lives depend on that infrastructure (through transportation; building heating, cooling, and refrigeration; lighting, etc.). Further, over 25 million jobs worldwide depend on the fossil-fuel infrastructure.

Given the great financial resources available to the fossil-fuel industry and the dependency of many people's jobs and livelihoods on the current infrastructure, great care is needed to ensure that a transition to WWS brings along the same or a better standard of living, the same or more jobs, and the same or better comfort to as many people as possible. This means encouraging shareholders of fossil-fuel companies to move their investments to clean, renewable energy and storage, setting policies to retrain workers, and setting policies to ensure that WWS electricity supply is at least as reliable as the current supply of electricity on the grid.

Movement is already occurring in all three areas. Investors are moving capital resources into WWS. The WWS infrastructure is creating more jobs per unit energy than is the fossil infrastructure. The WWS system is also proving to be even more reliable and less expensive than was the preceding fossil system in the few places that WWS has completely replaced conventional fuels.

> **Transition highlight**
> For example, after 1 year of construction, a large (100 megawatt, 129 megawatt-hour) battery system that was installed in South Australia to fill in gaps in electric power supply for a 317-megawatt wind farm saved $40 million (Australian dollars) in grid stabilization costs. This was due to the rapid speed (100 milliseconds) at which batteries can react to shortages.[332] In another example, during July 2019, NextEra energy determined that investing in a combined wind (250 megawatt), solar PV (250 megawatt), and battery storage (200 megawatt, 800 megawatt-hour) system was less expensive than investing in a natural gas peaker plant.[333]

14.2.2 Zoning Issues (NIMBYism)

The second obstacle that needs to be overcome to implement a 100 percent WWS system is the **not-in-my-backyard** syndrome (**NIMBYism**). NIMBYism is an objection to the siting of something that a person thinks is unpleasant or dangerous in their neighborhood while not objecting to the development of the same thing somewhere else. Classic examples of NIMBYism are the siting of a landfill or a hazardous waste site near a neighborhood. However, NIMBYism extends to most every type of infrastructure development. People generally do not want to see additional buildings or facilities, including energy facilities, near them.

Although the installation of onshore and offshore wind and solar farms and transmission lines will face opposition in many locations, they will likely face less opposition than the building of more coal and natural gas plants, nuclear plants, oil and gas wells, refineries, and pipelines. The reason is that most people realize that WWS technologies do not bring air pollution or catastrophic risk along with them, whereas the fossil technologies bring both, and nuclear plants bring the risk of catastrophic failure and exposure to radiation above background levels. A second reason is that, although no one wants to add anything to the landscape, the addition of WWS often reduces land use relative to a fossil-fuel infrastructure. Similarly, whereas land used for a coal or nuclear plant cannot be used for another purpose at the same time, land occupied by a wind farm can be used for multiple purposes simultaneously: agriculture, animal grazing, open space, or even housing solar PV and batteries.

In addition, many people are becoming accustomed to rooftop solar panels, so they object to them less than in the past. Some types of solar panels are also integrated into building design, so it is difficult to tell whether the panels are even there. Most wind farms are located away from buildings because winds are fastest where no obstacles exist on the ground. With the advent of floating offshore wind turbines, visual objections to siting offshore wind are virtually eliminated because people cannot see such turbines.

Probably the most difficult WWS infrastructure to site is aboveground transmission. Siting transmission lines and pipelines today already results in NIMBYism, for good reason. Aboveground transmission lines and pipelines are not pretty. In addition, sparks from aboveground transmission lines can trigger wildfires. Wildfires around the world caused by transmission-line malfunctions have resulted in enormous damage, including loss of life.

Wildfires triggered by transmission-line sparks were a problem long before WWS, but the issue still needs to be addressed. The best solution is to bury the lines underground. To that end, in 2021, the northern California utility Pacific Gas & Electric announced that it will bury, over several years, 16,000 kilometers of existing transmission lines to reduce fire hazard. Although underground transmission lines are more expensive upfront than are overhead lines, burying lines may cost less over the long term because it eliminates devastating fire damage due to transmission lines. Burying transmission lines also solves their view problem.

Another partial solution to reducing reliance on transmission lines is to install more rooftop solar PV and batteries in fire-prone regions and to use more storage in general instead of transmission. A third partial solution is to improve safety measures by clearing more brush, and by learning from the causes of previous fires.

Adding new overhead or underground transmission lines due to 100 percent WWS will enable the elimination of all oil and natural gas pipelines. Such pipelines leak and/or rupture, sometimes causing explosions that result in damage and death. Particularly dangerous are leaky natural gas distribution pipes, which exist almost everywhere in densely populated cities.

Finally, many new transmission lines will be built on the same pathways as existing transmission lines, reducing the need for new land.

Because of the delays caused by zoning requirements, the installation of many new transmission pathways for a 100 percent WWS system may be limited in some countries and states. Fortunately, though, an alternative to transmission is more energy storage. Because storage costs are declining rapidly, the future WWS system may be dominated by storage over transmission.

14.2.3 Countries Engaged in Conflict

A third potential obstacle to a large-scale WWS buildout is the difficulty in building new energy infrastructure in countries suffering from conflict, such as civil war, terrorism, or war with another country. In 2022, five countries or regions of the world had conflicts with more than 10,000 deaths per year (Afghanistan, Yemen, Myanmar, Ukraine, and Eritrea–Ethiopia–Sudan). Another 14 had conflicts with between 1,000 and 10,000 deaths per year (Somalia, Azerbaijan, Iraq, Mexico, Colombia–Venezuela, the Democratic Republic of the Congo, Syria,

Libya, Sudan, South Sudan, Mali, Algeria–Morocco–Tunisia–Libya–Mali–Niger–Chad, Nigeria–Cameroon–Niger–Chad, and Mozambique–Tanzania). Dozens of additional countries suffered less than 1,000 casualties during the year from conflict.

Because millions of people live in countries of ongoing conflict, it is even more essential to bring distributed, clean renewable energy to these countries. Oil pipelines and transmission lines are often targets for destruction or theft, so local microgrids may be the best way to provide energy safely to communities in countries engaged in conflict until the conflict is resolved. A **microgrid** is a grid isolated from or disconnected from outside sources of electricity. It may consist, in its simplest form, of solar panels plus batteries. The next step may be to add heat storage in a water tank. The electricity may be combined with a water filtration or desalinization system to provide clean water from wastewater or salt water, respectively. It may also be combined with a greenhouse or a food-growing container to provide food year-round.[18] In locations with suitable terrain, the microgrid may be combined with a small pumped-hydropower storage facility and small wind turbines.

Major problems in a war-torn country are famine and poverty. Microgrids, if set up and maintained properly, can help produce food, water, and energy together, mitigating both problems. One way to help alleviate some of the difficulty in setting up a microgrid in a war-torn country is for safer countries to provide aid in the form of clean, renewable WWS microgrid technology.

Some conflicts between countries arise over energy. For example, one country that provides natural gas to another may withhold the gas to extract concessions or use money from the sale of the gas to fund a war against another country. Alternatively, one country may need more energy and so invade a neighbor to take control of its oil and natural gas wells or coal mines. Transitioning a country entirely to clean, renewable energy, where most or all of the raw WWS resources are obtained from within the country, will make the country more energy-independent, reducing the reasons for conflict. However, many countries, even if they produce most of their own energy in the annual average, will benefit from trading energy with their neighbors to reduce the cost of matching power demand with supply over time. The reason is that, whereas the wind may not blow in one country at a given time, it is more likely to blow somewhere among several countries. The same applies to solar.

14.2.4 Countries with Substantial Poverty

Countries with a large segment of their population in poverty will benefit from transitioning to 100 percent clean, renewable energy as soon as possible. Millions of people die from indoor plus outdoor air pollution each year. Of these about 20 percent are children under the age of five. Almost all of the indoor air pollution deaths (about 2.6 million per year in 2016) are due to the burning of biomass and coal in developing countries for home heating and cooking.

The main step in eliminating indoor air pollution death is to eliminate indoor burning of fuel for cooking and heating. This can be accomplished most readily with the use of electric induction burners and electric heat pumps. For remote communities, the electricity would be obtained from a microgrid that combines solar PV or wind with batteries or pumped hydropower. Such an infrastructure costs money, and impoverished communities usually do not have access to such funds. However, costs of WWS generation and storage continue to decline. In addition, national governments may help financially. International aid will help with this solution too.

Impoverished countries will also benefit from a large-scale transition to WWS. Their end-use energy requirements will go down by an average of 56.4 percent; their annual cost of energy will consequently drop by over 60 percent. Their annual social costs of energy will decrease by near 90 percent.

However, the main barrier these countries face is the capital cost of their investment. WWS generators have no fuel cost, but they have an upfront capital cost. Although that capital cost pays for itself relatively quickly, raising upfront capital is not a trivial matter. To that end, wealthier countries need to work with more impoverished countries to help finance the upfront cost of a transition.

14.2.5 Transitioning Long-Distance Aircraft and Long-Distance Ships

Whereas 95 percent of the technologies needed for a transition to 100 percent WWS are currently commercialized, the rest are not yet available. The two most obvious technologies not yet commercialized are long-distance aircraft and long-distance ships. About 15 percent of worldwide commercial aircraft flights by number and 46 percent of such flights by distance are mid-haul flights (3 to 6 hours) or long-haul flights (longer than 6 hours).[61] The best WWS solution for such flights may be the development of hydrogen fuel-cell planes. However, for such aircraft to work cost-effectively, fuel-cell sizes and efficiencies need to improve. A transition to long-distance WWS aircraft may not occur until 2027 to 2035 because of the improvements and testing needed to commercialize hydrogen fuel-cell aircraft for mid- and long-distance travel. For short-haul flights (less than 3 hours), aircraft will be primarily battery-electric.

Long-distance ships have a similar issue to aircraft and are also proposed to be mostly hydrogen fuel-cell although some may be battery-electric. Long-distance ships can stop during their journey to refuel (if they are hydrogen fuel-cell) or recharge (if they are battery-electric). In addition, ships have less constraint on mass than aircraft, so ships are easier to design. As such, a transition of ships may be implemented between 2022 and 2030.

14.2.6 Competition among Solutions

One more obstacle facing a rapid transition is the competition among proposed solutions to the problems of air pollution, global warming, and energy security. Given the severity of the problems facing the world and the short time to fix them, competition among energy plans can result in poor technologies being implemented, thereby slowing down the solution. The main competitors for solutions to date have been a WWS solution and an **all-of-the-above** solution. An all-of-the-above solution includes WWS but also includes nuclear, fossil fuels with carbon capture, biofuels with or without carbon capture, direct air capture, and hydrogen from natural gas with or without carbon capture, among other technologies. However, those non-WWS technologies are opportunity costs. They result in more air pollution, more greenhouse gas emissions, more energy security risk, more land use risk, higher costs, or longer planning-to-operation times, or all of these, relative to WWS. As such, spending money on non-WWS technologies results in less benefit and a longer delay before enough WWS can be implemented to eliminate air pollution, global warming, and energy insecurity. The best way to overcome this obstacle is to educate the public and policymakers about it and guide them to the WWS solution.

14.3 What Can Individuals Do?

The solution to air pollution, global warming, and energy insecurity requires actions by individuals, businesses, nonprofits, and policymakers. This section discusses some of the actions that individuals can take to reduce energy use and to change the types and sources of the energy that they consume.

Most individuals consume energy that results in carbon dioxide and air pollution emissions in all of the main energy sectors (electricity, transportation, building heating and cooling, and industry). We consume electricity for lights, computers, refrigerators, and food production. We drive to work or for leisure. We desire buildings of a certain temperature for comfort. We consume products and food created by industry. Because we consume, we can control our emissions both by changing our habits and choices and by changing the type and sources of energy we use.

Because each person is in a different situation in terms of both the energy they consume and their financial and practical ability to transition their energy, each recommendation given here may apply to some people but not others.

14.3.1 New Home Construction

Those building a new home from scratch can plan their home not only to minimize energy use but also to produce their own WWS energy. The first step in planning such a home is to ensure that it uses only one energy source, electricity, rather than both electricity and natural gas. There is no reason to have two sources of energy in a building since every appliance that runs on natural gas has an electrical equivalent that is the same or better in quality. To the contrary, adding natural gas to a home or another building adds unnecessary costs, including for a gas hookup fee charged by the utility, for ditches to bury natural gas pipes, for natural gas pipes themselves, for a natural gas meter, for vents to provide air and ventilation for natural gas combustion, for carbon monoxide monitors, and for inspection of all of the above. These items sum to tens of thousands of dollars of unnecessary cost per home. Natural gas use in buildings also releases health-affecting air pollutants. Health problems from natural gas fumes and combustion products are eliminated by eliminating natural gas use.

The appliances in a building that run on natural gas often include an air heater, a water heater, a cooktop, a dryer, and a pool heater. In locations on district heating, air heating and water heating (and rarely, air cooling) are supplied from centralized boilers and chillers, so in-house appliances are not needed for these tasks.

For locations not on district heating loops, a natural gas air heater, water heater, dryer, and pool heater can be avoided by using an electric heat pump version of each. A natural gas cooktop can be avoided by using an electric induction cooktop. Because heat pump air heaters all automatically run in reverse as air conditioners, the use of a heat pump for air heating and air conditioning eliminates the need for a separate air conditioner, saving additional money. Because heat pumps consume one-quarter the energy of natural gas heaters, the installation of a heat pump saves a lot of money each year. Finally, the use of a heat pump air heater/air conditioner that has an indoor unit in each room of the house, and an outdoor unit to exchange air between the inside and outside, eliminates the need for ducts throughout the house, saving more money.

The third step in new home planning is to minimize energy use by maximizing energy efficiency. The first way to make a building energy efficient is to install thick insulation within walls, below the bottom floor, between floors, above the ceiling, around water pipes, and in and around windows. Concrete, which helps to keep the temperature relatively stable inside a building, can also be used as floor or wall material. Triple-paned windows minimize heat transfer through the windows. The greater the insulation, the less the energy needed to heat or cool a home, reducing energy consumption and extending heat pump life.

Similarly, smart glass windows reduce cooling and heating needs. They are translucent during summer, thereby blocking incoming heat from the outside, and transparent during winter, maximizing sunlight into your home during that time. The second way to minimize energy use is to install LED lights everywhere. The third way is to use only energy-efficient electric appliances (refrigerator, dishwasher, washing machine, television, vacuum, etc.).

The fourth step in new home planning is to install rooftop solar PV panels, garage batteries, an inverter to control both, and an electric car charging port. In some windy locations, a backyard wind turbine may be possible instead of solar PV. The solar PV or wind system should be sized, at a minimum, to meet 100 percent of your annual average electricity demand for your home, including for any electric vehicles you might use. Ideally, the PV system will be sized larger than this to account for any future growth in demand and to help the grid when the grid is not producing enough electricity to meet everyone's demand. One or two batteries are helpful to store electricity during times of low demand, such as during the morning, and to discharge electricity during times of peak demand, such as just after sunset. Such batteries help to relieve stress on the grid and save you money.

The fifth step is to use sustainable building materials for construction. One option is to use prefabricated, recycled steel for the structure.[97] Since the steel is produced precisely in a factory and is from recycled material, it minimizes waste. Wood homes typically result in wood waste during construction. Concrete can also be recycled and used as aggregate, either as a sub-base material or mixed as an aggregate with new concrete. Some sustainable building materials include composite roofing shingles and bamboo. Composite roofing materials require much less repair and replacement than do asphalt shingles, tile, and wood shake. Bamboo re-grows in 3 years versus 25 years for most trees.

14.3.2 Home Retrofits

Hundreds of millions of buildings already exist worldwide, and the turnover is only 1 percent per year. Given the substantial energy use in existing buildings, retrofitting such buildings to become all electric is essential for solving the air pollution, climate, and energy security problems we face.

Solutions for existing buildings are usually not technically challenging. However, unlike for new buildings, where one can save a lot of money by going to 100 percent WWS, retrofitting existing buildings usually saves less.

One of the lowest-cost ways to reduce energy use in your home, thus reduce energy cost, is to weatherize your home by sealing cracks in windows and doors and adding insulation around water pipes. Adding or changing the insulation below the floor or in the attic also helps. Even insulating an attached garage can minimize heat loss from the house to the garage.

A second step is to change lights to efficient LED lights and to change appliances to electric, energy-efficient appliances. Certainly, if a natural gas water heater needs replacing, it should be replaced with a heat pump water heater. Most natural gas air heaters in buildings are centralized units that provide heat that is sent through ducts to the rest of the building. When or before the heater breaks down, it should be replaced with a centralized heat pump air heater and air conditioner that also uses ducts. Similarly, a natural gas dryer should be replaced with a heat pump dryer. A natural gas cooktop should be replaced with an induction cooktop. Gradually, all appliances in a building can be changed from natural gas to electric.

A third step is to install rooftop PV or backyard wind, one or two batteries, an inverter, and a car charging port, just like with a new building.

Although some of these measures are low cost and others can be done when an existing appliance breaks down, the rest require new investments. Although some people can afford such investments, many others cannot. As such, policies are needed to encourage and help finance retrofits. Otherwise, only a small fraction of the needed retrofits will be performed.

14.3.3 Renting

Many people rent apartments, condominiums, or town houses, so they have little control over either their source of electricity or the appliances that they use. However, renters do have some control. First, renters pay electricity bills to a utility. Many utilities today, particularly community choice aggregation utilities, but also many others, offer to sell customers 100 percent WWS electricity. Second, even if an apartment has a natural gas cooktop, it is possible to purchase an individual induction cooktop burner for $40 to $100. These just plug into the wall and replace the natural gas cooktop. Third, weatherizing doors and windows in the rental can save a lot of money, as can changing lights to LED lights. If the building is on district heating, then natural gas is not used for any other purpose in the rental, and you are done. If your rental has a natural gas air or water heater, then it would be up to the landlord to change these. The first step in doing this is to request the landlord to invest in electric heat pump air and water heating, especially when existing units break down. The second step is to ask your local policymaker to incentivize or require such changes.

14.3.4　Transportation

People travel primarily by foot, bicycle, car, streetcar, bus, train, ship, or airplane. Many people own or lease cars. If a new car is needed, the next one should be electric. A variety of electric cars are currently available. Whereas only expensive electric cars were available until a few years ago, lower-cost models are now available most everywhere. Traveling by bus, train, ship, or plane usually results in less energy consumed per unit distance traveled than traveling by car.

14.3.5　Consumer Choices

Individuals make conscious or unconscious choices every day about how much energy they consume and, therefore, how much pollution they emit. Consumers decide what to eat, where to travel, whether to use a computer or television, and what goods to buy. Based on today's fossil-fuel use worldwide, each product purchased or action taken results in certain **carbon footprint**, which is the lifecycle CO_2-equivalent emissions of the product or action. However, as we go to 100 percent WWS and address non-energy emissions too, the carbon footprint of actions and products will go to zero, which is good news. In the meantime, though, certain actions can definitely reduce air pollution and carbon emissions.

Such actions include telecommuting for work instead of driving to and from work; holding more meetings virtually instead of in person; minimizing travel; using more public transportation; biking or walking instead of driving; shutting off lights when they are not needed; dimming the screen on a computer or phone slightly; reducing meat consumption; and purchasing fewer nonessential goods.

Transition highlight
In an extreme example of how consumer choices can affect energy use, on July 11, 2021, Richard Branson and three others flew into space 85.3 kilometers above the surface of the Earth as the world's first space tourists. However, the 1.5-hour, 11,260-kilometer flight resulted in 135,000 kilograms of carbon dioxide emissions per person along with other rocket fuel exhaust emissions.[334] The average person in the United States emits 3.4 kilograms of carbon dioxide per 1.5 hours. So, the emissions per person for that 1.5-hour flight were 40,000 times those of the average person. A better solution is not to travel to space for tourism, thereby avoiding wasteful energy use.

14.4 Policies

The policies necessary to transform to 100 percent WWS differ by country, depending largely on the willingness of the government
and people in each country to effect a rapid change. This section first
defines several types of policy options that each country can consider
implementing. Policy options for different energy sectors are then discussed in more detail.

14.4.1 Policy Options for a Transition

Below are some policy options that have been used in the past.
The list is by no means complete.

Renewable portfolio standards, also called renewable electricity standards, are policy mechanisms requiring a certain fraction of electric
power generation to come from specified renewable energy sources
by a certain date. Thus, for example, a mandate of 80 percent WWS
by 2030, and 100 percent no later than 2035, is a renewable portfolio standard. 100 percent renewable portfolio standards have been
enacted in many countries, states, cities, and towns to date.

Financial incentives and laws for increasing energy efficiency and
reducing energy use are policy methods to reduce the demand for
energy. For example, a law requiring the substitution of low-energy-
consuming LED light bulbs for high-energy-consuming incandescent ones reduces energy demand. Demand reduction reduces the
pressure on energy supply, which makes it easier for WWS supply
to match demand.

Laws requiring demand response are helpful because they force utilities to incentivize customers to shift the time of their electricity use
from a time of peak electricity demand to a time of lower demand
during the day or night. Demand response is usually accomplished
in one of two ways. The first way is for utilities to increase the price
of electricity during times of peak electricity use. The higher cost
of electricity incentivizes customers to use less electricity during
those times of day. The second way is for utilities to pay customers
to use less electricity during times of high electricity demand. In
both cases, the customer can react manually to the change in rate
or payment by reducing or stopping energy consumption during
times of peak demand. Alternatively, the customer can agree to

have internet- or radio-controlled switches installed on air conditioners or other devices to automatically reduce energy use during
times of peak demand.

Feed-in tariffs are subsidies to cover the difference between electricity generation cost (ideally including grid connection cost) and
wholesale electricity prices. Feed-in tariffs have been an effective policy tool for stimulating the market for renewable energy.
To encourage innovation and the large-scale implementation
of WWS, which will itself lower costs, feed-in-tariffs should be
reduced gradually. Otherwise, technology developers have little
incentive to improve.

Output subsidies are payments by governments to energy producers
per unit energy produced. For clean, renewable energy producers,
the justification of such subsidies is to correct the market because
fossil-fuel and bioenergy energy producers are not paying for the
pollution they emit, which has health and climate costs to society.
In other words, the subsidies attempt to address the tragedy of the
commons, which arises because the air is not privately owned. As
such, the air historically has been polluted without polluters paying
the health, climate, and environmental costs of the pollution.

Investment subsidies are direct or indirect payments by governments to
energy producers for research and development. Historically, such
subsides have been given mostly to conventional fuel producers
through legislation and clauses in the tax code, such as deductions
and credits for specified activities. Conventional generators have
benefited historically from such subsidies by not paying externality
costs of the pollution that their energy creates. Investment subsidies
are now available to WWS energy sources in many countries.

One type of investment subsidy is a **loan guarantee**, whereby the government guarantees a loan to a company for constructing a facility.
Without such guarantees, many large energy projects would not be
approved for a construction loan. Such loan guarantees have been historically provided to the conventional fuels industry. Guarantees will
benefit the WWS industry as well.

On a smaller scale, **municipal financing** of residential energy
efficiency retrofits and solar installations help to overcome the financial
barrier of the high upfront cost to individual homeowners. **Purchase
incentives and rebates** can also help stimulate the market for electric
vehicles.

A potential policy tool that has not been used widely to date is a **revenue-neutral carbon tax or pollution tax**. This is a tax on polluting energy sources, with the revenue transferred directly to nonpolluting energy sources. In this way, no net tax is collected, so the cost to the public is zero in theory. If the tax is a revenue-neutral carbon tax, it may not address air pollution. For example, a biomass or coal-with-carbon-capture electricity plant can claim it is low carbon (which itself is not accurate), thus avoid much of the tax, yet still emit substantial health-affecting air pollutants along with carbon. A second problem is that heavy polluters can choose to pay the tax and withstand a lower profit margin or raise their prices, yet still emit.

A related tool is straight **pollution tax**, such as a **carbon tax**. Such a tax is not so popular since it is perceived as a cost to the public, because the money collected from the tax does not necessarily go back to reducing energy prices. Instead, polluting companies merely increase their prices charged to customers to pay the tax.

Another noneconomic policy is **mandatory emission limits** (usually tailpipe or stack exhaust emission limits) for technologies. This is a **command-and-control** policy option implemented widely under the U.S. Clean Air Act Amendments and in many countries to reduce vehicle emissions. By tightening emission standards, including for carbon dioxide, policymakers can force the adoption of cleaner vehicles. Such emission limits can and have also been set for other pollution sources. A disadvantage of emission limits (unless they are zero) is that they allow the fossil-fuel industry and their upstream mining to persist while the industry pursues incremental reductions in tailpipe or stack emissions. However, a mandate of absolutely zero tailpipe pollution emissions would mean that only electric and hydrogen fuel-cell cars could meet these limits. That would lead to the phase out of internal combustion engine vehicles.

Related to mandatory emission limits is **cap and trade**. Under this policy mechanism, mandatory emission limits lower than current emissions levels are set for an entire industry, and pollution permits are issued corresponding to the total pollution emissions allowed. Polluters in the industry can then buy and sell pollution permits. The net result is lower overall emissions and a payment by the polluters for the remaining emissions. The problem with this mechanism is similar to that with emission limits. Unless the cap is zero, pollution will persist long into the future.

14.4.2   Policy Options by Sector

Current energy markets, institutions, and policies have been developed to support the production and use of fossil fuels, bioenergy, nuclear power, and clean renewable energy. New policies are needed to ensure that a 100 percent clean, renewable energy system develops quickly and broadly in each country, and that dirtier energy systems are not promoted. Below, several policy mechanisms are proposed for each energy sector to accomplish these goals. For each sector, the policy options are listed roughly in order of proposed priority.

14.4.2.1 Energy Efficiency and Building Energy Measures

- Expand energy efficiency standards.
- Incentivize the conversion from natural gas water and air heaters to electric heat pump heaters.
- Promote, through incentives, rebates, and municipal financing, energy efficiency measures in buildings.
- Revise building codes to incorporate "green building standards" based on best practices for building design, construction, and energy use.
- Incentivize landlord investment in energy efficiency. Allow owners of multi-family buildings to take a property tax exemption for energy efficiency improvements in buildings.
- Create energy performance rating systems with minimum performance requirements to assess energy efficiency levels and pinpoint areas of improvement.
- Create a green building tax credit program for the corporate sector.

14.4.2.2 Energy Supply Measures

- Increase the percentage of WWS energy supply required under renewable portfolio standards.
- Extend or create WWS production tax credits (a tax credit for every kilowatt-hour of electricity produced).
- Invest in job retraining from business-as-usual energy to WWS energy.
- Incentivize the expansion of building-scale electricity storage.
- Streamline the permit approval process for large-scale WWS power generators.

- Streamline the permit approval process for high-capacity transmission lines. In the United States, for example, 750 gigawatts of WWS electricity generation projects were sitting in an "interconnection queue" in 2021, waiting to connect to the transmission grid.[335] Once at the front of the queue, project owners are often charged high fees to cover the cost of transmission upgrades to support their project. One solution to this problem is to identify and pre-approve regions where large generation and transmission can be added simultaneously, reducing the cost of additional transmission.
- Work with local and regional governments to streamline zoning and permitting for new electricity generation and storage within existing planning efforts to reduce the cost and uncertainty of projects and to expedite their build-out.
- Streamline the small-scale solar and wind installation permitting process. Create common codes, fee structures, and filing procedures across a country.
- Lock in fossil-fuel and nuclear power plants to retire under enforceable commitments. Implement taxes on air pollution and carbon emissions by current utilities to encourage their phase-out.
- Incentivize home or community battery storage (through garage electric battery systems, for example) that accompanies rooftop solar.

14.4.2.3 Utility Planning and Incentive Structures

- Incentivize district heating and community seasonal heat and cold storage.
- Incentive the development of utility-scale grid electric power storage in CSP, pumped hydropower, more efficient large hydropower, and large battery systems.
- Require utilities to use demand response grid management to reduce the need for short-term energy backup on the grid.
- Incentivize the use of excess WWS electricity to produce and store hydrogen, heat, and cold to help manage the grid.
- Develop programs to use electric-vehicle batteries, after the end of their useful life in vehicles, for stationary storage.
- Implement **net metering**, whereby rooftop solar owners can sell electricity back to the grid to offset a part or all of the cost of the electricity that they buy from the grid.

14.4.2.4 Transportation Measures

- Mandate battery-electric vehicles for government transportation (mail delivery, service vehicles, military base vehicles) and use incentives and rebates to encourage the transition of commercial and personal vehicles to battery-electric and hydrogen fuel-cell vehicles.
- Promote more public transport by increasing its availability.
- Increase safe biking and walking infrastructure, such as dedicated bike lanes, sidewalks, crosswalks, and timed walk signals.
- Set up time-of-use electricity rates to encourage vehicle charging during off-peak hours.
- Use incentives or mandates to stimulate the growth of private fleets of electric and/or hydrogen fuel-cell buses.
- Incentivize battery-electric and hydrogen fuel-cell ferries, riverboats, tugboats, speed boats, dredgers, and other local watercraft.
- Adopt zero-emission standards for all new on-road and off-road vehicles for all purposes, with 100 percent of new production required to be zero-emission of every air pollutant by 2030. Incentivize the transition by gasoline and diesel superusers the fastest, since the top 10 percent of gasoline users, for example, consume more gasoline than the bottom 60 percent.[336]
- Ease permitting for installing electric charging stations in public parking lots, hotels, suburban metro stations, on streets, and in residential and commercial garages.
- Incentivize the electrification of freight rail and shift freight from trucks to rail.

14.4.2.5 Industrial Sector Measures

- Provide financial incentives for industry to convert to electricity for high-temperature manufacturing processes.
- Provide financial incentives to encourage industry to use WWS electric power generation.
- Encourage industry to take part in demand response measures to help match power demand with supply and storage on the grid.
- Encourage steel, concrete, and silicon manufacturing plants to shift to processes producing those products without releasing carbon dioxide chemically.

The measures listed above are a limited list. Yet, many are necessary to speed a transition, which would otherwise drag on long past 2050. Each town, city, state, province, or country must select its own policies based on what works best in their region. Yet, the reduction in cost that has already occurred combined with economies of scale due to further WWS expansion are a cause for optimism. If effective policies are put in place, an all-sector transition to 100 percent WWS, ideally by 2035, but no later than 2050, is possible.

15 MY JOURNEY

Since the early 2000s, substantial progress has been made toward raising awareness about the need for a rapid transition to WWS. A movement toward 100 percent clean, renewable energy has begun. Popular support for such a transition has grown, as have the numbers of laws and commitments furthering that goal. Such commitments have been made by cities, states, provinces, countries, international businesses, nonprofits, community groups, policymakers, and individuals. This chapter discusses my personal journey to develop and implement 100 percent WWS roadmaps, and how these roadmaps and collaborations with other scientists, cultural influencers, business leaders, and community leaders helped to shape the 100 percent WWS movement.

15.1 First Exposure to Severe Air Pollution

During the summer of 1978, as a 13-year-old tennis player, I travelled to San Diego from my home in Northern California to play in a tennis tournament. What struck me was the air pollution, which was visible on and off the freeways. I could see it, taste it, and smell it. The pollution hung like a morbid pall. It was one thing to sit inactively in a car in the middle of this pollution. It was another to run through it for hours while playing tennis. During play, my throat and lungs became irritated, and my eyes became scratchy. Taking deep breaths while lunging for tennis balls was a chore. If this soup of pollution was hurting me after only a few minutes, I imagined the damage it caused people who lived in it. Indeed, living in such pollution is equivalent to smoking two to three packs of cigarettes per day.

I took additional trips to Los Angeles and San Diego during the next few years. After only the second trip, I realized that the smog was not just a one-time event. After the third trip, I concluded this pollution was a way of life in these cities. I then began to ask myself, "*Why should anyone live like this?*" and "*Isn't there a solution to this problem?*" I decided then and there, that when I grew up, I wanted to understand and try to solve this avoidable air pollution problem, which affects so many people. I knew what I wanted to do for my career.

Later in my teens, I learned from popular magazine articles about the emissions of greenhouse gases since the Industrial Revolution creating a blanket over the Earth, trapping heat and increasing globally averaged temperatures. I also learned about acid rain and its impact on forests and lakes. I realized that these problems were intimately connected to the air pollution problem. The main sources of the chemicals that cause air pollution, which were fossil-fuel and bioenergy combustion, also produce chemicals that cause global warming and acid rain. If combustion is the problem, then eliminating combustion must be the solution.

Yet, at that time, I was busy trying to finish high school academically while playing tennis. I had neither the time nor the skills to help solve the problem. I did figure though, that if I kept studying and learned as much math and science (and later, at university, engineering and economics) as possible, I would equip myself as best I could to solve these problems.

15.2 Hungry for Knowledge

When I started my undergraduate degree program at Stanford University during the autumn of 1983, it was not obvious to me what courses or major to focus on. Indeed, there was no major available to study air pollution, climate, or acid rain. There was not even a general environmental major. The closest degree (not even an undergraduate degree) was a master's degree in Environmental Engineering, which centers around the study of groundwater pollution.

Because it was not evident to me what undergraduate major to focus on, I took five engineering courses in different engineering majors during the spring and autumn of 1984. These included courses in material science, electrical engineering, aeronautics and astronautics, statics, and environmental science and technology. The last course broadly covered water pollution, urban air pollution, acid rain, global climate, and energy. Professor Gil Masters was the instructor. His engaging teaching style consistently won him many teaching awards. This course was by far the most interesting to me among the five. It was taught through the Civil Engineering Department. I decided then to major in Civil Engineering although there were hardly any other related courses available in the department or the university.

One additional relevant course I did find was on small-scale energy systems. It was also taught by Professor Masters. In that course, he taught about the efficiencies and engineering characteristics of solar panels, wind turbines, and other types of renewable energy systems. I took that course during the winter of 1985, which was in the middle of my sophomore year. Although I was interested in the renewable energy information in that course, I was more interested at the time in understanding air pollution, climate, and acid rain problems before focusing on solutions. As such, the information in that course stayed dormant inside me until the year 2000 when I dusted off my notes from the course to come up with an *aha* moment.

In the meantime, I had taken a basic economics course during my freshman year. Since I had several credits from high school that I could transfer to Stanford, I had room to consider a second major. Although I was not so excited about taking economics courses, I felt it was important to do because I knew that if I wanted to understand and solve large-scale air pollution, climate, and acid rain problems, I should understand costs, financing, and economics. As a result, I selected Economics as a second major.

During the rest of my undergraduate career, I built up skills in engineering and economics. However, I was disappointed by the fact that there were simply no other courses that I could take that focused on my specific interests in air pollution, climate, and acid rain. Toward graduation, I thought deeply about what my next step would be. I learned about a program at Stanford, called the co-term program, that would allow me to complete a master's degree concurrently with my undergraduate degrees. I decided to apply to the co-term master's degree program in Environmental Engineering through the Department of Civil

Engineering. This program focused on groundwater pollution, which was still not my main interest. However, it was the closest I could come to a program that moved me toward my goal. Since the program took only one more year, I used the time as an opportunity to broaden my knowledge about the environment and obtain some research skills.

15.3   Lessons for Life

After completing my undergraduate degrees and master's degree at the end of winter quarter, 1988, I contemplated my next move. I knew I wanted to enter a Ph.D. program to study the atmosphere. However, I had also been playing tennis continuously for about 12 years, including 4 years on the Stanford University tennis team, and I felt that I should try to go onto the professional circuit for a period. I allocated myself a year and a half, at which point, during the autumn of 1989, I would enter a Ph.D. program. While I always hoped to break through the tennis ranks and rise to the top, I was under no illusion I was good enough to do that, so I planned my obsolescence ahead of time. Otherwise, like many of my tennis-player friends, I could have stayed out on the tennis circuit for years traveling and playing but eventually having to come back to real life. I also felt that the longer I procrastinated trying to understand the problems that I was interested in solving, the longer I would need to come to a solution. I felt passionate about finding a solution. As such, I compromised by giving myself a limited time to play.

Halfway through the one-and-a-half years of travel, I developed a fracture in my right knee that resulted in a piece of bone breaking loose and floating around my knee joint. This required surgery to screw the bone chip back in and suture up my meniscus, which the bone chip had shredded. During the surgery, the doctor mistakenly sutured my right peroneal nerve down. That nerve runs down from the knee to the foot. By compressing the nerve, he paralyzed my foot, giving me a drop foot. I could push down but not pull my foot back up. He realized this the next morning and took me back in for another surgery to undo the suture. Even though the nerve was not cut, it was compressed so severely that my foot remained paralyzed for 2 years and did not recover fully for 7 years. Feeling gradually came back to my leg and foot at the rate of 1 inch per month down my leg, starting at my knee. Needless to say, this was the end of my competitive tennis career, although I tried to use a prosthetic to run and play tennis with a drop foot. However, while I could play, the loss of speed made it almost impossible to win matches.

I never regret playing tennis. It taught me many lessons that I still use today.

One is to focus, regardless of all the distractions around you. Always keep your eye on the ball. Don't let outside noise or movement disturb you. I have applied this in my research countless times. There are many distractions that can take focus away from research on solutions.

Second, there is no substitute for practice and hard work. The harder one trains and practices tennis, the less chance that a loss will be due to not being in shape or not training sufficiently. The more one studies and practices a research subject, the deeper one's knowledge of the subject becomes and the lesser the chance of making an error.

Third is switching gears. While playing tennis through high school and university, I simultaneously took heavy course loads. This was difficult, but it taught me to be efficient in my current work, where I need to teach, research, advise students, write proposals, respond to emails, review applications, attend committee meetings, and attend conferences and seminars. Even though switching gears to work on different tasks has been stressful at times, I learned how to focus on each task based on years of training.

Fourth is being accurate and truthful. There is nothing more important than trying to be as accurate as possible in research and while reviewing other people's work. In most tennis tournaments growing up, players would call their own lines. Thus, if someone cheated, it would be known quickly. As such, most players tried to be honest despite the temptation to get an extra point here or there. Honesty is important in science as well. It is incredibly important to be accurate not only in one's own work but also when describing another's work. Too often, scientists criticize other studies either without fully understanding them, because they do not want to spend the time to understand them, or, in some cases, because they simply want to make other researchers look bad to make their own work look better in comparison. These scientists would have benefited from learning tennis etiquette.

15.4 Modeling Regional Pollution and the Weather

During my hiatus from education to play tennis, I looked at Ph.D. programs at the University of Washington and the University of California, Los Angeles (UCLA), both of which had strong Atmospheric Science programs. I ended up going to UCLA because I found an advisor there, Professor Richard Turco, who had a project available on the main topic I wanted to study, urban air pollution. In addition, I felt that, if I wanted to understand and solve air pollution problems, I needed to be in a living laboratory. Los Angeles was just the place. It was and still is ground zero for air pollution in the United States.

I started my Ph.D. at UCLA during the autumn of 1989 and completed my studies in June of 1994 with an M.S. and Ph.D. in Atmospheric Sciences. There, I learned how to develop physical equations describing phenomena in the atmosphere, how to solve the equations, how to write computer programs to represent the equations and their solutions, and how to compare model results with data. My project was to build an urban air pollution–weather prediction model, apply it to study air pollution in Los Angeles, and compare model results with data. I loved my project from beginning to end.

An **air pollution computer model** is really a four-dimensional (three dimensions in space and one in time) mathematical representation of atmospheric processes, including gas processes, particle processes, transport processes, solar and heat radiation processes, ground-surface processes, and ocean processes, among others. At the time, air pollution models were decoupled from weather prediction models. In other words, the sources of wind speed and direction data to move gases and particles from place to place in the air pollution model were either winds estimated from observations or winds from an electronic data set, where the data were produced by running a separate weather prediction model. No regional air pollution model at the time had a built-in weather prediction model, where the winds drove pollution movement *and* the pollutants themselves fed back to modify the weather.

In 1990, I began my computer modeling research by focusing on building computer algorithms for gas chemistry and particle physics and chemistry. Some of the codes I eventually developed for the air pollution model included three algorithms that simulated gas chemistry and others that simulated particle physics and chemistry. In particular, I focused on treating pollution particles of different size and composition.

I then coupled these gas and particle algorithms with existing transport and radiation algorithms developed by Dr. Owen Toon and my advisor, Professor Turco. Dr. Toon was a colleague of my advisor.

I worked during the summer of 1990 in Dr. Toon's lab at NASA Ames Research Center in Mountain View, California.

The resulting combination of gas, particle, radiation, and transport algorithms comprised an air pollution model. I set up the model to predict gas and particle concentrations in space and time in the Los Angeles basin. However, it was applicable anywhere that sufficient data were available. At the time, though, the only source of winds for the model was an electronic file with interpolated observations. This was not ideal because the observations were far apart in space and not very frequent in time. Fortunately, in our research group at UCLA, another graduate student, Rong Lu, was building a regional four-dimensional **weather prediction model**. Such a model predicts winds, temperatures, pressures, and humidity.

In 1993, I coupled Rong's weather model with my air pollution model in such a way that heating rates from the radiation algorithms in the air pollution model now fed back to the weather model to predict temperatures in the weather model. Heating rates are calculated by considering the transfer of sunlight and heat through the atmosphere, and the interactions of the sunlight and heat with gases, aerosol particles, and cloud particles. Simultaneously, I set up the coupled models so that winds from the weather model now moved gases and particles around in the air pollution model. Also, temperatures and air pressures from the weather model now fed back to affect gas and particle chemical reaction rates and physical interactions. Little did I know at the time, but this was the first air pollution–weather model, coupled with feedback for gases and aerosol particles, developed worldwide to study urban or regional air pollution. I gave a copy of the resulting model back to Rong for him to use as well, which he did. Sadly, Rong passed away from a brain tumor just a few years after he graduated.

I named the coupled model GATOR-MMTD (Gas, Aerosol, TranspOrt, Radiation-Mesoscale Meteorological and Tracer Dispersion model). I subsequently used the model to simulate gas, aerosol, radiative, and meteorological parameters and to compare results with hourly data in different locations in the Los Angeles basin.[337,338,339,340]

15.5 Modeling Global Pollution and Climate

In early 1994, before graduating from UCLA, I was fortunate to land a job as an Assistant Professor at Stanford University in what became the Department of Civil and Environmental Engineering. While still at UCLA, I wrote a large research proposal to the U.S. Environmental Protection Agency to expand my regional air pollution model to the global scale in order to study global climate and air pollution. I included on the proposal my advisor and Professor Akio Arakawa, primary developer of the UCLA General Circulation Model (GCM). The UCLA GCM predicts winds, temperatures, air pressures, cloud cover and moisture on a global scale but lacked detailed treatment of gases, particles, clouds, or radiation. The goal of the project was for me to couple, with feedback, the UCLA GCM to a globalized version of my air pollution model. I would perform the same type of interactive coupling as I did with the regional pollution and weather models. To my shock and that of everyone else on our proposal team, the proposal was funded. The hard part, work on the project, followed.

Shortly after the project funded, I left for my new job at Stanford. I then worked on the project while also teaching and advising students. I first stretched my regional air pollution model to the global scale. Owing to the intense treatments of air pollution chemistry, particle physics, and radiation, this resulted in a very detailed global air pollution model. I then coupled it, with feedback, to the UCLA GCM. The result, in 1995, was the first coupled global air quality–weather–climate model worldwide to treat the feedback of gases and size- and composition-resolved particles to weather and climate and vice versa.[341,342]

A regional air pollution model simulation requires inputs at its horizontal boundaries. Such inputs include wind speeds and directions, gas concentrations, and particle concentrations. These values must come from a larger scale, such as the global scale, than that of the regional model. At the time, boundary inflows into regional models were coming primarily from interpolated winds. Since I now had both a global and regional model, I realized I could improve the regional model by coupling the global model to it and feeding boundary conditions from the global model, not only horizontally, but also vertically, into the side and top edges of the regional model. I did this coupling and ensured atmospheric processes on all scales were solved consistently. The first version, completed in 1998, was a nested global-through-regional model.

A **nested model** starts at the global scale to produce meteorological, gas, and particle fields. These fields are then fed into one or more smaller, more finely resolved domains placed anywhere within the global domain. Within each smaller domain may lie one or more even smaller, more finely resolved domains that receive boundary conditions from the next-larger domain. In this way, air pollution or weather can be modeled anywhere in the world at high resolution while receiving boundary conditions that continuously vary in time and space. I eventually called the nested model GATOR-GCMOM (Gas, Aerosol, TransPOrt, Radiation-General Circulation, Mesoscale, and Ocean Model).[343,344,345] It was the first model worldwide to nest gases, particles, radiation, and meteorology from the global to local scale. Since then, many models have made efforts to nest pollution, weather, and climate in a similar manner.

15.6   Black Carbon, the Kyoto Protocol, and Wind versus Coal

I have used the global and nested models over the years since then to study several phenomena, including the impacts of aerosol particles on ultraviolet radiation, temperatures, and ozone in urban areas,[346,347] and the effects of black carbon on climate. One hypothesis I proposed was that black carbon, the main component of soot, may be the second-leading cause of global warming after carbon dioxide in terms of direct radiative forcing.[9,10] I further hypothesized that controlling the emissions of black carbon may be the fastest method of slowing global warming.[11,12] The contention that black carbon was the second-leading cause of warming was subsequently supported by others, including in a major review article.[13]

Up to the year 2000, I had focused on understanding air pollution and climate problems. In 2000, I started to delve into solutions. In particular, I became concerned about whether the United States would ratify the Kyoto Protocol. The **Kyoto Protocol** was an international climate agreement adopted by the United Nations Framework Convention on Climate Change on December 11, 1997. The protocol called for developed, but not less-developed, countries to reduce greenhouse gas emissions. Between the adoption of the protocol and 2022, 191 countries and the European Union ratified it. The only countries not to ratify it were the United States, Sudan, and Afghanistan. Canada renounced its ratification in 2012.

In 2000, I wondered what it would take for the United States to satisfy its share of the Kyoto Protocol, which was to reduce greenhouse gas emissions 7 percent below 1990 levels. I first estimated that this could be accomplished by reducing about 59 percent of 1999 U.S. coal emissions. I then wondered how many wind turbines this would require. I dusted off my notes from the Environmental Science and Technology course I took in 1984 from Professor Gil Masters. I vaguely remembered an equation that he derived that could help answer this question. The equation estimated the capacity factor of most any wind turbine given just three parameters: mean wind speed, rated power of the turbine, and the turbine's blade diameter. At the time, the equation did not exist anywhere except in his course notes, which I had fortunately saved.

After thinking about this a bit, I went to Professor Masters to discuss my goal. He was encouraging and agreed to help me on a paper. I then used the equation along with information about a recently developed efficient 1.5-megawatt wind turbine to calculate the number of such turbines that might be needed to replace enough coal for the United States to satisfy the Kyoto Protocol. This analysis resulted in a very short (three-quarters of one page) paper in *Science Magazine* called "Exploiting wind versus coal".[265] Despite its brevity, the paper received an enormous amount of pushback, primarily from coal supporters. One person kept sending negative comments about the paper to hundreds of people on a blind email list. He defended himself by saying that he was just a concerned citizen. However, an investigation by a reporter revealed he was a former senior vice president of the National Coal Association.

The United States signed the Kyoto Protocol on November 12, 1998. During May 2001, shortly after George W. Bush became U.S. President, he had to decide whether to submit the Kyoto Protocol for ratification to the U.S. Senate. At that time, I was working on a draft of the paper called "Control of fossil-fuel particulate black carbon and organic matter, possibly the most effective method of slowing global warming".[11] The main conclusion of the paper was:

> *Reductions in black carbon plus organic matter emissions from fossil-fuel sources will not only slow global warming, but also improve health.*

During May 2001, I sent a draft of the paper to some people I knew at the U.S. Environmental Protection Agency to provide comments. About the same time, coincidentally, the White House asked the Environmental Protection Agency if they were aware of any new papers on the causes of climate change. My paper was mentioned. On May 18,

2001, the White House then requested the Environmental Protection Agency to ask for my permission for the agency to send the draft paper to the White House. I agreed but asked that the paper not be cited or quoted as it was not yet published. The White House agreed, and the paper was sent.

Sure enough, on June 11, 2001, President Bush gave a speech in which he explained why he was not sending the Kyoto Protocol to the Senate for ratification. In his speech, he said, *"The Kyoto Protocol was fatally flawed in two fundamental ways."* First, the protocol would have *"a negative economic impact, with layoffs of workers and price increases for consumers."* Second, *"Kyoto also failed to address two major pollutants that have an impact on warming, black soot and tropospheric ozone. Both are proven health hazards. Reducing both would not only address climate change, but also dramatically improve people's health."*[348]

The first reason for President Bush's rejection of the Kyoto Protocol was not true because a transition to renewable energy creates more jobs than lost and reduces costs to consumers. While the second statement was true, it was not a reason to pull out of the Kyoto Protocol. I was flattered that the President thought my conclusions were important enough to mention almost verbatim. However, I was not happy that he used them to draw an incorrect inference, namely that the Kyoto Protocol should not be ratified. The correct implication of the conclusions was that the Kyoto Protocol should have been ratified but it could have been improved by including black carbon (and ozone). I do understand, though, that even if the President did not use this reason, he would have found another reason to prevent ratification. Nevertheless, I learned a lesson here. It is extremely important for scientists to ensure their results are used by policymakers in a way that the results are meant to be used. This is not easy since, once a paper is published, it is impossible to ensure that everyone applies the results properly. However, scientists can take steps, such as by disseminating follow-up statements or opinion-editorials, that clarify the implications of a paper.

15.7 Analyzing Wind Energy and Other Technologies

The substantial feedback I received from the 2001 paper on wind versus coal motivated me to examine wind energy in more detail. I asked a Ph.D. graduate student, Cristina Archer, who had substantial experience with and knowledge about meteorology, if she wanted to work on a wind-mapping project for the United States. She agreed to work on this along with her main project, which was to analyze a wind circulating around the Monterey Bay that she discovered while visiting the Santa Cruz beach. Over the next few years, she developed the world's first wind maps from data alone at 80 meters above ground level. One of the maps was for the United States and the other, for the world as a whole.[247,267] The first of these studies, along with another,[248] also showed that interconnecting geographically dispersed wind farms could turn completely intermittent wind power into partial baseload power.

Another student, Mike Dvorak, then performed high-resolution computer modeling of wind energy resources offshore of California and offshore of the United States east coast.[349,350,351] These studies found not only substantial wind resources offshore of the United States but also that peak offshore winds often coincide in time with the time of peak energy demand.

In the meantime, I was using the GATOR-GCMOM model to compare the impacts on air pollution and climate of different energy technologies. I compared the effects of diesel versus gasoline vehicles on air pollution;[352] the effects of gasoline versus hydrogen fuel-cell versus gasoline-electric hybrid vehicles on U.S. air quality and climate;[116,266] the effects of hydrogen fuel-cell versus gasoline vehicles on the global ozone layer and climate;[75] and the effects of ethanol versus gasoline vehicles on air pollution mortality.[180,181,182]

After performing the wind analyses and the comparisons of different fuels and electricity sources, I began to think that it would be useful to review systematically different proposed solutions to global warming, air pollution, and energy security. At the time, several technologies were being proposed as solutions to replace fossil fuels for electricity. Other technologies were being proposed as solutions for transportation. Aside from renewables, major proposed technologies for electricity included nuclear power and coal with carbon capture. For transportation, battery-electric vehicles, hydrogen fuel-cell vehicles, and biofuel vehicles were being proposed.

Given our previous work on evaluating some of these technologies, I felt I was ready to carry out such a review. The study[133] compared 11 different proposed energy solutions to global warming, air pollution, and energy security in terms of 11 different impact categories. Such categories included CO_2-equivalent emissions, air pollution mortality, land footprint on the ground, spacing area, water consumption, resource abundance, effects on wildlife, thermal pollution, water chemical pollution/radioactive waste, risk of energy supply disruption, and normal operating reliability.

With respect to reliability, in 2008, I asked a graduate student, Graeme Hoste, to see if it were possible to match California's hourly electricity demand with only wind, solar, geothermal, and hydroelectric power supply. Geothermal would provide constant electricity each hour; solar and wind would provide variable power; and hydroelectric power would fill in gaps in supply. Graeme completed a report on this topic.[271] He found that, in theory, California could meet 100 percent of its monthly averaged hour-by-hour power demand during the two months tested, April and July 2020. The upshot was that, when treated as a bundle with geothermal and hydroelectric, wind and solar are more reliable for meeting peak demand than they are when treated individually. This result, which was borne out in dozens of subsequent independent studies, was used in my 2009 review paper.

The overall conclusion of the review paper was that the best electric power generating technologies for minimizing the 11 impacts were onshore and offshore wind, solar PV, CSP, geothermal, tidal, wave, and hydroelectric power. These technologies were all **wind, water, and solar** (**WWS**) technologies. The paper also proposed the use of battery-electric and hydrogen fuel-cell vehicles for transportation.

15.8   One Hundred Percent Wind–Water–Solar and the TED Debate

The review paper garnered substantial interest from the climate and energy communities. Shortly after the paper was published, I was approached by *Scientific American* to consider writing a follow-up paper about the feasibility of powering the world with the best technologies identified in the review. While pondering this, I asked a colleague of mine, Dr. Mark Delucchi, who at the time was a research scientist at the University of California at Davis' Institute of Transportation Studies, whether he would be interested in partnering with me on such a study. He agreed. Together, we analyzed the technical and economic feasibility of transitioning the world's all-purpose energy to 100 percent WWS. We analyzed the change in energy demand upon a conversion to WWS, the numbers of WWS devices needed, the land areas required for footprint and spacing, the materials needed, reliability, and costs. The study concluded that, with aggressive policies, a transition to 100 percent WWS by 2030 was technically and economically possible, but for social and political reasons, a more likely transition end-goal was around 2050.

The paper[272] was published in November 2009. It was immediately attacked as pie-in-the-sky and an impossible dream. Yet, within 12 years, the 100 percent goal outlined in the paper had advanced significantly, and a movement had sprung up around it. Since then, many 100 percent laws and commitments have been put in place by cities, states, countries, and businesses. WWS costs have come down substantially, and the public is overwhelmingly supportive of the goal. Nevertheless, the solution is still a long way from being fully realized.

As a result of the review paper and the *Scientific American* article, I was asked to take part in a debate at a TED (Technology, Entertainment, Design) conference in Long Beach, California, on February 11, 2010. The debate was with Stewart Brand, former editor of the *Whole Earth Catalog*, turned nuclear advocate. The debate was on nuclear power versus renewables.[353] I had given lots of talks but had not experienced this debate format before, where we each had 6 minutes to lay out a case and a few minutes each for rebuttals.

At the beginning of the debate, the moderator, Mr. Chris Anderson, asked the audience of about 2,000 if they favored or opposed nuclear energy. To my surprise, about 75 percent favored it. Given the strong support for Mr. Brand's position from the get-go and the facts that he was an experienced speaker at TED events, had such a charming personality, and was the first to speak; I thought I was doomed. The only advantage I had was that the data I was about to show were based substantially on new, raw research our group had performed, and he had not seen them before. Such data included data from wind-mapping studies and comparisons among different energy sources. He was using information exclusively from third-party sources.

After his 6 minutes of speaking, I felt more in a hole because he just presented lots of information with a compelling style. I felt I needed to get my words out correctly from the beginning; otherwise, I would lose the audience. I focused on my first few words and managed to start strong. I then reeled off statistics and my own graphs. While I was presenting my 6 minutes of information, I could see out of the corner of my eye the concern growing on his face. At the same time, I could feel the audience resonating with my arguments and statistics. In fact, a few times, the audience started cheering me on. A vote after the debate indicated that I had switched about 10 percent of the audience, or 200 people, such that in the end, 65 percent instead of 75 percent favored nuclear.

This debate, coupled with the review paper that found that nuclear was better than some technologies but not so good as WWS technologies, rendered me a target for nuclear advocates. Most never understood or did not want to understand that my goal has always been focused on solving air pollution, climate, and energy security problems with the technologies that result in the fastest solution, the least cost, and the greatest benefit. Nuclear power, while beneficial for some processes on some timescales, is not so good as other solutions. Thus, it isn't that I am *against* nuclear power. Instead, from a scientific point of view, nuclear is not so good as other technologies, and I needed to honestly state that, as I have been trained to do. The fact that nuclear has problems has consistently been borne out over the years since the review paper and debate.

Subsequent to the *Scientific American* paper, Dr. Delucchi and I embarked on writing a more detailed version of the global 100 percent WWS roadmap and also a roadmap for the United States as a whole. These were published in the journal *Energy Policy*.[78,274] Concurrently, I engaged another Ph.D. student, Elaine Hart, to develop an optimization model to simulate rigorously the matching of electricity demand with a bundle of geothermal, wind, solar, and hydropower electricity supply in California. She completed that work,[257,287] confirming in more detail what Graeme Hoste had found. Another student, Bethany Frew, followed Elaine's work with an optimization model study that found that a combination of renewables would allow electricity grid reliability across the United States.[297] An additional Ph.D. student, Eric Stoutenburg, examined the impact of combining wind and wave power to increase grid reliability.[249]

15.9 The Solutions Project

In the midst of the flurry of research activity, I was invited to a dinner at the Axis Café and Gallery in San Francisco on July 10, 2011 to discuss the economic viability of renewable energy alternatives for the state of New York. At the time, I had no idea that this meeting would catalyze a series of events that would turn a scientific theory into a popular mass movement to transition the world to clean, renewable energy and storage. Nor did I envision at the time that this movement would result in country, state, and city laws and proposed laws, including the *Green New Deal*, along with business commitments.

The dinner was hosted by my now good friend, Marco Krapels. At the time, Marco worked in the banking business. He had also started a nonprofit, Empowered by Light, whose goal was to bring solar power to remote communities worldwide. Marco had invited several others to this dinner, both from the local area and from New York. In particular, he invited Mark Ruffalo and Josh Fox. Mark is an actor and activist who had been asked by the governor of New York, Andrew Cuomo, to participate in a New York renewable energy task force. Josh directs documentaries and was coming off the success of his 2010 documentary *Gasland*, which exposed the problem of natural gas fracking to a worldwide audience. The movie received an Academy Award nomination.

I had been invited to the dinner because Marco and several others had seen our *Scientific American* article about powering the world for all purposes with renewable energy. My contribution to the dinner would be to offer ideas about what New York could do to obtain its energy from something other than fracked natural gas. At the time, fracking was not legal in New York, but the governor was under significant pressure to legalize it. Fracking was legal in nearby Pennsylvania, and tens of thousands of wells had already been drilled there, causing damage and upheaval in many communities.

At the dinner, Marco, Josh, Mark, and I bonded together. It was as if we had known each other for years. We were all passionate about finding a sustainable solution to New York's energy problem and eliminating air pollution and climate problems in general. The interesting thing was that we all had different backgrounds. I was a scientist. Marco was a businessperson. Mark and Josh were cultural heroes – artists and entertainers. Later, this combination of science–business–culture would prove to be a powerful combination, much stronger than any of the individual parts.

At the dinner, we discussed the fracking issue in New York and whether there was an alternative energy solution to it. I speculated that New York could be powered entirely by clean, renewable energy for all purposes. This prompted the question as to whether I would be interested in developing a WWS plan for New York state. My immediate impulse was that this would be great to do, but I knew how much work it would take, so I wavered. Put on the spot, I told them I would be willing to write a one-paragraph summary but that it would be better for someone else to take that summary and turn it into a real roadmap given the effort involved.

Josh Fox then suggested I speak to two Cornell University professors, Tony Ingraffea and Robert Howarth, to get their perspective on a New York plan. Tony was a professor in Civil and Environmental Engineering who had worked a lot on methane leaks from cement casings in natural gas fracking wells. Robert was a professor in Ecology and Evolutionary Biology who had just published a seminal paper with Tony on methane leakage rates from the fracking of shale rock.[139] I spoke with them and Josh Fox by phone almost 2 weeks later. We discussed, among other topics, the different renewable energy resources available in New York.

Subsequent to the call, I sent Tony and Robert a wind resource analysis of New York that my Ph.D. student, Mike Dvorak, had put together. Josh Fox, Robert Howarth, Tony Ingraffea, and I then scheduled a follow-up call with a larger group, adding Gianluca Signorelli, a work associate of Marco Krapels; Stan Scobies, a retired researcher in New York familiar with the energy landscape; and Marcia Calicchia, a policy expert and researcher at Cornell University.

On the follow-up call, which was at 4 p.m. Pacific time on the afternoon of September 13, 2011, I was asked again if I could put together a 100 percent renewable energy plan for New York. Feeling the stress of a lot of other work commitments, I wavered again. I repeated that I could only write a brief one-paragraph summary and then help guide someone else to do the rest.

Later that same night, when I started writing the paragraph, I began to think more deeply about a state plan for New York, and something in me clicked. I thought to myself, "*If we really want to solve the problems of air pollution, global warming, and energy security, we need granular state and country level plans.*" I also realized that a New York plan would be a natural extension of Mark Delucchi's and my previously global and U.S. roadmaps.

My inspiration took over. I worked into the night. I found wind, solar, hydroelectric, and geothermal resource data and air pollution mortality data for New York State. I then did an energy, air pollution, and climate data cost analysis of fossil fuels versus WWS for the state. Finally, I identified a set of policy mechanisms that could be proposed to implement 100 percent WWS in the state.

In the morning, I woke up from my trance and happily emailed the paragraph-turned-14-page, single-spaced draft 100 percent New York energy roadmap to the group. They were as shocked as I was. Stan quipped, "*I figure about three more inspiring conference calls ought to have the whole thing done.*" We now had a starting point to change the New York energy infrastructure.

The first draft catalyzed a flurry of activity and edits among several people in the group that we just formed. Ultimately, the New York roadmap went through 40 drafts before it was completed and published[275] in the journal *Energy Policy* 18 months later, on March 13, 2013.

At the time, though, our immediate goal was to develop a draft paper that was clean enough to present to New York Governor Andrew Cuomo and his staff. I worked feverishly on this paper, engaging students and incorporating comments by several in the group. The group as a whole also started holding regular phone conversations. Marco, Mark, Josh, and I also began talking more, taking part in several middle-of-the-night phone conversations. We became close as we shared a common passion and goal to solve major problems.

The communication and written material from our overall group had become substantial enough that we decided we needed a name for the group. We wanted the name to represent something positive because we felt it is better to be for than against something. Since this all-volunteer group was focused on solutions for New York's energy future, we settled on **The Solutions Project**. On December 15, 2011, Marcia sent the group its first brochure, which summarized The Solutions Project as a collaboration among

> *...scientists, renewable energy industry pioneers, experts in renewable energy financing and investment, mission-oriented investors, businesses, labor organizations, inner city community groups, farmers, environmentalists, and many celebrities and cultural figures.*

Its initial main goals, laid out in a February 25, 2012 brochure, were to

- Raise mass awareness about the viability of renewable energy;
- Raise awareness about the 100 percent WWS roadmap we were developing for New York; and

- Leverage private capital to demonstrate the success and viability of some specific renewable energy projects.

On January 20, 2012, The Solutions Project, still behind the scenes, began recruiting members for an advisory council. Ultimately, the advisory council consisted of scientists, business leaders, and celebrities. Some of the celebrities who agreed to take part included Deepak Chopra, Leonardo DiCaprio, Jesse Eisenberg, Woody Harrelson, Ethan Hawke, Scarlett Johansson, Sean Lennon, Julianne Moore, Leilani Munter, Elon Musk, Edward Norton, Yoko Ono, Robert Redford, Eileen Rockefeller, Antonio St. Lorenzo, Wendy Schmidt, Martha Stewart, Channing Tatum, Michelle Williams, and Deborah Winger, among others. The purpose of the Advisory Council was not only to provide feedback, but also to help The Solutions Project meet its main goal, which was to bring awareness to the world about the potential of a 100 percent transition to clean, renewable WWS energy.

During early March 2012, a new integral member came on board The Solutions Project. Jon Wank had been working in advertising and learned about The Solutions Project from Marco Krapels. Jon's passion for making a difference resulted in him quitting his job and volunteering to work with us on media and social engagement. It was Jon's ingenuity that led to the development of state, country, and city infographics. Each infographic summarizes a 100 percent roadmap for a different locale. The infographics have been used worldwide and are still available online.[354]

15.10 Effects of New York State Roadmap on Policy

The original task of The Solutions Project was to bring our science-based 100 percent New York energy roadmap to the governor of New York. The hope was that the governor might see that a WWS plan is the best solution for the state's future and translate the plan into law. Such an action would also mean that hydraulic fracturing was not needed.

To that end, The Solutions Project began to engage other groups, including the National Resources Defense Council, which had a major presence in New York. After some discussion, they committed, on January 2, 2012, to support The Solutions Project goal of bringing 100 percent WWS to New York. This was an important first step.

The Solutions Project held its first all-hands meeting at Cornell University on May 21, 2012. It was there that many of the people involved in numerous phone calls could finally meet each other and focus further on a path forward. Such a path included the idea of getting information out to large numbers of people. Some of the ideas that came forth were to hold a concert, go on a speaking tour, hold rallies, and develop media content. Marcia introduced finger puppets that we could wave at each other during disagreements to lighten the mood.

Shortly after that meeting, on June 20, 2012, Mark Ruffalo, Marco Krapels, and I met at Stanford University, California, where we spoke with students, some of whom were helping to develop the New York energy transition roadmap. We then went over to Google in Mountain View, where the three of us gave a joint talk to a packed house.[355] We offered our experiences, through the lenses of culture (Mark Ruffalo), business (Marco Krapels) and science (me), about how it is possible and necessary to transition from fossil fuels to 100 percent WWS.

We then met with the Google energy and sustainability team led by its director, Rick Needham. Google had been making investments in renewable energy for several years. In 2007, they installed a 1.6-megawatt PV array on their Mountain View buildings. In 2010, they purchased two wind farms in North Dakota and contracted for additional wind in Iowa. Our meeting with them was a meeting of like minds. We discussed with them the importance of companies working together with scientists, cultural and community leaders, and policymakers to affect a transition. Ultimately, in 2017, Google became the first company in the world to provide at least 100 percent of its yearly energy needs with WWS.

On the same day as the Google visit, we visited Facebook, also in Mountain View, and their sustainability and energy efficiency team, led by Bill Weihl. We discussed the need for Facebook and other businesses to participate in a 100 percent WWS transition.

The talk at Google was received so well that the three of us (Mark, Marco, and me) were asked to give another joint talk, this time at an October 6, 2012 conference on Nantucket Island, Massachusetts. There, policymakers from across political parties and media were gathered. They included Secretary of State John Kerry, President of Americans for Tax Reform Grover Norquist, presidential advisor David Gergen, U.S. Treasury Secretary Larry Summers, and MSNBC commentator Chris Matthews.

The talk[356] led to an invitation to Chris Matthews' home in Washington D.C. for a dinner. On February 27, 2013, Mark, Marco, and I presented The Solutions Project vision at the dinner to U.S. Senator Kirsten Gillibrand, Secretary Kerry, some members of the U.S. House of Representatives, and some members of President Obama's staff. Although these meetings did not lead to direct policies at the time, they provided new information to policymakers, increased the familiarity of The Solutions Project name, and allowed us to hone our explanations and learn about the concerns of others.

In a parallel effort to raise awareness on a large scale, during August 2012, Jon Wank created an animated cartoon, called *Tommy and the Professor*. Mark Ruffalo came up with this idea at our May 12, 2012 retreat at Cornell. He and I were brainstorming outside, and he suggested a brilliantly crazy idea to produce a cartoon where he was an annoying kid and I was a pedantic professor trying to teach him about renewable energy. Months later, Jon Wank brought this cartoon idea to life. Tommy (voiced by Mark Ruffalo) was the annoying student who hung out with a girl (voiced by Zoe Saldana). After seeing that the Professor's house was the only one on the block with its lights on after a blackout, the two came to a class on renewable energy taught by the Professor (voiced by me). There, the kids learned about energy from wind, water, and sunlight. After the lecture, Tommy came to an epiphany about what 100 percent WWS meant and wanted to inform the world about it.

In the meantime, during the rest of the 2012, several students at Stanford plus Tony Ingraffea, Bob Howarth, Stan Scobies, economist Dr. Jannette Barth, and I continued to revise the New York energy roadmap. I also engaged Dr. Mark Delucchi to work on the New York paper. He has been a coauthor on most roadmaps since then.

On November 17, 2012, Mark Ruffalo, Marco Krapels, Josh Fox, Jon Wank, and I brought the latest version of the roadmap with us to a meeting with the National Resources Defense Council in New York City. The goal of the meeting was to discuss how they and The Solutions Project could bring 100 percent WWS to New York State. On February 5, 2013, the National Resources Defense Council and The Solutions Project wrote a joint letter to the governor of New York informing him about our roadmap. The letter, signed by Frances Beinecke (President of the National Resources Defense Council), Mark Ruffalo, and me, stated in part,

We are writing today to let you know about a clean energy framework for New York that has been developed by scientists, financial specialists, business leaders, and policy experts under the Solutions Project and reviewed and updated for practical implementation in New York State by the Natural Resources Defense Council...

The letter then went on to propose several first steps that the governor could take to reach 100 percent WWS. These steps included increasing the installation of large-scale wind and distributed solar, scaling up energy efficiency, removing barriers to the adoption of electric vehicles, creating a **green bank** (which leverages public and private financing to advance clean energy projects), and expanding the existing solar rooftop program.

Soon after (during March 2013) our New York roadmap paper was published and received attention in the press, particularly in New York. The roadmap spread like wildfire throughout New York. As a result, I was asked to give a talk at an antifracking rally about the potential of New York to go to 100 percent WWS. The rally was on the footsteps of the governor's office in Albany. I spoke on June 17, 2013 to a crowd of thousands. This was the first and only rally I ever spoke at. It was gratifying because the crowd was on board with transitioning to 100 percent WWS. Our roadmap for New York gave people hope that there was an alternative solution to natural gas.

Ultimately, on December 17, 2014, Governor Cuomo banned hydraulic fracturing in New York because of concerns over its health risks. A decision on fracking had been held in abeyance for 6 years while health effects data were gathered. Grass roots organizations throughout the state ensured that the health risks were considered. The ban was facilitated by the fact that an alternative to fracking now existed. That alternative, supported strongly by the grassroots organizations, was WWS.

Transition highlight
To that end, with the fracking issue behind him, Governor Cuomo submitted to the New York Public Service Commission a proposal for New York to obtain half of its electric power from renewable sources by 2030. On August 1, 2016, the Public Service Commission approved the proposal, giving rise to a mandatory and enforceable 50 percent WWS law by 2030 for the state of New York, called the Clean Energy Standard.

The law required half of the state's electricity to come from onshore and offshore wind, solar, and hydroelectric power. Our New York roadmap called for 80 percent WWS in all energy sectors by 2030 (and 100 percent by 2050), but the Clean Energy Standard requirement of 50 percent WWS in the electric power sector was a good first step. It was the first piece of proposed legislation for which the 100 percent WWS roadmaps provided a scientific basis.

On July 18, 2019, Governor Cuomo went further by signing a law requiring that New York reach 70 percent WWS by 2030 and up to 100 percent by 2040 in the electric power sector. The law also required sufficient emission reductions in other sectors for overall greenhouse gas emissions in the state to decrease 85 percent by 2050. It was nice to see that science, business, culture, and community could come together to motivate the passage of a law mandating a clean, renewable energy solution.

15.11   Effect of California Roadmap on City Policies

In the meantime, by September 2012, I wanted to develop a 100 percent all-sector WWS roadmap for California. Aside from the facts that California is the most populous U.S. state and its residents have an affinity for renewable energy and efficiency, our research group had already analyzed the electricity grid and offshore wind energy in California. I formalized a Solutions Project student research group at Stanford and engaged over 20 new students to help develop the California roadmap. I also worked on the California plan with several of the scientists who had helped with the New York roadmap.

We finished an early draft of the California roadmap on January 8, 2013. I sent a copy that day to the Sierra Club nonprofit environmental organization to review it for practical implementation. Jodie Van Horn, at the Sierra Club, became the point person, and she obtained internal suggestions for improvement. This was the first of many iterations of the California roadmap over the next year and a half.

In early 2013, The Solutions Project was still an all-volunteer group. However, the group was considering becoming a formal nonprofit. Before deciding to take the leap, we considered other options. Given that we were now interacting more with the Sierra Club, it was logical to discuss an option to become part of them as well. On February 6, 2013, Mark Ruffalo approached Michael Brune, the Executive Director of the Sierra Club, about the possibility of The Solutions Project and Sierra Club becoming *"one diverse coalition around science, business, and culture advancing 100 percent renewables."*

The Sierra Club was strongly interested and proposed to incorporate The Solutions Project as a campaign inside of the Sierra Club and to move resources from their *Beyond Coal Campaign* to a *100 Percent WWS Campaign* led by The Solutions Project. The Sierra Club was interested because The Solutions Project offered raw science, the media power of celebrity, and creativity. The offer was tempting to The Solutions Project because it negated the need for us to go through the pains of forming a nonprofit. However, it also would have reduced the autonomy of The Solutions Project, whose success to date was based on being free to pursue ideas without gaining the consensus of a large board.

After lengthy deliberations, including a meeting in New York City on April 20, 2013, The Solutions Project chose to become its own nonprofit. Its focus remained, for the time being, on educating the public and policymakers about the 100 percent roadmaps. The Sierra Club saw this as an important mission as well but shifted to focus on what it does best, grass-roots campaigning. In this case, the campaign centered around getting cities across America to commit to 100 percent WWS. This effort, called the *Ready for 100 Campaign*, started in earnest in late 2015.

On October 29, 2015, Jodie Van Horn from the Sierra Club contacted me again. We talked about the Sierra Club's pending cities campaign, the goal of which was to get 100 U.S. cities to commit to 100 percent WWS within 3 years. She asked if I could prepare 100 percent WWS roadmaps for some targeted cities. On the one hand, I knew that each roadmap took a lot of work. On the other hand, I knew they were important for analyzing the ability of cities to transition. I agreed but told her it would take some time. Indeed, the first roadmap paper we finally completed for 53 towns and cities in North America[279] was not published until 2½ years later, on June 30, 2018.

Transition highlight
The Sierra Club used the 100 percent WWS state roadmaps and resulting infographics,[354] to help provide confidence to communities and town and city leaders that 100 percent WWS was

possible. When the town and city roadmaps we were developing finally dribbled in, they became helpful as well. However, by the time the roadmaps were published in 2018, the Sierra Club had already obtained commitments from over 100 U.S. cities, meeting their goal of at least 100 cities within 3 years. By the middle of 2022, this number had increased to over 200 cities. The 100 percent movement had really accelerated with the help of many groups working together.

The Solutions Project finally became an independent nonprofit in mid-2013. Much of the rest of the year was spent hiring an Executive Director, Sarah Hope, and staff, setting up the organization, and preparing a launch event. The launch event finally took place on June 19, 2015 in New York City. Leonardo DiCaprio joined the rest of The Solutions Project team at the event.

During May 2013, while still updating the California roadmap, I embarked on an individual roadmap for Washington State. Washington was chosen because of the substantial renewable resources it has (wind and existing hydroelectric in particular). I thought that Washington might be a state that could reach 100 percent quickly with respect to electric power. During the autumn of 2013, I recruited a new set of over 20 students to work more intensely on the roadmap. The Washington State roadmap paper was published in 2016.[277]

15.12 The Letterman Show

On September 25, 2013, a television producer, Mike Buckiewicz, invited me to appear on the *Late Show with David Letterman*. He said that he invited me because he saw my interview in the documentary *Gasland Part II* by Josh Fox. Josh had interviewed me for his documentary on July 12, 2011, the day after we first met in San Francisco. His film is about the damage done by natural gas fracking. My short interview was about how it was possible to get off of gas, coal, and oil and move completely to renewables. The producer liked the positive message I related. He told me that when he showed the clip to Mr. Letterman, the latter asked, "*Why haven't we had that guy on before?*" Mr. Letterman himself had a serious concern about our future in the face of climate change, so he felt hopeful there might be a possible solution.

I was humbled about the thought of being on the *Late Show*. Only a handful of scientists had appeared on a late-night show on a major network. One who I deeply admired was the late Professor Stephen Schneider, who was a climate scientist and ironically had an office a few doors down from me at Stanford University. He had appeared on the Johnny Carson show in 1977 to talk about climate science. Ever since I was a young scientist and heard about Stephen Schneider doing this, I had thought that being able to speak on such a stage about science would be almost impossible, but amazing if it could happen. When I was offered an appearance on the show, I felt a weight and responsibility. I needed to communicate clearly and accurately given that Stephen was such an effective communicator, and I had nothing like his skills.

I appeared on the show[357] on the evening of October 9, 2013. I spent the whole day training in New York City with Mark Ruffalo and others. Mark came with me backstage to the Green Room to help prepare me further. I thought I had everything under control. My time slot was right after Lucy Liu. I was given 11 minutes to converse with Mr. Letterman. This was long for an interview on a show like this.

As I was walking toward the stage with about 30 seconds to go before appearing live, the music was playing and the lights were blinding. Even though I had been teaching for over two decades and had given hundreds of talks and had prepared all day, I panicked. I was overthinking what I wanted to say, and all of a sudden, I had too much running through my head. Then, the thought *"I'm going to be speaking to millions of people"* crept into my head. With seconds ticking by, all I could think was, *"This is going to be a disaster."* I kept walking toward the stage in a panic, as if I were walking to my death.

Then, with about 4 seconds to go, I remembered one of my tennis lessons. *Keep your eye on the ball.* I thought to myself, *"Ok, relax. Take a breath. Just focus on one thing to say. Say something about why I want to change the energy system."* Because my concern about air pollution is the reason that I started this career, I immediately knew I should just focus on this topic. I decided to say something about how many people die from air pollution. As morbid as that sounds, I felt that few people were aware of this fact, and it was important. I believed that if I said that, everything else would flow.

When I finally sat down and the music died down, I was still jittery but felt more under control. I also remembered what the producer had told me, *"Look toward Dave and keep your hands planted."*

Mr. Letterman made me feel at ease and threw me a softball statement to respond to: "...*tonight, you have something positive that you can present to all Americans.*" My response was, "*So Dave, we're developing science-based plans to eliminate global warming, air pollution, including the 2.5 to 4 million deaths that occur worldwide each year.*" His immediate response was, "*Due to air pollution? That many people are dying due to air pollution?*" Once he asked that, I knew I had made the right decision to focus on that topic, because it grabbed his attention and likely the attention of those watching. After that, I relaxed, and everything flowed for the rest of the interview.

The response to my being on the show was overwhelmingly positive. The New York governor's staff was watching, as were people interested in energy and the environment across America. Our Solutions Project student group had organized a viewing party, which humbled me further. The show brought 100 percent clean, renewable energy to the large-scale public sphere. One-hundred percent WWS was no longer just a niche idea. It had begun to capture the public's imagination and was accelerating into a movement.

Soon after the show, a new star entered The Solutions Project scene. On November 26, 2013, Marco Krapels hosted a dinner in San Francisco, where he introduced me and others to Leilani Munter. Leilani was a racecar driver who was passionate about clean, renewable energy. She enthusiastically embraced the 100 percent goal and wanted to bring it to 80 million U.S. racecar fans who normally would not care about energy. Over the years, since that meeting, she has been a spokesperson for electric racing and electric powered racing stadiums. She also graciously volunteered her time at events and to speak out whenever possible about 100 percent WWS.

15.13   Impact of California Roadmap on California Law

The California roadmap paper[276] was ultimately published on July 22, 2014. Once it became public, the press began writing articles about it, and some articles reached the office of California Governor Jerry Brown. On August 22, 2014, Mr. Brown's senior advisor on energy and environmental issues, Cliff Rechtschaffen, emailed me stating,

> *We read with interest about your recent paper on a 2050 renewables strategy for California. I'm wondering if you*

368 / 15 My Journey

*would be interested in coming up to brief a group of advisors
and policymakers who work on climate and energy issues in
the administration.*

Needless to say, I was ecstatic over the thought, because the whole goal
of our work was to perform science in order to inform policymakers
about solving air pollution and climate problems. Air pollution alone
in California caused at the time about 13,000 deaths per year. This
was also a chance to inform the staff of the governor of the fifth largest
economy in the world about a pressing global and local issue.

On October 27, 2014, Marco Krapels and I trekked to Sacra-
mento to present our California 100 percent WWS roadmap. The group
we presented to included Mr. Rechtschaffen, Mr. Brown's senior policy
advisor Ken Alex, California Energy Commissioner David Hochschild,
and others. I laid out the 100 percent roadmap for California in detail.
The roadmap called for an 80 percent transition of all energy sectors by
2030 and 100 percent by 2050. Marco and I were peppered with lots of
good questions. We left feeling like we gave it our best shot but had no
idea at the time what the impact might be. We did know that nothing
would happen right away, if at all, because Mr. Brown was running for
re-election, and the election was coming up in 11 days.

Mr. Brown was re-elected. During his inauguration speech,
on January 5, 2015, he pleasantly surprised us by proposing several
laws that followed logically from our discussion and our California
roadmap. He proposed that at least 50 percent of all electricity in
the state of California should come from WWS sources by 2030. He
also proposed reducing petroleum use in vehicles 50 percent by 2030,
which meant electrifying 50 percent of the vehicle fleet by 2030. We
had proposed 80 percent by 2030. He finally proposed doubling the
efficiency of existing buildings by 2030. We had proposed providing all
heating, cooling, and electricity in buildings with WWS and reducing
energy use through energy efficiency measures. Although the oil industry gutted his proposal to reduce petroleum use, California did pass the
50 percent WWS portfolio standard and the energy efficiency standard
through California SB 350.

In another surprise, on September 19, 2016, Governor Brown
signed California SB 1383, which required a 50 percent reduction in
black carbon (and other short-lived climate forcers) emissions below
2013 levels by 2030. After two decades of work on black carbon,

my scientific studies and results were finally paying off with specific, enforceable legislation.

Between 2016 and 2018, the 100 percent movement swelled nationally. The Solutions Project was central to this expansion. They, along with the Sierra Club and a few other key groups, mobilized almost 100 nonprofits in a group later called the 100 **Percent Network**, around the 100 percent WWS goal. The main purposes of The Solutions Project had shifted slightly, not only to raise awareness that 100 percent WWS is possible, but also to bring 100 percent WWS to 100 percent of the people.

To amplify the latter point, The Solutions Project created a subgroup called 100.org. The purpose of 100.org was to provide small grants for small organizations that had creative ideas about how to engage people to transition to 100 percent WWS, especially people who might not otherwise prioritize a transition. The group 100.org also provided awards to honor transition leaders and people with innovative ideas. It also sought an **equitable transition** to 100 percent WWS. This involved, for example, proposing that at least a certain fraction of new jobs during a transition would be reserved for disadvantaged communities. Through The Solutions Project and 100.org, communities in many inner cities suddenly became part of the 100 percent movement.

Owing to the groundswell of public support for 100 percent WWS in California following the passage of SB 350, State Senator Kevin DeLeon proposed a follow-up law, **SB 100**. The law called for 60 percent of all grid electricity in the state to come from eligible renewables by 2030 and the remaining 40 percent to come from either eligible renewables, large hydropower, or other zero-carbon technologies not yet invented, by 2045.

In California, an eligible renewable includes primarily WWS technologies (e.g., wind, solar, geothermal). Whereas small hydropower (run-of-the-river hydropower) is also an eligible renewable, large hydropower (with a dam) is not. However, because California has so much existing large hydropower and imports more from Washington State and Canada, and because state lawmakers wanted to maintain existing large hydropower as part of the 100 percent goal, it was necessary to create a separate category from eligible renewables to put large hydropower in. So, the last 40 percent of SB 100 allows for eligible renewables and large hydropower. It also allows for other zero-carbon technologies that have not yet been invented. The California

Senate states[358] that the reason for including other possible zero-carbon technologies is to leave *"the door open to new technologies we may not know about today."* The term was not intended for nuclear power or technologies, such as CCS, that are not zero-carbon. California Senate also states that nuclear is being phased out of California, and no new nuclear plants are being planned. The remaining 40 percent will likely be filled by hydropower plus other renewables.

> **Transition highlight**
> SB 100 was signed into law on September 10, 2018. This was a landmark legislation mandating the transition to effectively 100 percent WWS in the state's electric power section. The passage of this law was the kind of result hoped for when we embarked on the first 100 percent WWS roadmap in 2009. However, much more work is needed in other sectors and in other locations.

15.14   Fifty-State and 139-Country Roadmaps and the Paris Conference

While working on individual Washington State and California roadmaps during October 2013, I came to the realization that, at the rate we were creating one state roadmap at a time, it could take decades to finish all 50 states. I then became determined to automatize the process. I took the Washington State roadmap spreadsheets and expanded them to include all 50 U.S. states. Data could be gathered simultaneously for all states. This began the process of developing 50 individual state roadmaps, which were ultimately published 20 months later, in June 2015.[145]

Once we produced the first draft of the 50-state roadmaps during February 2014, Jon Wank created infographics for each state. These were ultimately posted on a clickable map on a website.[354] These maps were invaluable because they were simple and informative.

A ripe opportunity then arose for us to use the infographics. On August 27, 2014, Leilani Munter, Marco Krapels, Mark Ruffalo, Brandon Hurlbut (a Solutions Project board member), and Andres Lopez (a Solutions Project staff member), and I went to the White House to meet with Vice President Joe Biden. The purpose of our visit was to provide information to the Vice President about our 100 percent roadmaps. He was taking a leading role in issues related to the U.S. renewable energy infrastructure. Ahead of the meeting, we provided his staff with a document that included an infographic of our 100 percent WWS roadmap for Delaware, the state that Mr. Biden had represented for 36 years in the U.S. Senate. We were told we had only 30 minutes to speak with him, so we each prepared short remarks.

Mr. Biden came in 15 minutes late and apologized for this. The first thing he said, before we could put in a word, was that he could stay only a few minutes but that he fully supported our 100 percent WWS roadmaps. He cautioned, though, that the political landscape was such that it would be difficult for the federal government at the time to take any action on the roadmaps. We later chuckled, because Mr. Biden ended up staying an hour with us and talked almost the entire time. He was endearing and easy to listen to. Halfway through, he said he really had to go and started to leave. I asked a parting question, and he came back in the room and spoke for another 30 minutes. For me, the best thing that came out of it was the feeling he was on board with solving the problem in a big way.

On September 21, 2014, New York held a climate march, attended by about 400,000 people. I was not able to attend, but The Solutions Project had a large and central presence there. Leonardo DiCaprio joined Mark Ruffalo, Marco Krapels, Jon Wank, Leilani Munter, and Brandon Hurlbut from The Solutions Project. The United Nations building itself was even lit up with the sign "Solutions Exist."

Mr. DiCaprio, who was good friends with Mark Ruffalo, was passionate about what The Solutions Project stood for. He wanted to play a key role in helping to disseminate research results on our behalf. To that end, on September 23, 2014, he spoke in front of the United Nations General Assembly after being designated as a United Nations Messenger of Peace.

He spoke about climate change and about the solutions we had developed. He stated in front of the world body, *"New research shows that by 2050 clean, renewable energy could supply 100 percent of the world's energy needs using existing technologies, and it would create millions of jobs."*

A month later, on October 29, 2014, Leo DiCaprio came to Stanford to interview me and several students working with me on The Solutions Project roadmaps. He was producing a documentary called *Before the Flood*. He and Director Fisher Stevens spent the day with us filming. Marco Krapels was there as well. We all saw firsthand the passion that the students had for trying to solve the climate, pollution, and energy problems we face. The documentary ended up taking a different direction so did not focus on solutions, but some of the interviews were used as a postscript on the DVD version of the film. This event motivated our students for months after.

During 2015, I was not only completing the 50-state roadmaps but had also embarked on an even more ambitious project, which was to develop roadmaps for most countries of the world. Although we had developed roadmaps for the world as a whole in 2009 and 2011, individual countries could not practically implement a world plan, so country roadmaps were needed. The work on country roadmaps began during August 2014. Our group at Stanford first put together a sample roadmap for the Ukraine. We expanded this in earnest to 139 countries by March 2015. This was the number of countries for which we could find raw energy data from the International Energy Agency. Work on the country roadmaps continued all year but accelerated feverishly to meet a deadline we set for ourselves, the end of November 2015, in order to have the roadmaps ready for the Paris climate conference.

In December, Paris was hosting the United Nations Climate Change Conference, referred to as the Conference of the Parties 21 (COP 21). This was a big international event and a ripe opportunity to disseminate our country roadmaps to world leaders. The event lasted 2 weeks, but I could stay for only a few days. I met Marco Krapels there and gave several talks but only one major one. Because of a quirk in scheduling, I was fortunate to be able to give a recorded talk at the Petit Palais in between a talk by the U.N. Secretary General Ban Ki-moon and another one by U.S. Secretary of State John Kerry. I later quipped to myself that I was the only speaker who did not have a security detail. I had only a few minutes to talk but made the most of it. I laid out for the first time to the world our 100 percent all-sector WWS energy roadmaps for 139 countries. These countries emit more than 99 percent of all greenhouse gases and air pollutants worldwide.

In Paris, many nongovernmental organizations, including those that helped build the 100 percent network, supported the 100 WWS percent goal that The Solutions Project had been disseminating. Paris was filled with the spirit of 100 percent. The Eiffel Tower was lit up with the words "*100% Renewable*," and a group even formed a human sign stating "*100% Renewable*." I felt that the 100 percent movement had hit the world stage.

15.15 Impacts of Roadmaps on U.S. and Business Commitments

Shortly after completing and publishing the 50-state roadmaps in 2015, we published a companion paper that examined whether it was possible to match power demand among all energy sectors with 100 percent WWS supply continuously over time among the 48 contiguous U.S. states.[296] The new paper concluded that it was possible. The paper received an award, the Cozzarelli Prize, from the journal that it was published in, the *Proceedings of the National Academy of Sciences*.

The combination of the 50-state roadmaps and the grid reliability study gave confidence to lawmakers and politicians to legislate 100 percent clean, renewable energy at the national level in the United States.

On November 30, 2015, the first of several proposed U.S. federal laws or resolutions calling for 100 percent WWS was introduced into the U.S. Congress. U.S. House Resolution 540 called for the United States to transition to 100 percent clean, renewable energy for all energy sectors. The resolution summary stated:[359]

> *Expressing the sense of the House of Representatives that the policies of the United States should support a transition to near zero greenhouse gas emissions, 100 percent clean renewable energy...*

It was humbling to see that the text of the bill itself acknowledged our work as the scientific basis for this resolution:

> *Whereas a Stanford University study concludes that the United States energy supply could be based entirely on renewable energy by the year 2050 using current technologies;*

Policymakers were realizing on the U.S. national stage that a solution to the horrendous problems of air pollution, global warming, and energy security was possible. This realization continued. Between 2015 and 2019, seven more resolutions and bills were introduced into the U.S. House of Representatives and Senate proposing that the United States go to 100 percent clean, renewable WWS energy in one or more energy sectors (Senate Resolution 632 in 2016; Senate Bill 987 in 2017; House Bill 3314 in 2017; House Bill 3671 in 2017; House Bill 330 in 2019; House Resolution 109 in 2019; and Senate Resolution 59 in 2019). The last two of these resolutions were the Green New Deal proposals for the United States to go to 100 percent clean, renewable, zero-emission energy for all purposes by 2030. None of these resolutions or bills was ever voted on, but they educated the public and policymakers, fanning the 100 percent movement.

Meanwhile, during 2016, the United States was going through a presidential election process. On the Democratic side, the three major candidates were Governor Martin O'Malley, Senator Bernie Sanders, and Senator Hillary Clinton. The 100 percent movement had taken off to such a degree that all three candidates supported it.

Governor O'Malley was the first. On July 2, 2015, he issued a press release,[360] where the number one item stated on his platform was "*A complete transition to renewable energy... by 2050.*"

Next, Senator Sanders formulated a platform calling for a transition "*toward a completely nuclear-free clean energy system for electricity, heating, and transportation.*" To illustrate, he put The Solutions Project 50-state infographic map on his campaign website. Each state infographic showed what a 100 percent WWS system for all purposes would look like in the state in 2050.

Third, although Senator Clinton also supported the temporary use of natural gas, she stated publicly in a video[361] on October 16, 2015,

> We need to be moving as quickly as possible to 100 percent clean, renewable energy. We have a long way to go, but that should be our goal, and we should do nothing that interferes with or undermines our efforts to reach that goal as soon as possible.

The support for 100 percent WWS by the three Democratic presidential candidates culminated in the U.S. National Democratic Party platform mimicking this sentiment by stating,[362] "*We believe America must be running entirely on clean energy by mid-century.*" Thus, the 100 percent movement had grown to a point where the largest U.S. party had adopted the 100 percent goal.

Just 2 weeks before the November 7, 2016 presidential election, I received a phone call from Senator Sanders. He wanted me to provide him more details about our 100 percent roadmaps for the 50 states, which were prominent on his campaign website, because he was planning to submit a bill to the Senate calling for the United States to go to 100 percent renewables. He subsequently submitted the bill on

April 27, 2017. The bill (S.987) was co-sponsored by Senators Merkley, Markey, Booker, and Schatz.

The phone call with Senator Sanders lasted about 45 minutes. I asked him whether he wanted me to suggest possible policies that might help a 100 percent law be effective. His reply, in his characteristic voice, was, "*No, no, no. That's our job.*" We further discussed other action that could be taken to educate the public about 100 percent renewables. One such suggestion he had was to write a joint op-ed. Two days after he submitted his bill, S.987, in the Senate, we did publish a joint op-ed in the *Guardian*,[363] entitled "The American people – not big oil – must decide our climate future."

Meanwhile, the 139-country roadmaps were published[223] on September 6, 2017 and reported widely in the press. The Solutions Project then developed infographics for each country summarizing the roadmaps.[354]

> **Transition highlight**
> The international roadmaps helped to give confidence to countries that they can reach 100 percent clean, renewable energy in all energy sectors. As of 2020, at least 64 countries had committed to 100 percent renewables in the electric power sector.[364] Denmark had also committed to 100 percent by 2050 in all energy sectors. Although some of the countries propose 100 percent WWS by 2030, many other commitments are for 2050, far in the future. In all countries, it is necessary for effective policies to be put in place to reach the goal.

On November 13, 2017, results from an international public opinion poll about renewable energy were reported.[365] Twenty-six thousand people in 13 countries (Canada, China, Denmark, France, Germany, Japan, Netherlands, Poland, South Korea, Sweden, Taiwan, the United Kingdom, and the United States) were queried. The poll found that 82 percent of all the people who were asked responded that it is "*important to create a world fully powered by renewable energy.*" In the United States, 83 percent supported that statement.

Interestingly, only 66 percent of people in the same poll believed climate change was a significant international problem. The reasons that more people believed in renewable energy solutions than in the damaging effects of climate change could be extracted as follows: (1) 75 percent of people polled believed their country's renewable technology leadership gave them a sense of pride in their county;

(2) 73 percent believed that building and producing renewable energy increased economic growth and jobs; and (3) 69 percent believed that renewables made their country more energy independent. However, only 53 percent believed renewable energy reduced health problems. Ironically, that is their greatest benefit. The results of this poll suggest that people do not need to believe in climate change (although they should) to believe in clean, renewable energy.

In fact, because support for renewables is so strong, it crosses political party lines. In the United States, for example, 9 out of the 10 states with the most wind power installed have a population that generally votes for Republicans, whose base has traditionally been the most skeptical about climate change and renewables. Wind farms are still installed in those states because of their low cost and the abundance of wind.

Following the publication of our 139-country roadmap paper, we performed a follow-up study to determine whether matching continuous all-sector power demand with 100 percent WWS supply was possible in 20 world regions encompassing the 139 countries. This study was published in early 2018.[136] It supported the previous grid integration study[296] from 2015 for the United States, but it also showed that matching demand with supply and storage on the grid at low cost is possible under additional conditions beyond those shown in the earlier study, and throughout the world.

Both the 139-country roadmap study and the 20-world region grid integration study were subsequently updated in 2019 to 2021 with roadmaps and grid studies for 143 countries placed in 24 world regions.[278,245,252] Those studies were updated further in 2022 with roadmaps and grid studies for 145 countries grouped into 24 world regions.[5] All such studies support the potential for a low-cost transition throughout the world.

Similarly, in 2022, the 2015 U.S. 50-state roadmaps and grid studies were updated with a new study that examined whether the grid could stay stable in all individual U.S. grid regions and in six isolated states.[251] The study found that transitioning the United States to 100 percent WWS may more than double electricity use but reduce total end-use energy demand by about 57 percent versus business-as-usual. This energy reduction contributes largely to the 63 percent lower annual private energy cost and 86 percent lower annual social energy cost with WWS than with business-as-usual in the U.S. The study also found that costs per unit energy in California, New York, and Texas may be 11, 21 and

27 percent lower, respectively, and in Florida, 1.5 percent higher, when these states are interconnected regionally rather than islanded. Finally, it found that transitioning to WWS may create about 4.7 million more long-term, full-time jobs than lost and require only about 0.29 percent and 0.55 percent of new U.S. land for footprint and spacing, respectively, less than the 1.3 percent occupied by the fossil industry today.

During 2018, the policy momentum continued toward 100 percent WWS. For example, following the November 2018 U.S. elections, five new governors (in Colorado, Maine, Connecticut, Illinois, and Nevada) were elected whose platforms called for taking their respective states to 100 percent renewable electricity.

Subsequently, on December 18, 2018, Washington D.C. joined two states (Hawaii and California) by passing an enforceable 100 percent renewables law. Hawaii had previously passed a law on June 8, 2015, to go to 100 percent renewable electricity for the entire state by 2045. California had passed SB 100 to do the same on September 10, 2018. Washington D.C.'s law, though, was to go to 100 percent renewable electricity sooner, by 2032.

Then, on March 22, 2019, New Mexico passed a law to go to 80 percent renewable electricity by 2040 and up to 100 percent by 2045.

A few weeks later, on April 11, 2019, the U.S. territory Puerto Rico passed a law to go to 100 percent renewable electricity by 2050.

On April 19, 2019, Nevada passed a law to go to 50 percent renewable electricity by 2030 and up to 100 percent renewable (termed 'carbon-free') electricity by 2050.

On May 8, 2019, Washington State enacted a law to go to 100 percent renewables for all new electricity by 2045. The bill states (in Section 6.a.iii) that, to meet the law, electric utilities, when acquiring new resources, must "*rely on renewable resources and energy storage...*"

On May 23, 2019, New Jersey's governor signed an executive order requiring the development of a plan for the state to reach 50 percent renewable electricity by 2030 and "*100 percent clean energy by 2050.*" Clean energy can be renewable electricity or electricity from another technology defined by the state to be "clean."

On June 26, 2019, Maine passed a law to go to 80 percent renewable electricity by 2030 and 100 percent by 2050.

Not to be outdone, New York passed a law on July 18, 2019 to go to 70 percent renewable electricity by 2030 and effectively

100 percent by 2045. The law also calls for an 85 percent across-the-board reduction in greenhouse gas emissions in the state by 2050.

On August 16, 2019, the governor of Wisconsin signed an executive order expressing a goal for the state to go to 100 percent carbon free electricity, thus up to 100 percent renewable electricity, by 2050, and ordered state agencies to develop a plan to do this.

On September 3, 2019, the governor of Connecticut signed an executive order for the state to develop a plan to go to 100 percent zero-carbon electricity, and thus up to 100 percent renewable electricity, by 2040.

The domino effect continued. On September 17, 2019, the governor of Virginia signed an executive order expressing a goal for the state to reach 30 percent renewable electricity by 2030 and 100 percent carbon-free electricity, and thus up to 100 percent renewable electricity, by 2050.

On January 17, 2020, the governor of Rhode Island signed an executive order for the state to go to 100 percent renewable electricity by 2030.

On July 27, 2021, Oregon's governor signed a law for the state to go to 100 percent renewable electricity by 2040, with 80 percent by 2030.

On September 15, 2021, Illinois's governor signed a law for the state to provide 100 percent of its electricity from renewable or zero-carbon sources by 2045, with 40 percent from renewables by 2030.

Thus, as of the end of 2021, 16 U.S. states, districts, and territories had laws or executive orders to go up to 100 percent WWS in the electric power sector. One-hundred percent renewable laws have also been proposed, but not yet passed, in Florida, Minnesota, and Colorado. Over 200 U.S. towns and cities have passed 100 percent laws to date,[366] as have hundreds more international cities.[367]

Transition highlight

The 100 percent fever caught on not only with state and local governments, but also with utilities. Seven U.S. states (Massachusetts, Ohio, California, Illinois, New Jersey, New York, and Rhode Island) have laws permitting community choice aggregation utilities. These utilities take over the electricity generation portion of a utility bill. Another utility handles the transmission and distribution portion. Starting on March 7, 2010, the community choice utility Marin Clean Energy began offering the option for customers to purchase up to 100 percent of their electricity from clean, renewable WWS sources. Subsequently, the number of community choice utilities offering 100 percent WWS electricity accelerated rapidly across not only California, but also across most other states that allow community choice utilities.

Not to be outdone, many traditional utilities began offering 100 percent WWS options in 2019. For example, in 2019, Xcel Energy, which services Minnesota, Michigan, Wisconsin, North Dakota, South Dakota, Colorado, Texas, and New Mexico, set up an option for customers, called Windsource, that allows them to obtain *"all of their (electrical) energy from renewables."*

On August 20, 2019, Appalachian Power even opened a *"Wind, Water, and Sunlight (WWS) service"* to supply 100 percent renewable power to any customer in Virginia.[368] On August 21, 2019, Duke Energy obtained approval to allow military, University of North Carolina, and large nonresidential customers to procure 100 percent WWS electricity directly from WWS suppliers.[369]

On another front, during the week of the September 21, 2014, climate march in New York City, and 5 years after our *Scientific American* worldwide 100 percent WWS roadmap paper, a campaign called RE100 was launched by two nonprofits, The Climate Group and CDP. The original purpose of the campaign was to commit 100 companies to 100 percent renewable electricity in the annual average for their global operations by 2020. The first companies to commit were Ikea, Swiss Re, Mars, and H&M. By mid 2022, over 360 companies had joined, but with different compliance dates.[370] Among them are 11 out of the 20 largest companies in the world in 2021 in terms of total value of a company's shares of stocks: (in alphabetical order) Amazon, Apple, Facebook, Google, Johnson & Johnson, JP Morgan Chase, Mastercard, Microsoft, P&G, Visa, and Walmart.

Transition highlight
On yet another front, during September through November 2019, the cities of Berkeley, Menlo Park, and Mountain View, California, banned the use of natural gas in new residential and/or commercial buildings. Mountain View further required that 100 percent of parking spaces in multi-unit housing and commercial developments have the electrical infrastructure available

for electric-vehicle charging. Since then, dozens more cities in California have banned natural gas in new construction buildings. During December 2021, New York City banned new natural gas hookups in buildings seven stories or shorter by 2023 and all remaining buildings by 2027. During July 2021, Menlo Park went further by proposing an ordinance requiring residents to remove natural gas appliances (stoves, heaters, dryers) from their homes and offering financing to do this. Banning gas and facilitating vehicle charging are both necessary steps toward the full electrification of society. Full electrification is needed for WWS to supply all-purpose energy.

The first company to reach its 100 percent renewable goal for electric power generation was Google, in 2017. This was fitting, because Mark Ruffalo, Marco Krapels, and I had presented our first public Solutions Project vision of reaching 100 percent WWS, by combining science, business, and culture, at Google on June 20, 2012. Apple met the 100 percent goal for its global operations during April 2018.

Between September 2017 and January 2022, the Climate Group went further, galvanizing 120 international companies[371] to commit to transitioning their millions of company vehicles to 100 percent electric by 2030. The companies committed not only to changing their vehicle fleets, but also to installing charging stations. On September 19, 2019, Amazon, committed to replacing 100,000 delivery trucks with electric trucks. Transitioning company vehicles is important, given that up to half of all vehicles on the road in many countries are company-owned vehicles.

In 2019, a new law was proposed in the U.S. House of Representatives (Bill 330) calling for the U.S. electric power sector to go to 100 percent WWS by 2030. Shortly after, broader Green New Deal resolutions (rather than laws) were proposed in both the U.S. House (Resolution 109) and Senate (Resolution 59). The Green New Deal resolutions contained proposals related to energy, jobs, health care, education, and social justice. In terms of energy, the resolutions called for "*100 percent clean, renewable, and zero-emission energy sources*" for all energy sectors by 2030 throughout the United States. The 100 percent goal and the 2030 deadline both originated[372,373] from our *Scientific American* article[272] and New York State energy roadmap.[275] Both articles concluded that 100 percent WWS by 2030 is technically and economically feasible, but the 2009 article cautioned that, for social and political reasons, a complete transition by 2030 was unlikely and may take longer to achieve.

Prior to the 2009 *Scientific American* article, the public discussion centered on whether it was possible to shift from fossil fuels to even 20 percent renewables without collapsing the grid. Our subsequent country, state, city, and town roadmaps and grid studies, together with the 100 percent WWS movement spawned by The Solutions Project, its founders, members, and partner organizations, all reinforced by additional scientific papers by over a dozen independent, international research groups, changed the discussion to *when* and *how fast* 100 percent would be implemented.

I am under no illusion that we are finished solving the problem. Although progress has been made in the electric power sector, transitioning other sectors (transportation, buildings, industry, agriculture/forestry/fishing, and the military) must be addressed in full force as well. However, the goal posts have shifted closer to where they should be, and this was due to cooperation among people with diverse expertise but who also had a likeminded passion and goal. The progress made to date is much more than I thought it could ever have been when I first aspired to help solve air pollution problems in the late 1970s.

15.16   Conclusion: Where Do We Go from Here?

A solution to the problems of air pollution, global warming, and energy insecurity requires a large-scale conversion of the world's energy infrastructure. This book has described and analyzed many elements of such a conversion.

The main steps in a transition are to electrify and/or use direct heat for all energy sectors, then to provide that electricity and heat with clean, renewable wind, water, and solar energy and storage. The energy sectors that must be transitioned include the electricity, transportation, buildings, industry, agriculture/forestry/fishing, and military sectors. It is also necessary to reduce or eliminate non-energy emissions.

WWS generators proposed for such a transition include onshore and offshore wind turbines, solar PV, solar heat collectors, CSP plants, geothermal electricity plants, geothermal heat collectors, tidal turbines, wave devices, and hydroelectric facilities. Storage types includes electricity, heat, cold, and hydrogen storage. Short and long-distance transmission is also needed. By combining electricity, heat, cold, and hydrogen generation with storage and demand response, it appears possible to match power demand with supply on the electric power grid, thereby avoiding blackouts worldwide with 100 percent WWS.

The resulting 100 percent WWS system will use an average of about 56 percent less end-use energy than a business-as-usual system, and the annual energy cost to consumers will be about 63 percent lower, on average. Because WWS also eliminates health and climate costs of energy, which together are 4 times energy costs, on average, a WWS system will reduce the social (economic) cost that consumers pay by about 92 percent.

A worldwide transition to WWS will not only reduce up to 7 million air pollution deaths annually and slow, then reverse, global warming, but it will also create far more jobs than are lost, use less land than the current energy infrastructure, increase energy security and energy independence, and reduce terrorism risk to the power grid.

Given the limited time and funding available to solve climate, air pollution, and energy security problems, it is essential we focus on known, effective technologies for the solution. We should not waste money and allow more damage with inferior options. Such poorer options include new nuclear power, fossil fuels or bioenergy with or without carbon capture, biofuels, biomass with or without carbon capture, direct air capture, blue hydrogen, and geoengineering.

In sum, a concerted international effort can lead to a conversion of the world's energy infrastructure so that by between 2022 and 2030, the world will no longer build new fossil-fuel electricity generation power plants or new land- or marine-based transportation equipment with internal combustion engines. Rather, it will manufacture new wind turbines, solar power plants, and electric and fuel-cell vehicles. Long-distance aviation may take longer to transition. As the WWS infrastructure grows, the remaining existing infrastructure will retire, so that ideally by 2035 and no later than 2050, the world will be powered 100 percent by WWS.

The main barriers to a conversion to WWS worldwide are not technical, resource based, or economic. Instead, they are political and social. The most difficult places to transition may be countries beset by conflict and countries that have high poverty rates. NIMBYism is also an issue but not a barrier that cannot be overcome. Although recycling will be beneficial for keeping costs down, there is no materials limit to a 100 percent transition in all energy sectors.

Because most polluting energy technologies and resources currently receive government subsidies or tax breaks or are not required to eliminate their emissions, many such technologies can run inexpensively for a long time while competing economically with WWS. Wisely implemented policies can promote a fast conversion to WWS; however, useful policies can be implemented only if policymakers are willing to make changes.

Policymakers in democratic countries are elected, whereas those in autocratic countries make decisions dictated by one or a few leaders. Thus, in democratic countries, many people need to be convinced that changes will be beneficial; in autocratic countries, only a few need to be convinced. In both cases, those who need to be convinced require a social evolution in their thought. Such an evolution comes from a better understanding of global warming and air pollution science, consequences, and solutions. If the public and policymakers can become confident in understanding the problems and the large-scale solution needed to solve them, they will gravitate toward the solution. Thus, it is important for individuals, advocates, business leaders, scientists, and policymakers to come together to effect a change.

When the 100 percent WWS solution is finally implemented worldwide, the air pollution and climate problems outlined in this book will be relegated to the annals of history.

REFERENCES

1. Green, H., and W. Lane, *Particle Clouds*, Van Nostrand, 1969.
2. WHO (World Health Organization), Health statistics and information systems, 2021, www.who.int/health-topics/universal-health-coverage/health-statistics-and-information-systems (accessed November 20, 2021).
3. WHO (World Health Organization), Air pollution data portal, 2021, www.who.int/data/gho/data/themes/air-pollution (accessed November 20, 2021).
4. WHO (World Health Organization), Household air pollution and health, 2021, www.who.int/news-room/fact-sheets/detail/household-air-pollution-and-health (accessed January 8, 2022).
5. Jacobson, M.Z., A.-K. von Krauland, S.J. Coughlin, et al., Low-cost solutions to global warming, air pollution, and energy insecurity for 145 countries, *Energy Environ. Sci.*, https://doi.org/10.1039/d2ee00722c, 2022.
6. Jacobson, M.Z., Effects of biomass burning on climate, accounting for heat and moisture fluxes, black and brown carbon, and cloud absorption effects, *J. Geophys. Res.*, 119, 8980–9002, https://doi.org/10.1002/2014JD021861, 2014.
7. Canadell, J.G., P.M.S. Monteiro, M.H. Costa, et al., Global Carbon and other Biogeochemical Cycles and Feedbacks. In: *Climate Change 2021: The Physical Science Basis. Contribution of Working Group I to the Sixth Assessment Report of the Intergovernmental Panel on Climate Change*, Masson-Delmotte, V., P. Zhai, A. Pirani, et al., eds., Cambridge University Press, 2021.
8. Jacobson, M.Z., *100% Clean, Renewable Energy and Storage for Everything*, Cambridge University Press, 2020.
9. Jacobson, M.Z., A physically-based treatment of elemental carbon optics: Implications for global direct forcing of aerosols, *Geophys. Res. Lett.*, 27, 217–220, 2000.

10. Jacobson, M.Z., Strong radiative heating due to the mixing state of black carbon in atmospheric aerosols, *Nature*, 409, 695–697, 2001.

11. Jacobson, M.Z., Control of fossil-fuel particulate black carbon plus organic matter, possibly the most effective method of slowing global warming, *J. Geophys. Res.*, 107, 4410, https://doi.org/10.1029/2001JD001376, 2002.

12. Jacobson, M.Z., Short-term effects of controlling fossil-fuel soot, biofuel soot and gases, and methane on climate, Arctic ice, and air pollution health, *J. Geophys. Res.*, 115, D14209, https://doi.org/10.1029/2009JD013795, 2010.

13. Bond, T.C., S.J. Doherty, D.W. Fahey, et al., Bounding the role of black carbon in the climate system: a scientific assessment, *J. Geophys. Res.*, 118, 5380–5552, https://doi.org/10.1002/jgrd.50171, 2013.

14. Jacobson, M.Z, On the causal link between carbon dioxide and air pollution mortality, *Geophys. Res. Lett.*, 35, L03809, https://doi.org/10.1029/2007GL031101, 2008.

15. Jacobson, M.Z., The enhancement of local air pollution by urban CO_2 domes, *Environ. Sci. Technol.*, 44, 2497–2502, https://doi.org/10.1021/es903018m, 2010.

16. Ritchie, H., and M. Roser, Access to energy, 2020, https://ourworldindata.org/energy-access (accessed November 20, 2021).

17. Vavrin, J., Power and energy considerations at forward operating bases (FOBs). United States Army Corps of Engineers, Engineer Research and Development Center, Construction Engineering Research Laboratory, 2010 https://citeseerx.ist.psu.edu/viewdoc/download?doi=10.1.1.965.2873&rep=rep1&type=pdf (accessed November 20, 2021).

18. Sambor, D.J., M. Wilber, E. Whitney, and M.Z. Jacobson, Development of a tool for optimizing solar and battery storage for container farming in a remote Arctic microgrid, *Energies*, 13, 5143, https://doi.org/10.3390/en13195143, 2020.

19. Krane, J., and R. Idel, More transitions, less risk: how renewable energy reduces risks from mining, trade and political dependence, *Energ. Res. Soc. Sci.*, 82, 102311, 2021.

20. General Electric, Haliade-X offshore wind turbine platform, www.ge.com/renewableenergy/wind-energy/turbines/haliade-x-offshore-turbine, 2018 (accessed November 16, 2018).

21. Durakovic, A., Fixed bottom offshore wind farms 90 metres deep? Offshoretronics says yes, 2021, www.offshorewind.biz/2021/11/29/fixed-bottom-offshore-wind-farms-90-metres-deep-offshoretronic-says-yes/ (accessed November 29, 2021).

22. U.S. Department of the Interior, Reclamation: Managing water in the west; Hydroelectric power, 2005, www.usbr.gov/power/edu/pamphlet.pdf (accessed November 22, 2018).

23. Rahi, O.P., and A. Kumar, Economic analysis for refurbishment and uprating of hydropower plants, *Renew. Energy*, 86, 1197–1204, 2016.
24. Free Dictionary, Installed capacity: definition, 2019 https://encyclopedia2 .thefreedictionary.com/Installed+Capacity (accessed March 15, 2019).
25. Corbley, A., These underwater 'kites' are generating tidal electricity as they move, 2021, www.goodnewsnetwork.org/faroe-islands-looks-to-tidal-power-with-tis-underwater-kites/ (accessed December 3, 2021).
26. Lee, K., H.-D. Um, D. Choi, et al., The development of transparent photovoltaics, *Cell Rep. Phys. Sci.*, 1, 100143, https://doi.org/10.1016/j .xcrp.2020.100143, 2020.
27. Renovagen, Rollable solar panels, 2021, www.renovagen.com (accessed November 20, 2021).
28. Bellini, E., Flexible solar panel for vehicle-integrated applications, 2021, www.pv-magazine.com/2021/09/02/flexible-solar-panel-for-vehicle-integrated-applications/ (accessed November 20, 2021).
29. Kougias, I., K. Bodis, A. Jager-Waldau, et al., The potential of water infra-structure to accommodate solar PV systems in Mediterranean islands, *Solar Energy*, 136, 174–182, https://doi.org/10.1016/j.solener.2016.07.003, 2016.
30. Hussain, A., A. Batra, and R. Pachauri, An experimental study on effect of dust on power loss in solar photovoltaic module, *Renewables: Wind, Water, Solar* 4, 9, 2017.
31. U.S. Department of Energy, Concentrating solar power commercial appli-cation study: reducing water consumption of concentrating solar power electricity generation, Report to Congress, 2008, www1.eere.energy .gov/solar/pdfs/csp_water_study.pdf (accessed December 28, 2019).
32. Nonbol, E., Load-following capabilities of nuclear power plants, Technical University of Denmark, 2013, http://orbit.dtu.dk/files/64426246/Load_ following_capabilities.pdf (accessed November 22, 2018).
33. Utility Dive, Los Angeles considers $3B pumped storage project at Hoover Dam, 2018, www.utilitydive.com/news/los-angeles-considers-3b-pumped-storage-project-at-hoover-dam/528699/ (accessed December 15, 2021).
34. Blakers, A., M. Stocks, B. Lu, C. Cheng, and A. Nadolny, Global pumped hydro atlas, 2019, http://re100.eng.anu.edu.au/global/ (accessed March 31, 2019).
35. Willuhn, M., Sonnen battery still running after 28,000 full charge cycles, 2021, www.pv-magazine.com/2021/07/15/sonnen-battery-still-running-after-28000-full-charge-cycles/ (accessed July 15, 2021).
36. Morrison, R., New 'iron-air' battery stores electricity from renewables for days, 2021, www.msn.com/en-us/news/technology/new-iron-air-battery-stores-electricity-from-renewables-for-days/ar-AAMriqB (accessed July 22, 2021).

37. Aarhus University, Gridscale: Storing renewable energy in stones instead of batteries, 2021, https://scitechdaily.com/gridscale-storing-renewable-energy-in-stones-instead-of-lithium-batteries (accessed July 22, 2021).

38. Engineering with Rosie, Thermal energy storage tour with Stiesdal Gridscale battery, 2021, www.youtube.com/watch?v=72wkuIvUISs (accessed July 22, 2021).

39. Boukhalf, S., and N. Kaul, 10 disruptive battery technologies trying to compete with lithium-ion, 2019, www.solarpowerworldonline .com/2019/01/10-disruptive-battery-technologies-trying-to-compete-with-lithium-ion/ (accessed February 2, 2019).

40. Wang, Y., D. Zhou, V. Palomares, et al., Revitalizing sodium-sulfur batteries for non-high-temperature operation: a crucial review, *Energ. Environ. Sci.*, 13, 3848–3879, 2020.

41. Pilkington, B., Introducing an aluminum-ion battery that charges 60 times faster than lithium, 2021, w ww.azonano.com/article.aspx?ArticleID=5753 (accessed July 22, 2021).

42. Nithyanandam, K, and R. Pitchumani, Cost and performance analysis of concentrating solar power systems with integrated latent thermal energy storage, *Energy* 64, 793–810, 2014.

43. Denholm, P., Y.-H. Wan, M. Hummon, and M. Mehos, The value of CSP with thermal energy storage in the western United States, *Energy Procedia*, 49, 1622–1631, 2014.

44. Chakratec, Flywheel specifications and comparison, 2019, www.chakratec .com/technology/ (accessed March 18, 2019).

45. Crossley, I., Simplifying and lightening offshore turbines with compressed air energy storage, *Wind Power Monthly*, 2018, www.windpowermonthly.com/ article/1463030/simplifying-lightening-offshore-turbines-compressed-air-energy-storage (accessed April 11, 2022).

46. Harrabin, R., How liquid air could help keep the lights on, *BBC News*, 2019, www.bbc.com/news/business-50140110 (accessed October 30, 2019).

47. Allain, R., How much energy can you store in a stack of cement blocks, *Wired*, 2018, www.wired.com/story/battery-built-from-concrete/ (accessed April 10, 2019).

48. Hanley, S., Energy Vault proposes an energy storage system using concrete blocks, Cleantechnica, 2018, https://cleantechnica.com/2018/08/21/energy-vault-proposes-an-energy-storage-system-using-concrete-blocks/ (accessed April 10, 2019).

49. Gross, B., Efficiency of gravitational mass storage system, 2019, https:// twitter.com/Bill_Gross/status/1164617097927806976/photo/1 (accessed August 22, 2019).

50. Roberts, D., The train goes up, the train goes down: a simple way to store energy, *Vox*, 2016, www.vox.com/2016/4/28/11524958/energy-storage-rail (accessed April 10, 2019).

51. Hunt, J.D., B. Zakeri, G. Falchetta, et al., Mountain gravity energy storage: a new solution for closing the gap between existing short- and long-term storage technologies, *Energy*, https://doi.org/10.1016/j.energy.2019 .116419, 2019.

52. Holnicki, P., A. Kaluszko, Z. Nahorski, and M. Tainio, Intra-urban variability of the intake fraction from multiple emission sources, *Atmos. Pollut. Res.*, 9, 1184–1193, 2018.

53. Bistak, S., and S.Y. Kim, AC induction motors vs. permanent magnet synchronous motors, 2017, http://empoweringpumps.com/ac-induction-motors-versus-permanent-magnet-synchronous-motors-fuji/ (accessed January 5, 2018).

54. Rahman, D., A.J. Morgan, Y. Xu, et al., Design methodology for a planarized high power density EV/HEV traction drive using SiC power modules, IEEE Energy Conversion Congress and Exhibition, Sept. 18–22, 2016, https://doi.org/10.1109/ECCE.2016.7855018 (accessed March 2, 2019).

55. Evarts, E.C., The world's largest EV never has to be recharged, *Green Car Reports*, 2019, www.greencarreports.com/news/1124478_world-s-largest-ev-never-has-to-be-recharged (accessed August 20, 2019).

56. Wilson, K.A., Worth the Watt: a brief history of the electric car, 1830 to present, 2018, www.caranddriver.com/features/g15378765/worth-the-watt-a-brief-history-of-the-electric-car-1830-to-present/ (accessed July 9, 2021).

57. Matulka, R., The history of the electric car, 2014, www.energy.gov/articles/history-electric-car (accessed July 9, 2021).

58. Goodwin, A., S. Ewing, and K. Hyatt, Every electric vehicle on sale in the US for 2021 and its range, 2021, www.cnet.com/roadshow/news/electric-car-ev-range-audi-chevy-ford-tesla/

59. Tesla, Semi, 2021, www.tesla.com/semi (accessed September 8, 2021).

60. Grasso Macola, I., Electric ships: the world's top five projects by battery capacity, 2020, www.ship-technology.com/features/electric-ships-the-world-top-five-projects-by-battery-capacity/ (accessed July 12, 2021).

61. Wilkerson, J.T., M.Z. Jacobson, A. Malwitz, et al., Analysis of emission data from global commercial aviation: 2004 and 2006, *Atmos. Chem. Phys.*, 10, 6391–6408, 2010.

62. Jacobson, M.Z., J.T. Wilkerson, A.D. Naiman, and S.K. Lele, The effects of aircraft on climate and pollution. Part II: 20-year impacts of exhaust from all commercial aircraft worldwide treated individually at the sub-grid scale, *Faraday Discuss.*, 165, 369–382, https://doi.org/10.1039/C3FD00034F, 2013.

63. Narishkin, A., and A. Appolonia, Inside a $4 million electric plane, the first full-size all-electric passenger aircraft in the world, 2020, www.businessinsider.com/inside-alice-first-full-size-passenger-electric-plane-eviation-2020-10 (accessed July 13, 2021).

64. Harvey, L.D.D., Resource implications of alternative strategies for achieving zero greenhouse gas emissions from light-duty vehicles by 2060, *Appl. Energy*, 212, 663–679, 2018.

65. Petitt, J., Inside Redwood Materials, former Tesla CTO's effort to recycle batteries for rare components, 2021, www.cnbc.com/2021/04/10/tesla-jb-straubel-redwood-materials-battery-recycling.html (accessed June 23, 2021).

66. Carpenter, S., Salton Sea is key to CA's EV future, contains 1/3 of global lithium supply, 2021, https://spectrumnews1.com/ca/la-west/environment/2021/05/27/salton-sea-is-key-to-ca-s-ev-future--contains-1-3-of-global-lithium-supply (accessed June 23, 2021).

67. Richter, A., Lithium for batteries from the Upper Rhine's Graben's geothermal resources, 2020, www.thinkgeoenergy.com/lithium-for-batteries-from-the-upper-rhine-grabens-geothermal-resources/ (accessed June 23, 2021).

68. Martin, P., Part 3: Lithium and cobalt-risky materials, 2017, www.linkedin.com/pulse/part-3-lithium-cobalt-risky-materials-paul-martin/ (accessed June 23, 2021).

69. Rai-Roche, S., and L. Stoker, Renewables-plus-storage projects for mining operations in Australia, Madagascar for BHP, Rio Tinto, 2021, www.energy-storage.news/news/renewables-plus-storage-projects-for-mining-operations-in-australia-madagas (accessed August 2, 2021).

70. Parkinson, G., Potash mine to build wind, solar, and battery micro-grid for most of its power needs, 2021, https://reneweconomy.com.au/potash-mine-to-build-wind-solar-and-battery-micro-grid-for-most-of-its-power-needs/ (accessed September 7, 2021).

71. Wang, J.-P., Environment-friendly bulk $Fe_{16}N_2$ permanent magnet: review and prospective, *J. Magn. Magn. Mater.*, 497, 165962, https://doi.org/10.1016/j.jmmmm.2019.165962, 2020.

72. Bloom Energy, Specification Bloom electrolyzer, 2021, www.bloomenergy.com/wp-content/uploads/electrolyzer-data-sheet.pdf (accessed September 9, 2021).

73. Feng, Z., Stationary high-pressure hydrogen storage, 2018, www.energy.gov/sites/prod/files/2014/03/f10/csd_workshop_7_feng.pdf (accessed November 28, 2018).

74. Daimler, Fuel-cell truck: start of testing of the new GenH2 truck prototype, 2021, www.daimler.com/innovation/drive-systems/hydrogen/start-of-testing-genh2-truck-prototype.html (accessed June 23, 2021).

75. Jacobson, M.Z., Effects of wind-powered hydrogen fuel cell vehicles on stratospheric ozone and global climate, *Geophys. Res. Lett.*, 35, L19803, https://doi.org/10.1029/2008GL035102, 2008.

76. Katalenich, S.M., and M.Z. Jacobson, Toward battery electric and hydrogen fuel cell military vehicles for land, air, and sea, *Energy*, 254, 124355, https://doi.org/10.1016/j.energy.2022.124355, 2022.

77. Frangoul, A., Norway's Statkraft lined up to provide green hydrogen for 88-meter long zero-emission ship, 2021, www.cnbc.com/2021/06/24/statkraft-lined-up-to-provide-green-hydrogen-for-zero-emission-ship.html (accessed July 12, 2021).

78. Jacobson, M.Z., and M.A. Delucchi, Providing all global energy with wind, water, and solar power, Part I: Technologies, energy resources, quantities and areas of infrastructure, and materials, *Energy Policy*, 39, 1154–1169, https://doi.org/10.1016/j.enpol.2010.11.040, 2011.

79. Werner, S., International review of district heating, *Energy*, 15, 617–631, 2017.

80. Stagner, J., Stanford University's "fourth-generation" district energy system, District Energy, Fourth Quarter, 2016, https://sustainable.stanford.edu/sites/default/files/IDEA_Stagner_Stanford_fourth_Gen_DistrictEnergy.pdf (accessed November 27, 2018).

81. Stagner, J., Efficiency and environmental comparisons, Stanford Energy System Innovations, 2017, https://sustainable.stanford.edu/sites/default/files/SESI_Efficiency_Environmental_Comparisons.pdf, (accessed December 31, 2019).

82. Cornell University, How lake source cooling works, 2019, https://energyandsustainability.fs.cornell.edu/util/cooling/production/lsc/works.cfm (accessed May 1, 2019).

83. IRENA (International Renewable Energy Agency), Thermal energy storage. IEA-ETSAP and IRENA Technology Brief E17, IRENA, Abu Dhabi, 2013.

84. Sibbitt, B, D. McClenahan, R. Djebbar, et al., The performance of a high solar fraction seasonal storage district heating system – five years of operation, *Energy Procedia*, 30, 856–865, 2012.

85. Sorensen, P.A., and T. Schmidt, Design and construction of large scale heat storages for district heating in Denmark, 14th Int. Conf. on Energy Storage, April 25–28, Adana, Turkey, 2018, http://planenergi.dk/wp-content/uploads/2018/05/Soerensen-and-Schmidt_Design-and-Construction-of-Large-Scale-Heat-Storages-12.03.2018-004.pdf (accessed November 25, 2018).

86. Ramboll, World's largest thermal heat storage pit in Vojens, 2016, https://stateofgreen.com/en/partners/ramboll/solutions/world-largest-thermal-pit-storage-in-vojens/; https://ramboll.com/projects/re/south-jutland-stores-the-suns-heat-in-the-worlds-largest-pit-heat-storage (accessed November 25, 2018).

87. Arcon/Sunmark, Large-scale showcase projects, 2017, http://arcon-sunmark.com/uploads/ARCON_References.pdf (accessed November 25, 2018).

88. Damkjaer, L., Gram Fjernvarme 2016, 2016, www.youtube.com/watch?v=PdF8e1t7St8 (accessed November 25, 2018).
89. Ramboll, Pit thermal energy storage: update from Toftlund, 2020, www.heatstore.eu/documents/20201028_DK-temadag_Rambøll%20PTES%20project.pdf (accessed July 21, 2021).
90. International Energy Agency, Integrated cost-effective large-scale thermal energy storage for smart district heating and cooling, 2018, www.iea-dhc.org/fileadmin/documents/Annex_XII/IEA_DHC_AXII_Design_Aspects_for_Large_Scale_ATES_PTES_draft.pdf (accessed November 25, 2018).
91. Butti, K., and J. Perlin, *Solar Water Heaters in California*, California Energy Commission, 1980.
92. Dandelion, Geothermal heating and air conditioning is so efficient, it pays for itself, https://dandelionenergy.com, 2018 (accessed November 17, 2018).
93. Meyers, S., V. Franco, A. Lekov, L. Thompson, and A. Sturges, Do heat pump clothes dryers makes sense for the U.S. market? ACEEE Summer Study on Energy Efficiency in Buildings, 9-240–9-251, 2010, https://aceee.org/files/proceedings/2010/data/papers/2224.pdf (accessed October 29, 2019).
94. Fischer, D., and H. Madani, On heat pumps in smart grids: a review. *Renew. Sustain. Energ. Rev.*, 70, 342–357, 2017.
95. De Gracia, A., and L.F. Cabeza, Phase change materials and thermal energy storage for buildings, *Energy Build.*, 103, 414–419, 2015.
96. Consumer Reports, Electric lawn mowers that rival gas models, 2017, www.consumerreports.org/push-mowers/electric-lawn-mowers-that-rival-gas-models/ (accessed November 21, 2018).
97. BONE Structure, https://bonestructure.ca/en/ (accessed December 26, 2021).
98. U.S. Department of Energy, Quadrennial Technology Review, Chapter 6: Innovative clean energy technologies in advanced manufacturing: Technology assessment, 2015, www.energy.gov/sites/prod/files/2016/06/f32/QTR2015-6I-Process-Heating.pdf (accessed Nov. 17, 2018).
99. Bellevrat, E., and K. West, Clean and efficient heat for industry, 2018, www.iea.org/newsroom/news/2018/january/commentary-clean-and-efficient-heat-for-industry.html (accessed November 17, 2018).
100. Barber, H., Chapter 7, Electric heating fundamentals. In: *The Efficient Use of Energy*, 2nd ed., pp. 94–114, I.G.C. Dryden, ed., Butterworth-Heinemann, https://doi.org/10.1016/B978-0-408-01250-8.50016-7, 1982.
101. Hulls, P.J., Development of the industrial use of dielectric heating in the United Kingdom, *J. Microw. Power*, 17, 28–38, 2016.
102. Viking Heat Engines, Heat booster, 2019, www.vikingheatengines.com/news/vikings-industrial-high-temperature-heat-pump-is-available-to-order (accessed January 13, 2019).

103. Ramaiah, R., and K.S.S. Shekar, Solar thermal energy utilization for medium temperature industrial process heat applications, *IOP Conf. Ser. Mater. Sci. Eng.*, 376, 010235, 2018.

104. Vogl, V., M. Ahman, and L.J. Nilsson, Assessment of hydrogen direct reduction for fossil-free steelmaking, *J. Clean. Prod.*, 203, 736–745, 2018.

105. Allanore, A., L. Yin, and D. Sadoway, A new anode material for oxygen evolution in molten oxide electrolysis, *Nature*, 497, 353–356, 2013.

106. Wiencke, J., H. Lavelaine, P.-J. Panteix, C. Petijean, and C. Rapin, Electrolysis of iron in a molten oxide electrolyte, *J. Appl. Electrochem.*, 48, 115–126, 2018.

107. Andrew, R.M., Global CO_2 emissions from cement production, 1928–2018, *Earth System Science Data*, 2019, https://doi.org/10.5194/essd-2019-152.

108. Choate, W.T., Energy and emission reduction opportunities for the cement industry, 2003, www1.eere.energy.gov/manufacturing/industries_technologies/imf/pdfs/eeroci_deco3a.pdf (accessed July 21, 2021).

109. Singh, N.B., M. Kumar, and S. Rai, Geopolymer cement and concrete: properties, *Mater. Today Proc.*, 29, 743–748, 2020.

110. Stone, D., Ferrock basics, 2017, http://ironkast.com/wp-content/uploads/2017/11/Ferrock-basics.pdf (accessed November 20, 2018).

111. Build Abroad, Ferrock: a stronger, more flexible and greener alternative to concrete, 2016, https://buildabroad.org/2016/09/27/ferrock/ (accessed November 20, 2018).

112. Carbon Cure, Carbon Cure, 2018, www.carboncure.com (accessed November 20, 2018).

113. Maldonado, S., The importance of new "sand-to-silicon" processes for the rapid future increase of photovoltaics, *ACS Energy Lett.*, 5, 3628–3632, 2020.

114. Dong, Y., T. Slade, M.J. Stolt, et al., Low-temperature molten salt production of silicon nanowires by the electrochemical reduction of $CaSiO_3$, *Angew. Chem.*, 56, 14453–14457, 2017.

115. Jacobson, M.Z., The short-term cooling but long-term global warming due to biomass burning, *J. Clim.*, 17, 2909–2926, 2004.

116. Colella, W.G., M.Z. Jacobson, and D.M. Golden, Switching to a U.S. hydrogen fuel cell vehicle fleet: the resultant change in emissions, energy use, and global warming gases, *J. Power Sources*, 150, 150–181, 2005.

117. Howarth, R.W., and M.Z. Jacobson, How green is blue hydrogen? *Energy Sci. Eng.*, 9, 1676–1687, 2021, https://doi.org/10.1002/ese3.956.

118. Ussiri, D., and R. Lal, Global Sources of Nitrous Oxide. In: *Soil Emission of Nitrous Oxide and Its Mitigation*, Springer, pp. 131–175, 2012.

119. Nitric Acid Climate Action Group, Nitrous oxide emissions from nitric acid production, 2014, www.nitricacidaction.org/about/nitrous-oxide-emissions-from-nitric-acid-production/ (accessed December 1, 2018).

120. Boiocchi, R., K.V. Gemaey, and G. Sin, Control of wastewater N_2O emission by balancing the microbial communities using a fuzzy-logic approach, *IFAC-PapersOnLine*, 49, 1157–1162, 2016.

121. Santin, I., M. Barbu, C. Pedret, and R. Vilanova, Control strategies for nitrous oxide emissions reduction on wastewater treatment plants operation, *Water Res.*, 125, 466–477, 2017.

122. Hong, S., J.-P. Candelone, C.C. Patterson, and C.F. Boutron, Greenland ice evidence of hemispheric lead pollution two millennia ago by Greek and Roman civilizations, *Science*, 265, 1841–1843, 1994.

123. Nriagu, J.O., A history of global metal pollution, *Science*, 272, 223–224, 1996.

124. Hong, S., J.-P. Candelone, C.C. Patterson, and C.F. Boutron, History of ancient copper smelting pollution during Roman and Medieval times recorded in Greenland ice, *Science*, 272, 246–248, 1996.

125. Brimblecombe, P., Air pollution and health history. In *Air Pollution and Health*, S.T. Holgate, J.M. Samet, H.S. Koren, and R.L. Maynard, eds., Academic, pp. 5–18, 1999.

126. Hughes, J.D., *Pan's Travail: Environmental Problems of the Ancient Greeks and Romans*, The Johns Hopkins University Press, 1994.

127. Brimblecombe, P., *The Big Smoke*, Methuen, 1987.

128. McNeill, J.R., *Something New Under the Sun*, W. W. Norton & Company Ltd., 2000.

129. Rosenberg, N., and L. E. Birdzell, Jr., *How the West Grew Rich*, Basic Books, Inc., 1986.

130. Union Gas, Chemical composition of natural gas, 2018, www.uniongas .com/about-us/about-natural-gas/chemical-composition-of-natural-gas (accessed December 5, 2018).

131. De Boer, J.Z., J.R. Hale, and J. Chanton, New evidence of the geological origins of the ancient Delphic oracle (Greece), *Geology*, 29, 707–710, 2001.

132. Intergovernmental Panel on Climate Change, *Special Report: Global Warming of 1.5°*, 2018, www.ipcc.ch/sr15/ (accessed June 26, 2019).

133. Jacobson, M.Z., Review of solutions to global warming, air pollution, and energy security, *Energy Environ. Sci.*, 2, 148–173, https://doi .org/10.1039/b809990c, 2009.

134. Frangoul, A., Scandinavia's biggest offshore wind farm is officially open, 2019, www.cnbc.com/2019/08/23/scandinavias-biggest-offshore-wind-farm-is-officially-open.html?__source=sharebar%7Ctwitter&par=share bar (accessed July 12, 2021).

135. Jacobson, M.Z., and C.L. Archer, Saturation wind power potential and its implications for wind energy, *Proc. Natl Acad. Sci.*, 109, 15679–15684, https://doi.org/10.1073/pnas.1208993109, 2012.

136. Jacobson, M.Z., M.A. Delucchi, M.A. Cameron, and B.V. Mathiesen, Matching demand with supply at low cost among 139 countries within 20 world regions with 100 percent intermittent wind, water, and sunlight (WWS) for all purposes, *Renewable Energy*, 123, 236–248, 2018.

137. Intergovernmental Panel on Climate Change, *IPCC Special Report on Carbon Dioxide Capture and Storage*. Prepared by Working Group III, Metz. B., O. Davidson, H. C. de Coninck, M. Loos, and L.A. Meyer, eds., Cambridge University Press, http://arch.rivm.nl/env/int/ipcc/, 2005 (accessed June 26, 2019).

138. U.S. Energy Information Administration, Hydraulically fractured wells provide two-thirds of U.S. natural gas production, 2016, www.eia.gov/todayinenergy/detail.php?id=26112 (accessed December 2, 2018).

139. Howarth, R.W., R. Santoro, and A. Ingraffea, Methane and the greenhouse gas footprint of natural gas from shale formations, *Clim. Change*, 106, 679–690, 2011.

140. Howarth, R.W., Is shale gas a major driver of recent increase in global atmospheric methane? *Biogeosciences*, 16, 3033–3046, 2019.

141. Alvarez, R.A., D. Zavalao-Araiza, D.R. Lyon et al., Assessment of methane emissions from the U.S. oil and gas supply chain, *Science*, 361, 186–188, 2018.

142. Schneising, O., M. Buchwitz, M. Reuter, et al., Remote sensing of methane leakage from natural gas and petroleum systems revisited, *Atmos. Chem. Phys.*, 20, 9169–9182, 2020.

143. Massachusetts Institute of Technology, *The Future of Natural Gas*, 2011, https://energy.mit.edu/wp-content/uploads/2011/06/MITEI-The-Future-of-Natural-Gas.pdf (accessed December 2, 2018).

144. U.S. Environmental Protection Agency, 2008 U.S. National Emissions Inventory (NEI), 2011, www.epa.gov/air-emissions-inventories/2008-national-emissions-inventory-nei-data (accessed December 2, 2018).

145. Jacobson, M.Z., M.A. Delucchi, G. Bazouin, et al., 100 percent clean and renewable wind, water, sunlight (WWS) all-sector energy roadmaps for the 50 United States, *Energy Environ. Sci.*, 8, 2093–2117, https://doi.org/10.1039/C5EE01283J, 2015.

146. Geuss, M., Florida utility to close natural gas plants, build massive solar-powered battery, *Ars Technica*, 2019, https://arstechnica.com/information-technology/2019/03/florida-utility-to-close-natural-gas-plants-build-massive-solar-powered-battery/ (accessed April 1, 2019).

147. Allred, B.W., W.K. Smith, D. Twidwell, et al., Ecosystem services lost to oil and gas in North America, *Science*, 348, 401–402, 2015.

148. Reuters, Special Report: Millions of abandoned oil wells are leaking methane, a climate menace, 2020, www.reuters.com/article/us-usa-drilling-abandoned-specialreport/special-report-millions-of-abandoned-

oil-wells-are-leaking-methane-a-climate-menace-idUSKBN23N1NL
(accessed September 6, 2020).

149. Jacobson, M.Z., Short-term impacts of the Aliso Canyon natural gas
blowout on weather, climate, air quality, and health in California and
Los Angeles, *Environ. Sci. Technol.*, 53, 6081–6093, https://doi
.org/10.1021/acs.est.9b01495, 2019.

150. Jaramillo, P., W.M. Griffin, and S.T. McCoy, Life cycle inventory of CO_2
in an enhanced oil recovery system, *Environ. Sci. Technol.*, 43, 8027–
8032, 2009.

151. Roberts, D., Turns out the world's first "clean coal" plant is a backdoor
subsidy to oil producers, 2015, https://grist.org/climate-energy/turns-out-
the-worlds-first-clean-coal-plant-is-a-backdoor-subsidy-to-oil-producers/
(accessed September 7, 2021).

152. Government of Alberta, Quest CO_2 Capture Ratio Performance,
Quest Carbon Capture and Storage (CCS) Project, Government of
Alberta, Edmonton, Alberta, 2020, https://open.alberta.ca/dataset/
f74375f3-3c73-4b9c-af2b-ef44e59b7890/resource/c36cf890-3b27-4e7e-
b95b-3370cd0d9f7d/download/energy-quest-co2-capture-ratio-perfor-
mance-2019.pdf (accessed July 16, 2021).

153. U.S. Department of Energy, W.A. Parish post-combustion CO_2 capture
and sequestration demonstration project, Final Scientific and Technical Report, Project DE-FE0003311, 2020, www.osti.gov/servlets/
purl/1608572 (accessed July 16, 2021).

154. Jacobson, M.Z., The health and climate impacts of carbon capture and
direct air capture, *Energy Environ. Sci.*, 12, 3567–3574, https://doi
.org/10.1039/C9EE02709B, 2019.

155. Schlissel, D., Boundary Dam 3 coal plant achieves goal of capturing 4
million metric tons of CO_2 but reaches the goal two years late, 2021,
http://ieefa.org/wp-content/uploads/2021/04/Boundary-Dam-3-Coal-
Plant-Achieves-CO2-Capture-Goal-Two-Years-Late_April-2021.pdf
(accessed July 12, 2021).

156. Morton, A., "A shocking failure": Chevron criticized for missing
carbon capture target at WA gas project, 2021, www.theguardian
.com/environment/2021/jul/20/a-shocking-failure-chevron-criticised-for-
missing-carbon-capture-target-at-wa-gas-project (accessed August 3, 2021).

157. U.S. Energy Information Administration, Today in energy, 2017, www
.eia.gov/todayinenergy/detail.php?id=33552 (accessed December 4,
2018).

158. ScottMadden, Billion dollar Petra Nova coal carbon capture project a
financial success but unclear if it can be replicated, 2017, www
.scottmadden.com/insight/billion-dollar-petra-nova-coal-carbon-
capture-project-financial-success-unclear-can-replicated/ (accessed
December 3, 2018).

159. Karam, P.A., How do fast breeder reactors differ from regular nuclear power plants? *Sci. Am.*, October 2006.

160. IAEA (International Atomic Energy Agency), World's uranium resources enough for foreseeable future says NEA and IAEA in new report, 2021, www.iaea.org/newscenter/pressreleases/worlds-uranium-resources-enough-for-the-foreseeable-future-say-nea-and-iaea-in-new-report (accessed August 5, 2021).

161. Lazard's levelized cost of energy analysis – Version 15.0, 2021, www.lazard.com/media/451881/lazards-levelized-cost-of-energy-version-150-vf.pdf (accessed November 1, 2021).

162. Koomey, J., and N.E. Hultman, A reactor-level analysis of busbar costs for U.S. nuclear plants, 1970–2005, *Energy Policy* 35, 5630–5642, 2007.

163. Berthelemy, M., and L.E. Rengel, Nuclear reactors' construction costs: The role of lead-time, standardization, and technological progress, *Energy Policy*, 82, 118–130, 2015.

164. Morris, C., French nuclear power history – the unknown story, 2015, https://energytransition.org/2015/03/french-nuclear-power-history/ (accessed June 16, 2019).

165. Bruckner, T., I.A. Bashmakov, Y. Mulugetta, et al., Energy Systems. In: *Climate Change 2014: Mitigation of Climate Change. Contribution of Working Group III to the Fifth Assessment Report of the Intergovernmental Panel on Climate Change*, Edenhofer, O., R. Pichs-Madruga, Y. Sokona, et al., eds., Cambridge University Press, 2014.

166. Garthwaite, J., What should we do with nuclear waste? Stanford Earth, 2018, https://earth.stanford.edu/news/qa-what-should-we-do-nuclear-waste#gs.1sfxox (accessed March 20, 2019).

167. Ten Hoeve, J.E., and M.Z. Jacobson, Worldwide health effects of the Fukushima Daiichi nuclear accident, *Energy Environ. Sci.*, 5, 8743–8757, 2012.

168. Denyer, S., Eight years after Fukushima's meltdown, the land is recovering, but public trust is not, *Washington Post*, 2019, www.washingtonpost.com/world/asia_pacific/eight-years-after-fukushimas-meltdown-the-land-is-recovering-but-public-trust-has-not/2019/02/19/0bb29756-255d-11e9-b5b4-1d18dfb7b084_story.html?utm_term=.8344c816d5bb (accessed December 21, 2019).

169. Cebulla, F., and M.Z. Jacobson, Alternative renewable energy scenarios for New York, *J. Cleaner Prod.*, 205, 884–894, 2018.

170. De Coninck, H., A. Revi, M. Babiker, et al., Chapter 4: Strengthening and Implementing the Global Response. In *Special Report: Global Warming of 1.5°*, Intergovernmental Panel on Climate Change, 2018.

171. Fuhrmann, M., Spreading temptation: proliferation and peaceful nuclear cooperation agreements (March 9, 2009). *Int. Secur.*, 34, Summer 2009. Available at SSRN: https://ssrn.com/abstract=1356091 (accessed September 9, 2019).

172. IranWatch, Iran's nuclear potential before the implementation of the nuclear agreement, 2015, www.iranwatch.org/our-publications/articles-reports/irans-nuclear-timetable (accessed December 9, 2018).

173. Johnson, G., When radiation isn't the real risk, *New York Times*, 2015, www.nytimes.com/2015/09/22/science/when-radiation-isnt-the-real-risk.html (accessed December 8, 2018).

174. BBC News, Japan confirms first Fukushima worker death from radiation, 2018, www.bbc.com/news/world-asia-45423575 (accessed June 9, 2019).

175. Henshaw, D.L., J.P. Eatough, and R.B. Richardson, Radon as a causative factor in induction of myeloid leukemia and other cancers, *Lancet*, 335, 1008–1012, 1990.

176. Lagarde, F., G. Pershagen, G. Akerblom, et al., Residential radon and lung cancer in Sweden: risk analysis accounting for random error in the exposure assessment, *Health Physics*, 72, 269–276, 1997.

177. Center for Disease Control and Prevention, Research on long-term exposure: Uranium miners, 2000, www.cdc.gov/niosh/pgms/worknotify/uranium.html (accessed December 9, 2018).

178. Hampson, S.E., J.A. Andres, M.E. Lee, et al., Lay understanding of synergistic risk: the case of radon and cigarette smoking, *Risk Analysis*, 18, 343–350, 1998.

179. Kadiyala, A., R. Kommalapati, and Z. Huque, Evaluation of the lifecycle greenhouse gas emissions from different biomass feedstock electricity generation systems, *Sustainability*, 8, 1181–1192, 2016.

180. Jacobson, M.Z., Effects of ethanol (E85) versus gasoline vehicles on cancer and mortality in the United States, *Environ. Sci. Technol.*, 41, 4150–4157, https://doi.org/10.1021/es062085v, 2007.

181. Ginnebaugh, D.L., J. Liang, and M.Z. Jacobson, Examining the temperature dependence of ethanol (E85) versus gasoline emissions on air pollution with a largely-explicit chemical mechanism, *Atmos. Environ.*, 44, 1192–1199, https://doi.org/10.1016/j.atmosenv.2009.12.024, 2010.

182. Ginnebaugh, D.L., and M.Z. Jacobson, Examining the impacts of ethanol (E85) versus gasoline photochemical production of smog in a fog using near-explicit gas- and aqueous-chemistry mechanisms, *Environ. Res. Lett.*, 7, 045901, https://doi.org/10.1088/1748-9326/7/4/045901, 2012.

183. Searchinger, T., R. Heimlich, R.A. Houghton, et al., Use of U.S. cropland for biofuels increases greenhouse gases through emissions from land-use change, *Science*, 319, 1238–1240, 2008.

184. Delucchi, M., A conceptual framework for estimating the climate impacts of land-use change due to energy crop programs, *Biomass Bioenergy*, 35, 2337–2360, 2011.

185. Chen, W.-H., M.-R. Lin, J.-J. Lu, Y. Chao, and T.-S. Leu, Thermodynamic analysis of hydrogen production from methane via autothermal reforming and partial oxidation followed by water gas shift reaction, *Int. J. Hyd. Energy*, 35, 11,787–11,797, 2010.

186. U.S. Drive, Hydrogen production tech team roadmap, 2017, www .energy.gov/sites/prod/files/2017/11/f46/HPTT%20Roadmap%20 FY17%20Final_Nov%202017.pdf (accessed August 25, 2021).

187. Duan, Y., and D.C. Sorescu, CO_2 capture properties of alkaline earth metal oxides and hydroxides: a combined density functional theory and lattice phonom dynamics study, *J. Chem. Phys.*, 133, 074508, 2010.

188. Jacobson, M.Z., and J.E. Ten Hoeve, Effects of urban surfaces and white roofs on global and regional climate, *J. Clim.*, 25, 1028–1044, https:// doi.org/10.1175/JCLI-D-11-00032.1, 2012.

189. King, G., Edison vs. Westinghouse: a shocking rivalry, *Smithsonian Mag.*, 2011, www.smithsonianmag.com/history/edison-vs-westinghouse-a-shocking-rivalry-102146036/ (accessed July 28, 2021).

190. Sadovskaia, K., D. Bogdanov, S. Honkapuro, and C. Breyer, Power transmission and distribution loses – a model based on available empirical data and future trends for all countries globally, *Int J. Electr. Power Energy Syst.*, 107, 98–109, 2019.

191. International Electrotechnical Commission, Efficient electrical energy transmission and distribution, 2007, https://basecamp.iec.ch/download/efficient-electrical-energy-transmission-and-distribution (accessed December 31, 2018).

192. ABB, HVDC – an ABB specialty, 2004, https://library.e.abb.com/public/ d4863a9b0f77b74ec1257b0c00552758/HVDC%20Cable%20Transmission.pdf (accessed December 31, 2018).

193. ABB, HVDC technology for energy efficiency and grid reliability, 2005, www02.abb.com/global/abbzh/abbzh250.nsf/0/27c2fdbd96a879a4c12 575ee00487a77/$file/HVDC+-+efficiency+and+reliability.pdf (accessed December 31, 2018).

194. World Bank, Electric power transmission and distribution losses (% of output), 2018, https://data.worldbank.org/indicator/EG.ELC.LOSS .ZS?end=2014&start=2009 (accessed January 1, 2019).

195. Becquerel, A.E., Memoire sur les effects d'electriques produits soul l'influence des rayons solairee, *Annal. Phys. Chem.*, 54, 35–42, 1841.

196. Adams, W.G., and R.E. Day, The action of light on selenium, *Proc. R. Soc. Lond.*, A25, 113, 1877.

197. Fritts, C.E., On a new form of selenium photocell, *Am. J. Sci.*, 26, 465, 1883.

198. Siemens, W., On the electromotive action of illuminated selenium, discovered by Mr. Fritts of New York, *Van Nostrands Engineering Magazine*, 32, 514–516, 1885.

199. Grondahl, L.O., The copper-cuprous-oxide rectifier and photoelectric cell, *Rev. Mod. Phys.*, 5, 141, 1933.

200. Green, M.A., E. Dunlop, J. Hohl-Ebinger, et al., Solar cell efficiency tables (Version 58), *Prog. Photovolt.*, https://doi.org/10.1002/pip.3444, 2021.

201. Bremner, S.P., M.Y. Levy, and C.B. Honsberg, Analysis of tandem solar cell efficiencies under {AM1.5G} spectrum using a rapid flux calculation method, *Prog. Photovolt. Res. Appl.*, 16, 225–233, 2008.

202. Breyer, C., *Economics of Hybrid Photovoltaic Power Plants*, Pro Business, 2012.

203. Jacobson, M.Z., V. Jadhav, World estimates of PV optimal tilt angles and ratios of sunlight incident upon tilted and tracked PV panels relative to horizontal panels, *Solar Energy*, 169, 55–66, 2018.

204. U.S. Department of Energy, How do wind turbines work? 2019, www.energy.gov/eere/wind/wind-energy-technologies-office (accessed March 27, 2019).

205. Pavel, C.C., R. Lacal-Arantegui, A. Marmier, et al., Substitution strategies for reducing the use of rare earths in wind turbines, *Resources Policy*, 52, 349–357, 2017.

206. Green Car Congress, Niron Magnetics raises $21.3M to commercialize rare-earth-free iron-nitride magnets, 2021, www.greencarcongress.com/2021/07/20210730-niron.html (accessed October 25, 2021).

207. Pires, O., X. Munduate, O. Ceyhan, M. Jacobs, and H. Snel, Analysis of high Reynolds numbers effects on a wind turbine airfoil using 2D wind tunnel test data, *J. Phys. Conf. Series* 753, 022047, 2016.

208. Schubel, P.J., and R.J. Crossley, Wind turbine blade design, *Energies*, 5, 3425–3449, 2012.

209. Hu, S.-y., and J.-h. Cheng, Performance evaluation of pairing between sites and wind turbines, *Renewable Energy*, 32, 1934–1947, 2007.

210. Sirnivas, S., W. Musial, B. Bailey, and M. Filippelli, Assessment of offshore wind system design, safety, and operation standards, NREL/TP-5000-60573, 2014.

211. Wiser, R., M. Bolinger, B. Hoen, et al., Land-based wind market report: 2021 edition, U.S. Department of Energy, 2021, www.energy.gov/eere/wind/articles/land-based-wind-market-report-2021-edition-released (accessed December 6, 2021).

212. Faulstich, S., B. Hahn, and P.J. Tavner, Wind turbine downtime and its importance for offshore deployment, *Wind Energy*, 14, 327–337, 2011.

213. Zhang, J., S. Chowdhury, and J. Zhang, Optimal preventative maintenance time windows for offshore wind farms subject to wake losses, AIAA 2012-5435, 2012.

214. Monitoring Analytics, Quarterly state of the market report for PJM: January through June, 2015, www.monitoringanalytics.com/reports/

PJM_State_of_the_Market/2015/2015q2-som-pjm-sec5.pdf (accessed January 19, 2019).

215. Enevoldsen, P., and M.Z. Jacobson, Data investigation of installed and output power densities of onshore and offshore wind turbines worldwide, *Energy Sustain. Dev.*, 60, 40–51, 2021.

216. Zhou, L., Y. Tian, S.B. Roy, et al., Impacts of wind farms on land surface temperature, *Nat. Clim. Change* 2, 539–543, 2012.

217. Jacobson, M.Z., C.L. Archer, and W. Kempton, Taming hurricanes with arrays of offshore wind turbines, *Nat. Clim. Change*, 4, 195–200, https://doi.org/10.1038/NCLIMATE2120, 2014.

218. American Bird Conservancy, https://abcbirds.org 2019 (accessed January 4, 2019).

219. Sovacool, B.K., Contextualizing avian mortality: a preliminary appraisal of bird and bat fatalities from wind, fossil-fuel, and nuclear electricity, *Energy Policy*, 37, 2241–2248, 2009.

220. National Wind Coordinating Collaborative, Wind turbine interactions with birds, bats, and their habitats, 2010, www1.eere.energy.gov/wind/pdfs/birds_and_bats_fact_sheet.pdf (accessed January 4, 2018).

221. Smallwood, K.S., Comparing bird and bat fatality rate estimates among North American wind energy projects, *Wildlife Society Bulletin*, 37, 19–33, 2013.

222. May, R., T. Nygard, U. Falkdalen, et al., Paint it black: efficacy of increased wind turbine rotor blade visibility to reduce avian fatalities, *Ecology and Evolution*, 10, 8927–8935, 2020.

223. Jacobson, M.Z., M.A. Delucchi, Z.A.F. Bauer, et al., 100 percent clean and renewable wind, water, and sunlight (WWS) all-sector energy roadmaps for 139 countries of the world, *Joule*, 1, 108–121, https://doi.org/10.1016/j.joule.2017.07.005, 2017.

224. von Krauland, A.-K., F.-H. Permien, P. Enevoldsen, and M.Z. Jacobson, Onshore wind energy atlas for the United States accounting for land use restrictions and wind speed thresholds, *Smart Energy*, 3, 100046, https://doi.org/10.1016/j.segy.2021.100046, 2021.

225. Moore, M.A., A.E. Boardman, A.R. Vining, D.L. Weimer, and D.H. Greenberg, Just give me a number! Practical values for the social discount rate, *J. Policy Anal. Manage.* 23, 789–812, 2004.

226. National Center for Environmental Economics, *Guidelines for Preparing Economic Analyses*, U.S. Environmental Protection Agency, 2014.

227. U.S. Office of Management and Budget, Circular A-4, Regulatory Analysis, The White House, Washington D.C., September 17, 2003, www.whitehouse.gov/sites/whitehouse.gov/files/omb/circulars/A4/a-4.pdf (accessed January 16, 2019).

228. Drupp, M., M. Freeman, B. Groom, and F. Nesje, Discounting disentangled: an expert survey on the determinants of the long-term social

discount rate. The Centre for Climate Change Economics and Policy Working Paper No. 195 and Grantham Research Institute on Climate Change and the Environment Working Paper No. 172, 2015.

229. Moore, C., S. Brown, U. Alparslan, E. Cremona, and G. Alster, European electricity review: 6-month update H1-2021, Ember, 2021, https://ember-climate.org/app/uploads/2022/02/European-Electricity-Review-H1-2021.pdf (accessed July 28, 2021).

230. Batjer, M., S. Berberich, and D. Hochschild (2020), Letter to Governor Gavin Newsom, www.gov.ca.gov/wp-content/uploads/2020/08/8.17.20-Letter-to-CAISO-PUC-and-CEC.pdf (accessed July 23, 2021).

231. Swenson, A., and A. Lajka, Texas blackouts fuel false claims about renewable energy, 2021, https://apnews.com/article/false-claims-texas-blackout-wind-turbine-f9e24976e9723021bec21f9a68afe927 (accessed July 23, 2021).

232. International Energy Agency, Data and statistics, 2021, www.iea.org/statistics/ (accessed July 23, 2021).

233. Scottish Government, Annual energy statement and quarterly statistics bulletin, December, 2021, www.gov.scot/binaries/content/documents/govscot/publications/statistics/2018/10/quarterly-energy-statistics-bulletins/documents/energy-statistics-summary---december-2021/energy-statistics-summary---december-2021/govscot%3Adocument/Scotland%2BEnergy%2BStats%2BQ3%2B2021.pdf (accessed January 1, 2022).

234. Kuhudzai, R.J., Renewables provided 92.3% of Kenya's electricity generation in 2020, Cleantechnica, 2021, https://cleantechnica.com/2021/11/04/renewables-provided-92-3-of-kenyas-electricity-generation-in-2020/ (accessed November 18, 2021).

235. Cockburn, H., Scotland generating enough wind energy to power two Scotlands, 2019, www.independent.co.uk/environment/scotland-wind-power-on-shore-renewable-energy-climate-change-uk-a9013066.html (accessed July 26, 2019).

236. Renewables Now, Chile's Coquimbo region nears 100% renewables share in H1 2019, 2019, https://renewablesnow.com/news/chiles-coquimbo-region-nears-100-renewables-share-in-h1-2019-663136/ (accessed July 26, 2019).

237. Vorrath, S., South Australia sets smashing new renewables record in final days of 2021, Renew Economy, 2022, https://reneweconomy.com.au/south-australia-winds-up-2021-with-smashing-new-renewables-record/ (accessed January 12, 2022).

238. Energy Information Administration, Renewables became the second-most prevalent U.S. electricity source in 2020, 2021, www.eia.gov/todayinenergy/detail.php?id=48896 (accessed July 28, 2021).

239. Linnerud, K., T. Mideksa, and G.S. Eskeland, The impact of climate change on nuclear power supply, *Energy J.*, 32, 149–168, 2011.

240. Ahmad, A., Increase in frequency of nuclear power outages due to climate change, *Nat. Energy*, 6, 755–762, 2021.

241. Elliott, D., M. Schwartz, and G. Scott, Wind Resource Base, *Encyclopedia of Energy*, 465–479, Elsevier, 2004.

242. Jacobson, M.Z., and Y.J. Kaufmann, Wind reduction by aerosol particles, *Geophys. Res. Lett.*, 33, L24814, 2006.

243. Pryor, S.C., R.J. Barthelmie, MS. Bukovsky, L.R. Leung, and K. Sakaguchi, Climate change impacts on wind power generation, *Nat. Rev. Earth Environ.*, 1, 627–643, 2020.

244. Beyer, H.G., and B.A. Niclasen, Assessment of the option 'wind power to heat for buildings' with respect to meteorological conditions, *Meteorol. Z.*, 28, 79–85, 2019.

245. Jacobson, M.Z., On the correlation between building heat demand and wind energy supply and how it helps to avoid blackouts, *Smart Energy*, 1, 100009, https://doi.org/10.1016/j.segy.2021.100009, 2021

246. Kahn, E., The reliability of distributed wind generators, *Electr. Power Syst. Res.*, 2, 1–14, 1979.

247. Archer, C.I., and M.Z. Jacobson, Spatial and temporal distributions of U.S. winds and wind power at 80 m derived from measurements, *J. Geophys. Res.*, 108, 4289, 2003.

248. Archer, C.L., and M.Z. Jacobson, Supplying baseload power and reducing transmission requirements by interconnecting wind farms, *J. Appl. Meteorol. Climatol.*, 46, 1701–1717, https://doi.org/10.1175/2007JAMC1538.1, 2007.

249. Stoutenburg, E.D., N. Jenkins, and M.Z. Jacobson, Power output variations of co-located offshore wind turbines and wave energy converters in California, *Renew. Energy*, 35, 2781–2791, https://doi.org/10.1016/j.renene.2010.04.033, 2010.

250. Stoutenburg, E.K., and M.Z. Jacobson, Reducing offshore transmission requirements by combining offshore wind and wave farms, *IEEE J. Ocean. Eng.*, 36, 552–561, https://doi.org/10.1109/JOE.2011.2167198, 2011.

251. Jacobson, M.Z., A.-K. von Krauland, S.J. Coughlin, F.C. Palmer, and M.M. Smith, Zero air pollution and zero carbon from all energy at low cost and without blackouts in variable weather throughout the U.S. with 100% wind-water-solar (WWS) and storage, *Renew. Energy*, 184, 430–444, 2022.

252. Jacobson, M.Z., The cost of grid stability with 100% clean, renewable energy for all purposes when countries are islanded versus interconnected, *Renew. Energy*, 179, 1065–1075, 2021.

253. Kempton, W., and J. Tomic, Vehicle-to-grid power fundamentals: calculating capacity and net revenue, *J. Power Sources*, 144, 268–279, 2005.

254. Kempton, W., and J. Tomic, Vehicle-to-grid power implementation: from stabilizing the grid to supporting large-scale renewable energy, *J. Power Sources*, 144, 280–294, 2005.

255. Budischak, C., D. Sewell, H. Thompson, et al., Cost-minimized combinations of wind power, solar power, and electrochemical storage, powering the grid up to 99.9% of the time, *J Power Sources*, 225, 60–74, 2013.

256. Child, M., A. Nordling, and C. Breyer, The impacts of high V2G participation in a 100% renewable Aland energy system, *Energies*, 11, 2206, https://doi.org/10.3390/en11092206, 2018.

257. Hart, E.K., and M.Z. Jacobson, A Monte Carlo approach to generator portfolio planning and carbon emissions assessments of systems with large penetrations of variable renewables, *Renew. Energy*, 36, 2278–2286, https://doi.org/10.1016/j.renene.2011.01.015, 2011.

258. Kirby, B.J., Frequency regulation basics and trends, ORNL/TM-2004/291, 2004, www.consultkirby.com/files/TM2004-291_Frequency_Regulation_Basics_and_Trends.pdf (accessed January 28, 2019).

259. U.S. National Research Council, *Real Prospects for Energy Efficiency in the United States*, National Academies Press, p. 251, www.nap.edu/read/12621/chapter/6#251, 2010 (accessed February 2, 2019).

260. Erlich, I., and M. Wilch, Frequency control by wind turbines, IEEE PES General Meeting, July 25–29, 2010, https://doi.org/10.1109/PES.2010.5589911, https://ieeexplore.ieee.org/document/5589911 (accessed March 2, 2019).

261. Roselund, C., Inertia, frequency regulation and the grid, *PV Magazine*, 2019, https://pv-magazine-usa.com/2019/03/01/inertia-frequency-regulation-and-the-grid/ (accessed March 2, 2019).

262. Sorensen, B., A plan is outlined to which solar and wind energy would supply Denmark's needs by the year 2050, *Science*, 189, 255–260, 1975.

263. Lovins, A.B., Energy strategy: the road not taken, *Foreign Affairs*, 55, 65–96, 1976.

264. Sorensen, B., Scenarios of greenhouse warming mitigation, *Energy Convers. Manag.*, 37, 693–698, 1996.

265. Jacobson, M.Z., and G.M. Masters, Exploiting wind versus coal, *Science*, 293, 1438–1438, 2001.

266. Jacobson, M.Z., W.G. Colella, and D.M. Golden, Cleaning the air and improving health with hydrogen fuel cell vehicles, *Science*, 308, 1901–1905, 2005.

267. Archer, C.L. and M.Z. Jacobson, Evaluation of global wind power, *J. Geophys. Res.*, 110, D12110, https://doi.org/10.1029/2004JD005462, 2005.

268. Czisch, G., *Szenarien zur zukünftigen Stromversorgung, kostenoptimierte Variationen zur Versorgung Europas und seiner Nachbarn mit Strom aus erneuerbaren Energien*, Ph.D. Dissertation, University of Kassel, 2005, https://kobra.uni-kassel.de/handle/123456789/200604119596 (accessed February 24, 2019).

269. Czisch, G., and G. Giebel, Realisable scenarios for a future electricity supply based 100% on renewable energies, Riso-R-1608 (EN), 2007, www.researchgate.net/publication/238787763_Realisable_Scenarios_for_a_Future_Electricity_Supply_based_100_on_Renewable_Energies (accessed April 11, 2022).

270. Lund, H., Large-scale integration of optimal combinations of PV, wind, and wave power into the electricity supply, *Renew. Energy*, 31, 503–515, 2006.

271. Hoste, G.R.G., M.J. Dvorak, and M.Z. Jacobson, Matching hourly and peak demand by combining different renewable energy sources, Stanford University Technical Report, 2009, web.stanford.edu/group/efmh/jacobson/Articles/I/CombiningRenew/HosteFinalDraft (accessed January 27, 2019).

272. Jacobson, M.Z., and M.A. Delucchi, A path to sustainable energy by 2030, *Sci. Am.* 301, 58–65, November 2009.

273. Lund, H., and B.V. Mathiesen, Energy system analysis of 100% renewable energy systems – the case of Denmark in years 2030 and 2050, *Energy*, 34, 524–531, 2009.

274. Delucchi, M.Z., and M.Z. Jacobson, Providing all global energy with wind, water, and solar power, Part II: Reliability, system and transmission costs, and policies, *Energy Policy*, 39, 1170–1190, https://doi.org/10.1016/j.enpol.2010.11.045, 2011.

275. Jacobson, M.Z., R.W. Howarth, M.A. Delucchi, et al., Examining the feasibility of converting New York State's all-purpose energy infrastructure to one using wind, water, and sunlight, *Energy Policy*, 57, 585–601, 2013.

276. Jacobson, M.Z., M.A. Delucchi, A.R. Ingraffea, et al., A roadmap for repowering California for all purposes with wind, water, and sunlight, *Energy*, 73, 875–889, https://doi.org/10.1016/j.energy.2014.06.099, 2014.

277. Jacobson, M.Z., M.A. Delucchi, G. Bazouin, et al., A 100 percent wind, water, sunlight (WWS) all-sector energy plan for Washington State, *Renewable Energy*, 86, 75–88, 2016.

278. Jacobson, M.Z., M.A. Delucchi, M.A. Cameron, et al., Impacts of Green-New-Deal energy plans on grid stability, costs, jobs, health, and climate in 143 countries, *One Earth*, 1, 449–463, https://doi.org/10.1016/j.oneear.2019.12.003, 2019.

279. Jacobson, M.Z., M.A. Cameron, E.M. Hennessy, et al., 100 percent clean, and renewable wind, water, and sunlight (WWS) all-sector energy roadmaps for 53 towns and cities in North America, *Sustain. Cities Soc.*, 42, 22–37, https://doi.org/10.1016/j.scs.2018.06.031, 2018.

280. Jacobson, M.Z., A.-K. von Krauland, Z.F.M. Burton, et al., Transitioning all energy in 74 metropolitan areas, including 30 megacities, to 100% clean and renewable wind, water, and sunlight, *Energies*, 13, 4934, https://doi.org/10.3390/en13184934, 2020.

281. Mason, I.G., S.C. Page, A.G. Williamson, A 100% renewable energy generation system for New Zealand utilizing hydro, wind, geothermal, and biomass resources, *Energy Policy* 38, 3973–3984, 2010.

282. Connolly, D., H. Lund, B.V. Mathiesen, and M. Leahy, The first step to a 100% renewable energy-system for Ireland, *Appl. Energy*, 88, 502–507, 2011.

283. Connolly, D., and B.V. Mathiesen, Technical and economic analysis of one potential pathway to a 100% renewable energy system, *Int. J. Sustain. Energy Plan. Manag.*, 1, 7–28, 2014.

284. Connolly, D., H. Lund, and B.V. Mathiesen, Smart energy Europe: the technical and economic impact of one potential 100% renewable energy scenario for the European Union, *Renew. Sustain. Energy Rev.*, 60, 1634–1653, 2016.

285. Mathiesen, B.V., H. Lund, and K. Karlsson, 100% renewable energy systems, climate mitigation, and economic growth, *Appl. Energy*, 88, 488–501, 2011.

286. Mathiesen, B.V., H. Lund, D. Connolly, et al., Smart energy systems for coherent 100% renewable energy and transport solutions, *Appl. Energy*, 145, 139–154, 2015.

287. Hart, E.K., E.D. Stoutenburg, and M.Z. Jacobson, The potential of intermittent renewables to meet electric power demand: a review of current analytical techniques, *Proc. IEEE*, 100, 322–334, https://doi.org/10.1109/JPROC.2011.2144951, 2012.

288. Hart, E.K., and M.Z. Jacobson, The carbon abatement potential of high penetration intermittent renewables, *Energy Environ. Sci.*, 5, 6592–6601, https://doi.org/10.1039/C2EE03490E, 2012.

289. Elliston, B., M. Diesendorf, and I. MacGill, Simulations of scenarios with 100% renewable electricity in the Australian National Electricity Market, *Energy Policy*, 45, 606–613, 2012.

290. Elliston, B., I. MacGill, and M. Diesendorf, Least cost 100% renewable electricity scenarios in the Australian National Electricity Market, *Energy Policy*, 59, 270–282, 2013.

291. Elliston, B., I. MacGill, and M. Diesendorf, Comparing least cost scenarios for 100% renewable electricity with low emission fossil fuel scenarios in the Australian National Electricity Market, *Renew. Energy*, 66, 196–204, 2014.

292. Rasmussen, M.G., G.B. Andresen, and M. Greiner, Storage and balancing synergies in a fully or highly renewable pan-European power system, *Energy Policy*, 51, 642–651, 2012.

293. Steinke, F., P. Wolfrum, and C. Hoffmann, Grid vs. storage in a 100% renewable Europe, *Renew. Energy*, 50, 826–832, 2013.

294. Becker, S., B.A. Frew, G.B. Andresen, et al., Features of a fully renewable U.S. electricity-system: optimized mixes of wind and solar PV and transmission grid extensions, *Energy*, 72, 443–458, 2014.

295. Becker, S., B.A. Frew, G.B. Andresen, et al., Renewable build-up pathways for the U.S.: generation costs are not system costs, *Energy*, 81, 437–445, 2015.

296. Jacobson, M.Z., M.A. Cameron, and B.A. Frew, A low-cost solution to the grid reliability problem with 100 percent penetration of intermittent wind, water, and solar for all purposes, *Proc. Natl Acad. Sci.*, 112, 15060–15065, https://doi.org/10.1073/pnas.1510028112, 2015.

297. Frew, B.A., S. Becker, M.J. Dvorak, G.B. Andresen, and M.Z. Jacobson, Flexibility mechanisms and pathways to a highly renewable U.S. electricity future, *Energy*, 101, 65–78, 2016.

298. Bogdanov, D., and C. Breyer, North-east Asian super grid for 100% renewable energy supply: optimal mix of energy technologies for electricity, gas, and heat supply options, *Energy Convers. Manag.*, 112, 176–190, 2016.

299. Bogdanov, D., J. Farfan, K. Sadovskaia, et al., Radical transformation pathway towards sustainable electricity via evolutionary steps, *Nat. Commun.*, 10, 1077, https://doi.org/10.1038/s41467-019-08855-1, 2019.

300. Child, M., and C. Breyer, Vision and initial feasibility analysis of a decarbonized Finnish energy system for 2050. *Renew. Sustain. Energy Rev.*, 66, 517–536, 2016.

301. Aghahosseini, A., D. Bogdanov, and C. Breyer, A techno-economic study of an entirely renewable energy-based powers supply for North America for 2030 conditions, *Energies*, 10, 1171, https://doi.org/10.3390/en10081171, 2016.

302. Aghahosseini, A., D. Bogdanov, L.S.N.S. Barbosa, and C. Breyer, Analyzing the feasibility of powering the Americas with renewable energy and inter-regional grid interconnections by 2030, *Renew. Sustain. Energy Rev.*, 105, 187–205, 2019.

303. Aghahosseini, A., D. Bogdanov, and C. Breyer, Towards sustainable development in the MENA regions: Analysing the feasibility of a 100% renewable electricity system in 2030, *Energy Strategy Rev.*, 28, 100466, 2020.

304. Blakers, A., B. Lu, and M. Socks, 100% renewable electricity in Australia, *Energy*, 133, 471–482, 2017.

305. Blakers, A., B. Lu, M. Stocks, K. Anderson, and A. Nadolny, Pumped hydro storage to support 100% renewable power, *Energy News*, 36, 11–14, 2018.

306. Barbosa, L.S.N.S., D. Bogdanov, P. Vainikka, and C. Breyer, Hydro, wind, and solar power as a base for a 100% renewable energy supply for South and Central America, *PloS ONE*, https://doi.org/10.1371/journal .pone.0173820, 2017.

307. Lu, B., A. Blakers, and M. Stocks, 90–100% renewable electricity for the South West Interconnected System of Western Australia, *Energy*, 122, 663–674, 2017.

308. Gulagi, A., D. Bogdanov, and C. Breyer, A cost optimized fully sustainable power system for Southeast Asia and the Pacific Rim, *Energies*, 10, 583, https://doi.org/10.3390/en10050583, 2017.

309. Gulagi, A., P. Choudhary, D. Bogdanov, and C. Breyer, Electricity system based on 100% renewable energy for India and SAARC, *PLoS ONE*, https://doi.org/10.1371/journal.pone.0180611, 2017.

310. Esteban, M., J. Portugal-Pereira, B.C. Mclellan, et al., 100% renewable energy system in Japan: smoothening and ancillary services, *Appl. Energy*, 224, 698–707, 2018.

311. Zapata, S., M. Casteneda, M. Jiminez, et al., Long-term effects of 100% renewable generation on the Colombian power market, *Sustain. Energy Technol. Assess.*, 30, 183–191, 2018.

312. Sadiqa, A., A. Gulagi, and C. Breyer, Energy transition roadmap towards 100% renewable energy and role of storage technologies for Pakistan by 2050, *Energy*, 147, 518–533, 2018.

313. Barasa, M., D. Bogdanov, A.S. Oyewo, and C. Breyer, A cost optimal resolution for sub-Saharan Africa powered by 100% renewables in 2030, *Renew. Sustain. Energy Rev.*, 92, 440–457, 2018.

314. Caldera, U., and C. Breyer, Role that battery and water storage play in Saudi Arabia's transition to an integrated 100% renewable energy power system, *J. Energy Storage*, 17, 299–310, 2018.

315. Liu, H., G.B. Andresen, and M. Greiner, Cost-optimal design of a simplified highly renewable Chinese network, *Energy*, 147, 534–546, 2018.

316. Teske, S., D. Giurco, T. Morris, et al., *Achieving the Paris Climate Agreement*, J. Miller, ed., 2019, https://oneearth.app.box.com/s/hctp4qlk34yg-domw3yjdtctsymsdtaqs (accessed September 6, 2019).

317. Ram, M., D. Bogdanov, A. Aghahosseini, et al., Global energy system based on 100% renewable energy – power, heat, transport, and desalination sectors, Lappeenranta University of Technology Research Reports 91, Lappeenranta, 2019, http://energywatchgroup.org/wp-content/uploads/EWG_LUT_100RE_All_Sectors_Global_Report_2019.pdf (accessed September 6, 2019).

318. Hansen, K., B. Mathiesen, and I.R. Skov, Full energy system transition towards 100% renewable energy in Germany in 2050, *Renew. Sustain. Energy Rev.*, 102, 1–13, 2019.

319. Oyewo, A.S., A. Aghahosseini, M. Ram, and C. Breyer, Transition towards decarbonized power systems and its socio-economic impacts in West Africa, *Renew. Energy*, 154, 1092–1112, 2020.

320. Marczinkowski, H.M., and L. Barros, Technical approaches and institutional alignment to 100% renewable energy system transition of Madeira Island – electrification, smart energy and the required flexible market conditions, *Energies*, 13, 4434, 2020.

321. Li, M., M. Lenzen, D. Wang, and K. Nansai, GIS-based modelling of electric-vehicle-grid integration in a 100% renewable electricity grid, *Appl. Energy*, 262, 114577, 2020.

322. Alves, M., R. Segurado, and M. Costa, On the road to 100% renewable energy systems in isolated islands, *Energy*, 198, 117321, 2020.

323. Kiwan, S., and E. Al-Gharibeh, Jordan toward a 100% renewable electricity system, *Renew. Energy*, 147, 423–436, 2020.

324. Zozmann, E., L. Goke, M. Kendziorski, et al., 100% renewable energy scenarios for North America – spatial distribution and network constraints, *Energies*, 14, 658, 2021.

325. Cole, W.J., D. Greer, P. Denholm, et al., Quantifying the challenge of reaching a 100% renewable energy power system for the United States, *Joule*, 5, 1732–1748, 2021.

326. Brown, T.W., T. Bischof-Niemz, K. Blok, et al., Response to 'Burden of proof: A comprehensive review of the feasibility of 100% renewable electricity systems,' *Renew. Sustain. Energy Rev.*, 92, 834–847, 2018.

327. Diesendorf, M., and B. Elliston, The feasibility of 100% renewable electricity systems: a response to critics, *Renew. Sustain. Energy Rev.*, 93, 318–330, 2018.

328. Hansen, K., C. Breyer, and H. Lund, Status and perspectives on 100% renewable energy systems, *Energy*, 175, 471–480, 2019.

329. Edelstein, S., Report: EV battery costs hit a new low in 2021, but they might rise in 2022, Green Car Reports, 2021, www.greencarreports .com/news/1134307_report-ev-battery-costs-might-rise-in-2022 (accessed January 2, 2022).

330. Irvine, M., and M. Rinaldo, Tesla's battery day and the energy transition, DNV, 2020, www.dnv.com/feature/tesla-battery-day-energy-transition .html (accessed January 2, 2022).

331. World Bank, Agricultural and rural development, https://data.worldbank .org/indicator/, 2017 (accessed September 16, 2019).

332. Macdonald-Smith, A., South Australia's big battery slashes $40m from grid control costs in first year, 2018, www.afr.com/business/energy/

solar-energy/south-australias-big-battery-slashes-40m-from-grid-control-costs-in-first-year-20181205-h18ql1 (accessed January 25, 2019).

333. Spector, J., 'Cheaper than a peaker': NextEra inks massive wind+solar+storage deal in Oklahoma, 2019, www.greentechmedia.com/articles/read/nextera-inks-even-bigger-windsolarstorage-deal-with-oklahoma-cooperative#gs.s8lb02 (accessed July 26, 2019).

334. Thomas, H., Richard Branson's disappointing space jaunt, 2021, www.cnn.com/2021/07/12/opinions/richard-branson-jeff-bezos-space-flight-is-a-waste-thomas/index.html (accessed December 8, 2021).

335. Wilson, M., FERC launches first transmission reforms in a decade, 2021, www.eenews.net/articles/ferc-launches-first-transmission-reforms-in-a-decade/ (accessed July 16, 2021).

336. Metz, M., J. London, and P. Rosler, Gasoline superusers, 2021, https://static1.squarespace.com/static/5888d6bad2b857a30238e864/t/60f65b1cbd95132ca728869f/1626757920037/Superusers+White+Paper+Final.pdf (accessed July 26, 2021).

337. Jacobson, M.Z., *Developing, Coupling, and Applying a Gas, Aerosol, Transport, and Radiation Model to Study Urban and Regional Air Pollution*. Ph.D. Dissertation, University of California, Los Angeles, 1994.

338. Jacobson, M.Z., R. Lu, R.P. Turco, and O.B. Toon, Development and application of a new air pollution modeling system. Part I: Gas-phase simulations, *Atmos. Environ.*, 30B, 1939–1963, 1996.

339. Jacobson, M.Z., Development and application of a new air pollution modeling system. Part II: Aerosol module structure and design, *Atmos. Environ.*, 31A, 131–144, 1997.

340. Jacobson, M.Z., Development and application of a new air pollution modeling system. Part III: Aerosol-phase simulations, *Atmos. Environ.*, 31A, 587–608, 1997.

341. Jacobson, M.Z., Simulations of the rates of regeneration of the global ozone layer upon reduction or removal of ozone-destroying compounds. *EOS Suppl.*, Fall, 1995, F119.

342. Jacobson, M.Z., Global direct radiative forcing due to multicomponent anthropogenic and natural aerosols, *J. Geophys. Res.*, 106, 1551–1568, 2001.

343. Jacobson, M.Z., GATOR-GCMM: A global through urban scale air pollution and weather forecast model. 1. Model design and treatment of subgrid soil, vegetation, roads, rooftops, water, sea ice, and snow, *J. Geophys. Res.*, 106, 5385–5401, 2001.

344. Jacobson, M.Z., GATOR-GCMM: 2. A study of day- and nighttime ozone layers aloft, ozone in national parks, and weather during the SARMAP field campaign, *J. Geophys. Res.*, 106, 5403–5420, 2001.

345. Jacobson, M.Z., Y.J. Kaufmann, Y. Rudich, Examining feedbacks of aerosols to urban climate with a model that treats 3-D clouds with

aerosol inclusions, *J. Geophys. Res.*, 112, D24205, https://doi
.org/10.1029/2007JD008922, 2007.

346. Jacobson, M.Z., Studying the effects of aerosols on vertical photoly-
sis rate coefficient and temperature profiles over an urban airshed,
J. Geophys. Res., 103, 10593–10604, 1998.

347. Jacobson, M. Z., Isolating nitrated and aromatic aerosols and nitrated
aromatic gases as sources of ultraviolet light absorption, *J. Geophys.
Res.*, 104, 3527–3542, 1999.

348. *New York Times*, Text of President Bush's remarks on global climate,
2001, www.nytimes.com/2001/06/11/world/text-of-president-bushs-
remarks-on-global-climate.html (accessed December 8, 2021).

349. Dvorak, M., C.L. Archer, and M.Z. Jacobson, California offshore wind
energy potential, *Renew. Energy*, 35, 1244–1254, https://doi
.org/10.1016/j.renene.2009.11.022, 2010.

350. Dvorak, M.J., E.D. Stoutenburg, C.L. Archer, W. Kempton, and M.Z.
Jacobson, Where is the ideal location for a U.S. East Coast offshore grid,
Geophys. Res. Lett., 39, L06804, https://doi.org/10.1029/2011GL050659,
2012.

351. Dvorak, M.J., B.A. Corcoran, J.E. Ten Hoeve, N.G. McIntyre, and M.Z.
Jacobson, U.S. East Coast offshore wind energy resources and their
relationship to peak-time electricity demand, *Wind Energy*, 16, 977–997,
https://doi.org/10.1002/we.1524, 2013.

352. Jacobson, M.Z., J.H. Seinfeld, G.R. Carmichael, and D.G. Streets, The
effect on photochemical smog of converting the U.S. fleet of gasoline
vehicles to modern diesel vehicles, *Geophys. Res. Lett.*, 31, L02116,
https://doi.org/10.1029/2003GL018448, 2004.

353. TED (Technology, Education, Development), Does the world need nuclear
energy? Public debate, Mark Z. Jacobson and Stewart Brand, Long Beach,
California, February 11, 2010, www.ted.com/talks/debate_does_the_
world_need_nuclear_energy?language=en (accessed December 9, 2021).

354. Solutions Project, Our 100% Clean Energy Vision, 2019, www
.thesolutionsproject.org/why-clean-energy/ (accessed December 8, 2021).

355. Ruffalo, M.A., M. Krapels, and M.Z. Jacobson, A plan to power
the world with wind, water, and sunlight, Talks at Google, Google,
Inc., Mountain View, California, June 20, 2012, www.youtube.com/
watch?v=N_sLt5gNAQs (accessed December 8, 2021).

356. Ruffalo, M.Z., M. Krapels, and video from J. Fox, Powering the world,
U.S., and New York with wind, water, and sunlight, The Nantucket
Project, Nantucket, Massachusetts, October 6, 2012, http://vimeo
.com/52038463 (accessed December 8, 2021).

357. Letterman, D., *The Late Show with David Letterman*, New York City,
October 9, 2013, www.youtube.com/watch?v=AqIu2J3vRJc (accessed
December 8, 2021).

358. California Senate, SB 100 FAQs, 2018, https://focus.senate.ca.gov/sb100/faqs (accessed December 8, 2021).

359. U.S. House of Representatives, H.Res.540, 2015, www.congress.gov/bill/114th-congress/house-resolution/540/text (accessed December 8, 2021).

360. O'Malley, M., A jobs agenda for our renewable energy future, 2015, www.p2016.org/omalley/omalley070215climate.html (accessed December 8, 2021).

361. Clinton, H., Hillary is ready for 100, October 16, 2015, www.c-span.org/video/?c4557641/hillary-readyfor100 (accessed December 8, 2021).

362. Democratic National Committee, 2016 Democratic Party Platform, July 9, 2016, https://democrats.org/wp-content/uploads/2018/10/2016_DNC_Platform.pdf (accessed December 8, 2021).

363. Sanders, B., and M. Jacobson, The American people, not big oil, must decide our climate future, *The Guardian*, April 29, 2017, www.theguardian.com/commentisfree/2017/apr/29/bernie-sanders-climate-change-big-oil (accessed December 8, 2021).

364. REN21 (Renewable Energy Policy Network for the 21st Century), Renewables 2020 global status report, 2020, www.ren21.net/gsr-2020/tables/table_06/table_06/ (accessed December 8, 2021).

365. Lillian, B., Orsted survey: Eight out of 10 support 100% global renewable energy, November 13, 2017, https://nawindpower.com/orsted-survey-eight-10-support-100-global-renewable-energy (accessed December 8, 2021).

366. Sierra Club, Ready for 100, 2021, www.sierraclub.org/ready-for-100/commitments (accessed January 2, 2022).

367. REN21 (Renewable Energy Policy Network for the 21st Century), Renewables in cities: 2019 global status report, 2019, www.ren21.net/wp-content/uploads/2019/05/REC-2019-GSR_Full_Report_web.pdf (accessed December 8, 2021).

368. Boyer, G., Appalachian power rolls out 100% renewable option for customers, *WFXR News*, 2019, www.wfxrtv.com/news/appalachian-power-rolls-out-100-renewable-option-for-va-customers/ (accessed December 8, 2021).

369. Duke Energy, More renewable energy options available under Duke Energy's Green Source Advantage, 2019, https://news.duke-energy.com/releases/more-renewable-energy-options-available-under-duke-energys-green-source-advantage?_ga=2.88266651.1875174277.1566405614-658711925.1566405614 (accessed December 8, 2021).

370. RE100, The world's most influential companies committed to 100% renewable power, 2022, http://there100.org (accessed January 2, 2022).

371. Climate Group, EV100 members, 2019, www.theclimategroup.org/ev100-members (accessed December 8, 2021).

372. Shepherd, M., The climate science behind the green new deal – a layperson's explanation, 2019, www.forbes.com/sites/marshallshepherd/2019/02/24/the-climate-science-behind-the-green-new-deal-a-laypersons-explanation/?sh=4e0260e36f2a (accessed December 8, 2021).

373. Green Party US, Green New Deal – Full Language, 2018, www.gp.org/gnd_full (accessed December 8, 2021).

INDEX

Printed in the United States
by Baker & Taylor Publisher Services